Peter Lax
Samuel Burstein
Anneli Lax

Calculus
with Applications
and Computing

Volume I

Springer-Verlag

New York   Heidelberg   Berlin

1976

Peter Lax
New York University
Courant Institute
New York, New York 10012

Samuel Burstein
New York University
Courant Institute
New York, New York 10012

Anneli Lax
New York University
Courant Institute
New York, New York 10012

AMS Subject Classifications
26-01, 26A06

Library of Congress Cataloging in Publication Data

Lax, Peter D
  Calculus with applications and computing.

  (Undergraduate texts in mathematics)
  Includes index.
  1. Calculus.   I. Burstein, Samuel Z., joint author.
II. Lax, Anneli, joint author.   III. Title.
AQ303.L35   515   76-4070

The cover design is based upon a page of Sir Isaac Newton's *Opera*,
reproduced through the courtesy of the New York Public Library.

© 1976 by Springer-Verlag New York Inc.

Printed in the United States of America

ISBN 0-387-90179-5   Springer-Verlag   New York

ISBN 3-540-90179-5   Springer-Verlag   Berlin Heidelberg

Undergraduate Texts in Mathematics

Preface

Mathematics is vigorously and brilliantly pursued in our time on a very broad front; yet the authors of this text feel that not enough mathematical talent is devoted to furthering the interaction of mathematics with other sciences and disciplines. This imbalance is harmful to both mathematics and its users; to redress this imbalance is an educational task which must start at the beginning of the college curriculum. No course is more suited for this than the calculus; there students can learn at first hand that mathematics is the language in which scientific ideas can be precisely formulated, that science is a source of mathematical ideas which profoundly shape the development of mathematics, and last but not least that mathematics can furnish brilliant answers to important scientific problems.

Our purpose in writing this text has been to emphasize this relation of calculus to science. We hope to accomplish this by devoting whole connected chapters to single—or several related—scientific topics, letting the reader observe how the notions of calculus are used to formulate the basic laws of science and how the methods of calculus are used to deduce consequences of those basic laws. Thus the student sees calculus at work on worthwhile tasks. The traditional course too often resembles the inventory of a workshop; here we have hammers of different sizes, there saws, yonder planes; the student is instructed in the use of each instrument, but seldom are they all put together in the building of a truly worthwhile object.

Finding numerical answers is a very important part of an application of mathematics. Even when qualitative rather than quantitative understanding is the aim, the calculation of a well-chosen special case can take the role of a crucial experiment in confirming an old speculation, or pointing to a new one. Also, the design of effective numerical methods is one of the finest applications of the ideas of calculus. Numerical methods are presented in this text as organic parts of calculus, not as a mere list of recipes appended as an afterthought.

Almost all numerical examples and exercises presented in this text can be calculated with the aid of programmable hand calculators. The use of a computer is advocated for those problems whose output is more than a few numbers and must be displayed in tabulated or graphical form to extract the essential features.

The educational value of numerical examples worked out by the students themselves, individually or as members of small teams, cannot be over-estimated. A good student is likely to be a better and more enterprising computer programmer than his instructor, and this will enable him to experiment on his own, instead of being bound to his text or the instructor's apronstrings. This active participation is a welcome alternative to merely sitting back and absorbing knowledge.

Our fairly radical outlook on what purpose calculus serves is echoed by a critical revision of some purely mathematical notions of calculus. Our treatment is rigorous without being pedantic; we do not hesitate to break with tradition where we feel that a change is called for. We pinpoint these changes in the following chapter-by-chapter account of our treatment:

Chapter 1 deals with real numbers; we teach the student to think of them in three complementary ways: (i) as entities which can be added, multiplied, etc., subject to the usual rules of algebra; (ii) as points on the number line; and (iii) as infinite decimals.

Infinite decimals are Platonic ideals of which we mortals only see shadows that appear as finite digits on the register, or printout, of calculators. The enormous advantage of thinking of real numbers as infinite decimals is that we can recognize at a glance when two numbers are close to each other. The notion of a convergent sequence can be explained without resorting to Greek letters, merely by noting that a sequence converges if more and more of the digits of its members are identical.

In Chapter 2 on functions we emphasize the role of functions in describing the relation of two quantities. We explain that complicated functions can be built out of very simple ones by composing, adding, multiplying, inverting, etc., functions repeatedly. Our definition of continuity is uniform continuity on a given interval; this is far more appropriate than the notion of continuity at each point. We define uniform convergence of a sequence of functions, a natural, useful, and elementary concept that Victorian prudishness usually reserves for mature audiences. We explain the notion of an algorithm for calculating a function, and give examples of distinct algorithms which calculate the same function, where one algorithm is faster and more accurate than the other.

In Chapter 3 on differentiation, the derivative is defined as the uniform limit of difference quotients. This makes it evident that a function whose derivative is positive on an interval is an increasing function. This observation is used as a workhorse throughout the book; in this chapter it is used to prove the mean value theorem, Taylor's theorem, and the characterization of maxima and minima. We give many illustrations of the notion of derivative and devote a section to one-dimensional mechanics.

The definite integral is defined in Chapter 4 in terms of two basic properties which are amply illustrated and motivated. We show that all properties of the integral follow from these two, including its relation to the derivative, the approximation of integrals by sums, and the rules of integration such as changing variables or integration by parts. These techniques change a given integral into another integral; we explain to the student that although in a few spectacular instances the application of these techniques leads to the explicit evaluation of an integral in terms of known functions, this is not so in most cases. Nevertheless an intelligently chosen change of variables or integration by parts can change an integral into another one that is far easier to approximate numerically than the original integral. Simpson's rule is introduced and applied lovingly.

In Chapter 5 the exponential function is defined as modeling growth; the functional equation of the exponential function is derived from this model, and from it the differential equation. We emphasize that all properties of the exponential function, including our ability to find accurate approximations to it, follow from this differential equation. The logarithmic function is defined as the inverse of the exponential function, and its usual properties are explored.

Chapter 6 is an introduction to probability theory, both discrete and continuous. The information content of a probability distribution is defined. Gauss' law of error is derived and applied to the diffusion process.

In Chapter 7 we derive the addition formulas for sine and cosine from the arithmetic of complex numbers. The differential equations for sine and cosine are derived; we emphasize that all properties of sine and cosine follow from these differential equations. There is a brief discussion of two-dimensional mechanics in terms of complex numbers. The basic facts of gravitational motion are derived.

In Chapter 8 on vibration the emphasis is placed on the law of conservation of energy. There is an elementary but nontrivial discussion of nonlinear vibrations which is a mixture of theory and numerical experimentation.

The last chapter on populations dynamics discusses the differential equations governing the growth of populations and the differential equations of chemical reactions. We emphasize that properties of solutions, qualitative and quantitative, can be deduced directly from the differential equations themselves without relying on explicit formulas for their solution.

Our experience in teaching this material convinces us that more is included in this text than can be covered in two semesters. The following has been taught successfully to an average freshman class at Washington Square College of New York University.

First semester: All of Chapter 1, including infinite sums; Chapter 2, excluding functions of several variables and partial fractions; Chapter 3, excluding Taylor's theorem; Chapter 4, omitting the section on the existence of the integral but including improper integrals.

Second semester: Chapter 5, including numerical methods for calculating exponential and logarithmic functions; all of Chapter 6 including the section on diffusion; Chapter 7, except for the section on isometry, and merely touching on complex valued functions; all of Chapter 8 minus nonlinear vibrations. In Chapter 9 we suggest covering either population dynamics or chemical kinetics, but not necessarily both. We would like to emphasize that Chapters 6, 8 and 9 are independent applications of the rest of the text.

About a week was spent in the first semester on a crash course in computing.

Volume II is in preparation; it will deal with functions of several variables in the same spirit as Volume I treated functions of a single variable. Vectors will be used from the outset.

It is a pleasure to thank friends and colleagues, within the Courant Institute and elsewhere, for critical review of parts of the manuscript and for general good advice. We particularly thank Robert Walker for an overall review of the first version of our manuscript, which appeared in 1972 in the Courant Institute Lecture Notes series. We also thank Paul Gans for a critical review of the section on chemical kinetics. One of the authors (SB) would like to thank his wife, Elaine, for her patience and understanding over many months of preparation and preoccupation. Finally, a bouquet of thanks to Gloria Lee for expert typing.

Each section was submitted to trial by fire, i.e., in the classroom. The students' reactions were taken into account in shaping the final version. We thank them for their cooperation and enthusiasm.

Contents

Chapter 3

Differentiation

Chapter 4

Integration

Chapter 5

Growth and decay

Chapter 6

Probability and its applications

Chapter 7

Rotation and the trigonometric functions

Chapter 8

Vibrations

Chapter 9

Population dynamics and chemical reactions

FORTRAN programs and instructions for their use

Index

Real numbers

1

There are at least three different ways of thinking about numbers, all valuable and almost indispensable. One can think of them *algebraically*, *geometrically*, or as *decimals*; we shall describe each in turn.

1.1 The algebra of numbers; a review

The set of all numbers is a collection of *symbols* which can be combined in two ways called *adding* and *multiplying*. Both operations are *commutative* and *associative*, and multiplication is *distributive* with respect to addition. In terms of symbols a, b, c representing three arbitrary numbers, our rules are:

Commutative rule
$$a + b = b + a$$
$$ab = ba$$

Associative rule
$$(a + b) + c = a + (b + c)$$
$$a(bc) = (ab)c$$

Distributive rule $\qquad a(b + c) = ab + ac.$

Two numbers, 0 and 1, are distinguished by the properties that adding 0 and multiplying by 1 does not alter a number; every number c has its *negative* $-c$, and every number c except 0 has its *reciprocal* $1/c$, also denoted by c^{-1}. We shall use the customary notations

$$a + (-b) = a - b$$

1

and

$$ab^{-1} = \frac{a}{b} = a/b.$$

The properties of negative and reciprocal are

$$a + (-a) = 0 \quad \text{and} \quad a \cdot \frac{1}{a} = aa^{-1} = 1.$$

The following rules ought to be familiar to all:

$$-(a + b) = -a + (-b),$$

and

$$(ab)^{-1} = a^{-1}b^{-1},$$

and

$$(-a)b = -(ab).$$

They can be deduced from the commutative, associative and distributive properties; but since we are not interested in analyzing logical interrelations, we are content with merely recapitulating all properties of numbers which actually will be used.

All numbers other than zero can be divided into two kinds, the *positive* and *negative* ones. If a is positive, $-a$ is negative, and conversely. Of the two numbers a and $-a$, the positive one is called the *absolute value* of a and is denoted by $|a|$. The absolute value of 0 is defined to be 0.

The sum, product, and reciprocal of positive numbers are positive. We call this the *basic property* of positivity.

Comparing numbers can be as important as adding or multiplying them. We can define the relation *less than* as follows: a *is less than* b means that $(b - a)$ is positive ; this is written

$$a < b \quad \text{and also} \quad b > a.$$

The statement $b > a$ is read "b is greater than a" and means the same as "a is less than b."

The symbol

$$a \le b \quad \text{or} \quad b \ge a$$

means that a *is not greater than* b, i.e., either $a < b$ or $a = b$.

Inequalities play a central role in calculus, and it is important to learn to handle them with ease. The following elementary rules will be used over and over:

1. If $a < b$ and $b < c$, then $a < c$.
   This is called the *transitivity* of $<$.
2. If $a < b$ and $c < d$, then $a + c < b + d$.
   This is called *adding inequalities*.

3. If $a < b$ and p is positive, then $pa < pb$.

   This is called *multiplying an inequality by a positive number*

4. If a and b are positive numbers and $a < b$, then $1/a > 1/b$.

   This is called *inverting an inequality.*

These rules for combining inequalities all follow from the basic property of positivity. Transitivity follows from the observation that $c - a$, being the sum of two positive numbers $b - a$ and $c - b$ is positive. The addition of inequalities holds since $(b + d) - (a + c)$ is the sum of $b - a$ and $d - c$; and so, being the sum of positive numbers, is positive. The rule for multiplying inequalities rests on the observation that $pb - pa = p(b - a)$, the product of two positive numbers, is positive. Finally the rule for inverting inequalities is obtained by multiplying $a < b$ by the positive number $1/ab$.

Two inequalities are called equivalent if each implies the other. For example, the inequalities

$$a < b \quad \text{and} \quad a + c < b + c$$

are equivalent. Similarly

$$a < b \quad \text{and} \quad pa < pb, \quad p > 0$$

are equivalent.

EXERCISES

1.1 Prove that if $a < b$ then $-a > -b$.

1.2 (a) Prove that the square of any number cannot be negative.

   (b) Prove that for any two numbers a and b

$$0 \le a^2 - 2ab + b^2.$$

   (c) Prove that for any two numbers a and b,

$$2ab \le a^2 + b^2.$$

1.3 Verify the statements in Exercises 1.1, 1.2(b), and 1.2(c) by setting

   (a) $a = 3, \quad b = -5$.

   (b) $a = -7, \quad b = -2$.

1.4 (a) For what value of m is it true that

$$\left(\frac{1}{2}\right)^m < 10^{-4}?$$

   (b) How large must m be in order that

$$\left(\frac{1}{2}\right)^m < 10^{-k}?$$

1.5 Prove the following inequalities:

(a) $\dfrac{1}{a} \geq \dfrac{1}{b}$ if $b \geq a > 0$

(b) $a^2 < b^2$ if $b > a > 0$

(c) $a + a^{-1} \geq 2, \quad a > 0$

(d) $\dfrac{a+b}{2} \geq (ab)^{1/2}, \quad a, b > 0$

(e) $a^2 \geq 2a - 1$.

1.6 Which of the following inequalities are valid for *all* a, b satisfying $0 < a < b < 1$?

(a) $ab > 1$

(b) $\dfrac{1}{a} < \dfrac{1}{b}$

(c) $\dfrac{1}{b} > 1$

(d) $a + b < 1$

(e) $a + b > 1$

(f) $a - b < 1$

(g) $a^2 < 1$

(h) $a^2 + b^2 < 1$

(i) $a^2 + b^2 > 1$

(j) $a^2 + b^2 > 2$

(k) $|a - b| < 1$

(l) $|b - a| > 1$.

1.2 The number line

In this section we describe a geometrical way of looking at numbers. We represent numbers as points on a line (see Figure 1.1). The point marked O is called the *origin*.

Figure 1.1

The importance of the number line is that it makes it easy to visualize certain relations between numbers and certain operations with them. For example, adding a and b can be described as follows: Shift the number line so that the point O is moved to where a was originally. The point b is thereby moved to where $a + b$ was originally (see Figure 1.2).

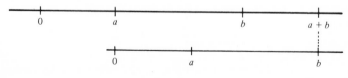

Figure 1.2

The natural geometrical representation of a product is as an area.

Inequalities are particularly easy to visualize on the number line: $a < b$ means that *a is to the left of b*. In particular, all negative numbers are to the left of the origin, all positive numbers to the right.

The *negative of a number* is obtained by reflecting it across the origin; thus, if a number a is represented by point A (see Figure 1.3, where $a < 0$), its mirror image A' (on the other side of, and as far from O) represents $-a$.

Figure 1.3

A number x lies *between* two numbers a and b if it is greater than one of them and less than the other. This relation is easily visualized on the number line; a and b are end-points of an interval containing x in its interior (see Figure 1.4, where $a < b$, so that $a < x < b$).

Figure 1.4

Suppose $a < b$. The *open interval* (a, b) is defined as the set of all points which lie between a and b, i.e., all x which satisfy

$$a < x < b.$$

The *closed interval* $[a, b]$ consists of the open interval plus the endpoints, i.e., of all x which satisfy

$$a \leq x \leq b.$$

Half-open (or *half-closed*) intervals are defined similarly and are denoted by $(a, b]$ and $[a, b)$.

The *absolute value* of a number is its distance from the origin O. The distance of b from a is $|b - a|$. Hence the *length* of an interval with endpoints a and b, open or closed, is defined as $|b - a|$. It is geometrically evident that, *if two numbers x and y lie in an interval of length l, then their distance cannot exceed l*:

$$|x - y| \leq l.$$

The following inequality, called the *triangle inequality*, is as important as it is simple:

$$|x + y| \leq |x| + |y|.$$

As the reader may easily convince himself, the sign of equality holds only when x and y are both positive, both negative, or when one of them is 0.

We recall the notion of *rational number*, defined as a quotient of two integers:

$$r = \frac{p}{q}, \quad q \neq 0.$$

These can be easily visualized on the number line: fix q as any positive integer and let p run through all the integers; the numbers p/q then divide the number line into intervals of length $1/q$. Each real number x lies in one of these intervals. Therefore given any positive integer q, for each real number x there is an integer p such that

$$\frac{p}{q} \leq x < \frac{p+1}{q}.$$

These inequalities can be rewritten in the equivalent form

$$\frac{p}{q} \leq x < \frac{p}{q} + \frac{1}{q}.$$

If we subtract p/q from all three members, we obtain

$$0 \leq x - \frac{p}{q} < \frac{1}{q},$$

so

$$\left| x - \frac{p}{q} \right| < \frac{1}{q}.$$

This inequality shows that every real number can be approximated arbitrarily closely by rational ones; for this reason the rationals are said to be *dense* on the number line. Yet it is an astounding fact of mathematics, discovered by the Greeks, that there are numbers which are not rational. It can be shown, for example, that there is no rational number whose square is 2. (A proof is given in Appendix 1.1.) On the other hand, there is a real number, denoted by $\sqrt{2}$, whose square is 2; namely the length of a diagonal of the unit square (see Figure 1.5).

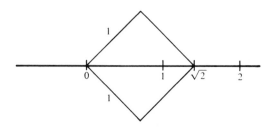

Figure 1.5

We present now another argument, not based on Euclidean geometry, in support of the existence of a real number whose square is two. Divide all positive rational numbers into two classes: into L go all those rationals whose square is less than 2, into U all those whose square is greater than 2. Clearly every number in L is less than any number in U. If the number line has no holes then there ought to be exactly one number larger than all numbers in L and less than all numbers in U. In Section 1.4 we shall formulate rigorously what it means that the number line has no holes.

EXERCISES

2.1 Prove that $|a| = \sqrt{a^2}$. [*Hint*: Show that the difference $|a|^2 - a^2$ vanishes.]

2.2 Prove that $|-a| = a$ if $a \geq 0$.

2.3 Prove that $|a/b| = |a|/|b|$.

2.4 (a) Prove that for any 3 numbers a, b, c,

$$|a + b + c| \leq |a| + |b| + |c|.$$

(b) Prove that for any finite set of n numbers $a_1, \ldots, a_n$,

$$|a_1 + \cdots + a_n| \leq |a_1| + \cdots + |a_n|.$$

[*Hint*: First prove the corresponding inequality for 4 numbers. Then show that, if it holds for k numbers, it holds for $k + 1$ numbers.]

2.5 Prove that $|ab| = |a||b|$.

2.6 Prove that if $a > b, c > 0$, then

$$ca > cb.$$

[*Hint*: Instead of an inequality of the form $x > y$, use the equivalent inequality $x - y > 0$.]

2.7 Modify the above inequality as well as your proof in Exercise 2.6 for the case where $c < 0$.

1.3 Infinite decimals

We start by reviewing the process of constructing the decimal expansion of an arbitrary number a. At the end of this section we shall take the position that a real number is *defined* as an infinite decimal.

The integers divide the number line into infinitely many half-open intervals $[n, n + 1)$, each of unit length. These intervals are mutually exclusive, i.e., each number belongs to exactly one of these intervals; in order to achieve this exclusiveness we have included the left endpoint n as part of the interval $[n, n + 1)$ but excluded the right endpoint $n + 1$.

7

Suppose the number a is *positive* and belongs to some interval $[n, n + 1)$; then the *integer part* of the decimal expansion of a is n, and the fractional part is $.a_1 a_2 a_3 \dots$:

$$a = n.a_1 a_2 a_3 \dots \ .$$

To determine the digits a_1, a_2 etc. after the decimal point, we divide the interval $[n, n + 1)$ into ten mutually exclusive half-open intervals, each of length $\frac{1}{10}$ (see Figure 1.6). As before we agree to count the left endpoint but

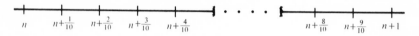

Figure 1.6

not the right as part of each interval, to achieve exclusiveness. Our number a belongs to exactly one of these ten intervals, say to

$$\left[n + \frac{a_1}{10}, n + \frac{a_1 + 1}{10} \right).$$

This determines the first digit a_1 of a. The second digit a_2 is determined similarly, by subdividing the above interval into 10 equal subintervals, etc. Thus using the representation of a number a as a point on the number line and the procedure just described, we can find a_k in $a = n.a_1 a_2 \dots a_k \dots$ by determining the appropriate half-open interval in the k-th step of this process.

At this point we free ourselves of any intuitive geometrical background and simply say that *real numbers are infinite decimals*. We point out that the infinite decimal representation for real numbers extends positional notation from integers to all real numbers. For example,

$$749 = 9 \cdot 10^0 + 4 \cdot 10^1 + 7 \cdot 10^2,$$

and

$$0.749 = 7 \cdot 10^{-1} + 4 \cdot 10^{-2} + 9 \cdot 10^{-3}.$$

The importance of the infinite decimal representation of real numbers lies in the ease with which numbers can be compared. It is instructive to look at the following questions: Which of the two numbers

$$\frac{17}{20} \quad \text{and} \quad \frac{45}{53}$$

is larger? How close are they? Only people with uncanny arithmetic ability can answer these without going through the laborious process of bringing the two numbers to a common denominator and writing them as

$$\frac{901}{1060} \quad \text{and} \quad \frac{900}{1060};$$

now we can tell that the first one is bigger than the second, but not by much, by less than a thousandth. Suppose we represent the two numbers as decimals:

$$\frac{17}{20} = 0.85000$$

$$\frac{45}{53} = 0.84905\ldots.$$

Now a glance discloses that they have the same first digit, and that the second digit of $\frac{17}{20}$ is greater than the second digit of $\frac{45}{53}$; therefore $\frac{17}{20}$ is greater than $\frac{45}{53}$.

Comparing the digits of the two numbers and noting at which point they begin to disagree tells us which is bigger. We shall see that it also tells us by how much. At this point we may conclude: *If a and b have the same integer parts and the same digits up to the m-th, then they differ by less than* 10^{-m},

$$|a - b| < 10^{-m}.$$

The converse is *not* true; e.g., in the previous example, $a = \frac{17}{20}$ and $b = \frac{45}{53}$ differ by less than 10^{-3}, yet only 1 and not 3 of their digits are equal. An even more extreme example is the following pair of numbers:

$$a = 0.29999\ldots \quad \text{and} \quad b = 0.30000\ldots.$$

Clearly, whatever the digits of a and b are beyond the fifth, a and b differ by less than 2×10^{-5}. It follows from the inequalities

$$0.299990 \leq a \leq b \leq 0.300010$$

that $b - a \leq 0.00002 = 2 \cdot 10^{-5}$. Thus the numbers a and b are very close. Nevertheless, the first 5 digits of a and b are different. Thus our digit comparison immediately answers the question "which number is bigger?," but gives a somewhat misleading answer to the question "by how much"?

The reason why two numbers may have different digits, yet be very close, lies in the manner of constructing the digits. If a and b lie on opposite sides of the point $p/10^k$, p an integer (see Figure 1.7), then the k-th digits of a and b are different no matter how close a and b are to each other.

Figure 1.7

How, then, do we tell how close two numbers are by looking at their digits? Given any number a, denote by $a_{(m)}$ the number with first m digits identical to those of a and with all other digits zero; $a_{(m)}$ is called the *truncation* of a to m digits. The process of truncating is also called *rounding down*. The number $a_{(m)}$ is the lower endpoint of the interval of length 10^{-m} containing a. The upper endpoint of this interval, denoted by $a^{(m)}$, is

$$a^{(m)} = a_{(m)} + 10^{-m}.$$

The process which assigns $a^{(m)}$ to a is called *rounding up*. It clearly follows from the construction of the decimal expansion of a that

$$a_{(m)} \le a < a^{(m)}.$$

Since the interval $[a_{(m)}, a^{(m)})$ has length 10^{-m}, *both* $a_{(m)}$ *and* $a^{(m)}$ *approximate a within* 10^{-m}:

(3.1) $$|a - a_{(m)}| \le 10^{-m}, \qquad |a - a^{(m)}| \le 10^{-m}.$$

When comparing two numbers a and b, we can now tell if they lie within $2 \cdot 10^{-m}$ of each other by comparing the approximations $a_{(m)}$, $a^{(m)}$ of a with those of b:

If one of the approximations $a_{(m)}$, $a^{(m)}$ *of a is equal to one of the approximations* $b_{(m)}$, $b^{(m)}$ *of b, then*

$$|b - a| \le 2 \cdot 10^{-m}.$$

PROOF. This statement, translated into geometry, says: if at least one of the endpoints of the interval of length 10^{-m} containing a coincides with at least one of the endpoints of the interval of length 10^{-m} containing b, then a and b lie in the same or in adjacent intervals of length 10^{-m}. Hence $|b - a| \le 2 \cdot 10^{-m}$. An analytic proof consists of using the triangle inequality, see Exercise 3.4. □

On the other hand, *if a and b differ by less than* 10^{-m}, *then one of the numbers* $a_{(m)}$, $a^{(m)}$ *is equal to one of the numbers* $b_{(m)}$, $b^{(m)}$.

PROOF. Indeed, if $|b - a| \le 10^{-m}$ the intervals $[a_{(m)}, a^{(m)})$ and $[b_{(m)}, b^{(m)})$ containing a and b, respectively, are either identical or adjacent. In either case, they have at least one endpoint in common. □

Examples

$$a = 0.2999937, \qquad a_{(5)} = 0.29999, \qquad a^{(5)} = 0.30000$$

$$b = 0.3000078, \qquad b_{(5)} = 0.30000, \qquad b^{(5)} = 0.30001$$

$$c = 0.7919928, \qquad c_{(4)} = 0.7919, \qquad c^{(4)} = 0.7920.$$

The approximations $a^{(5)}$ and $b_{(5)}$ in the example above are identical. According to inequalities (3.1) governing these approximations, $a^{(5)}$ differs from a by $\le 10^{-5}$, $b_{(5)}$ differs from b by $\le 10^{-5}$. Therefore a and b differ by $\le 2 \cdot 10^{-5}$. Of the two approximations $a_{(m)}$ and $a^{(m)}$, one lies closer to a than the other. To determine which, we look at the $(m + 1)$-st digit of a; if it is under 5, $a_{(m)}$ is the m-digit number closest to a; if the $(m + 1)$-st digit is 5 or greater, $a^{(m)}$ is the closest approximation to a.

The construction of the m-digit number closest to a is called *rounding*; the better approximation is called *a rounded to m digits*.

Example. The following table gives the values of a rounded to m digits, $m = 1, \ldots, 7$, where $a = 0.35997074$

m	a
1	0.4
2	0.36
3	0.360
4	0.3600
5	0.35997
6	0.359971
7	0.3599707

EXERCISES

3.1 Round the following numbers to m digits, $m = 1, \ldots, 5$:
   (a) 0.599708
   (b) 0.4952970
   (c) 0.0994972.

3.2 What is the largest value of m for which a and b agree up to m digits?
   (a) $a = 0.35497\ldots,$ $b = 0.35508\ldots$
   (b) $a = 0.23489\ldots,$ $b = 0.24978\ldots$
   (c) $a = 0.314159\ldots,$ $b = 0.31438\ldots$
   (d) $a = 0.289937\ldots,$ $b = 0.29002\ldots$.

3.3 Prove that the following operation produces a rounded to m digits: Add $\frac{1}{2} 10^{-m}$ to a, and round down the result to m digits.

3.4 Prove with the aid of the triangle inequality that $|b - a| \leq 2 \cdot 10^{-m}$ if $a_{(m)} = b^{(m)}$.

1.4 Convergent sequences

A sequence of numbers is, as the word indicates, one number after the other, such as

$$1, 2, 3, \ldots, n, \ldots$$

or

$$1, \frac{1}{2}, \frac{1}{4}, \ldots, \left(\frac{1}{2}\right)^n, \ldots .$$

It is customary to denote the n-th term of a sequence by a symbol with subscript n, such as a_n, in

$$a_1, a_2, \ldots, a_n, \ldots .$$

The dots indicate that the sequence goes on and on.

11

We shall study very particular kinds of sequences, called *convergent* sequences. Roughly speaking, these have the property that as n gets larger, the terms a_n get arbitrarily close to some number a, called the *limit* of the sequence. This is expressed symbolically by

$$\lim_{n \to \infty} a_n = a.$$

Here is a modest example of a convergent sequence:

$$a_1 = 0.30000\ldots$$
$$a_2 = 0.33000\ldots$$
$$\cdots\cdots\cdots\cdots$$
$$a_n = 0.\underbrace{3\ldots3}_{n\ 3\text{'s}}\underbrace{00}_{\text{all 0's}}\ldots$$

Clearly the limit of this sequence is $a = \frac{1}{3}$.

We now give a precise

Definition of convergence. A sequence a_n converges to the limit a if, no matter how large we choose m, there is an integer N such that

$$|a_n - a| < 10^{-m} \qquad \text{for } n > N;$$

N depends on m.

In the example just given, the first m digits of a_n and a are the same. For instance, suppose m is given as 7; all numbers beyond $a_7 = 0.3333333$ are within 10^{-7} of $a = \frac{1}{3}$. Similarly, for any given m, we may take $N = m$. We now present more examples of convergent sequences.

Example 1. $a_n = 1/n$; this sequence converges to zero. For any arbitrary given value of m, let $N = 10^m$; then $a_N = 1/N = 10^{-m}$, and for $n > N$, all digits of a_n up to the m-th are zero.

Example 2. $a_n = (\frac{1}{2})^n$. We claim that this sequence converges to zero; for

$$\left(\frac{1}{2}\right)^4 = \frac{1}{16} < 10^{-1};$$

therefore

$$\left(\frac{1}{2}\right)^{4m} < 10^{-m},$$

i.e.,

$$a_n < 10^{-m} \qquad \text{for } n > N = 4m.$$

Notice that N in Example 2 is much smaller than in Example 1; this means that the seond sequence converges faster than the first.

Example 3. $a_n = r^n$. We claim that this sequence tends to zero if the absolute value of r is less than 1. We present a proof for the case when $|r| < \frac{1}{2}$. By the result of the previous example

$$|a_n| = |r|^n < \left(\frac{1}{2}\right)^n < 10^{-m} \qquad \text{for } n > N = 4m.$$

A proof valid for any r, $|r| < 1$, will be given at the end of Section 1.6.

In all the examples of convergent sequences we have presented so far, we were able to guess what the limit would be. This is not typical. The most important convergent sequences arise as better and better approximations to quantities which cannot be guessed. Solutions of equations, maxima and minima of functions, derivatives, integrals, and solutions of differential equations are all quantities of this kind. The study and calculation of such quantities is the central problem of calculus; therefore the reader who wishes to master the subject must get used to dealing with convergent sequences. We now present two interesting examples of convergent sequences, typical of the kind one encounters in calculus.

Example 4. We have remarked earlier that $\sqrt{2}$ is not rational. We now construct a rapidly converging sequence of approximations to $\sqrt{2}$. The construction is based on two ideas:

(i) that the average $\frac{1}{2}(u + v)$ of two numbers u and v lies between them, and
(ii) that, if $\sqrt{2} < u$, then $v = 2/u < \sqrt{2}$, and conversely. To see this, note that $\sqrt{2} < u$ implies $2 < u^2$, which implies $2^2 < 2u^2$, which implies $2^2/u^2 < 2$, hence $2/u < \sqrt{2}$. The converse holds because the direction of each implication can be reversed.

Combining these two ideas, we show next that, if u_0 is an approximation to $\sqrt{2}$, then

$$u_1 = \frac{1}{2}\left(u_0 + \frac{2}{u_0}\right)$$

is a better approximation; that is, u_1 differs less from $\sqrt{2}$ than u_0 did. We compute

$$u_1 - \sqrt{2} = \frac{1}{2}\left(u_0 + \frac{2}{u_0}\right) - \sqrt{2} = \frac{1}{2u_0}\left[u_0^2 + 2 - 2u_0\sqrt{2}\right].$$

The expression in brackets is a perfect square, so we may write

(4.1) $$u_1 - \sqrt{2} = \frac{1}{2u_0}(u_0 - \sqrt{2})^2.$$

13

From this we conclude

(i) u_1 is always an overestimate of $\sqrt{2}$, since $u_1 - \sqrt{2}$ is positive (provided our first guess, u_0, was positive).

(ii) $u_1 - \sqrt{2} \leq \frac{1}{2}|u_0 - \sqrt{2}|$ because $u_0 > |u_0 - \sqrt{2}|$, and so

$$\frac{u_0}{|u_0 - \sqrt{2}|} > 1.$$

If we multiply the right member of (4.1) by

$$\frac{u_0}{|u_0 - \sqrt{2}|},$$

we increase it and obtain the inequality (ii), which says that the deviation from $\sqrt{2}$ of an approximation $u_1 = (u_0 + 2/u_0)/2$ is at most half the deviation from $\sqrt{2}$ of the previous approximation.

It follows that repeated application of step (4.1) leads to the sequence $u_0, u_1, u_2, \ldots$, where $u_0 > 0$ is arbitrary and

(4.2) $$u_{n+1} = \frac{1}{2}\left(u_n + \frac{2}{u_n}\right).$$

This sequence converges to $\sqrt{2}$ because

$$u_{n+1} - \sqrt{2} \leq \frac{1}{2}|u_n - \sqrt{2}| \leq \frac{1}{2^2}|u_{n-1} - \sqrt{2}| \leq \cdots \leq \frac{1}{2^{n-1}}|u_0 - \sqrt{2}|,$$

and for n sufficiently large, $|u_0 - \sqrt{2}|/2^{n+1}$ can be made as small as we please; that is, *the sequence $u_0, u_1, \ldots$ defined by (4.2) converges to $\sqrt{2}$.*

Let's see how well this works in practice. Starting with $u_0 = 2$, we get

$$u_0 = 2.0$$
$$u_1 = 1.5$$
$$u_2 = 1.4166\ldots$$
$$u_3 = 1.41421566\ldots$$
$$u_4 = 1.41421356\ldots$$
$$u_5 = 1.41421356\ldots .$$

If we could judge the convergence of a sequence from the behavior of the first six terms (which we never can) we would be strongly tempted to conclude just from the numerical evidence that the sequence converges to a number u whose first 8 digits are

$$u \simeq 1.41421356 .$$

The square of this number is

$$u^2 \simeq 2.00000005,$$

gratifyingly close to 2.

In the last part of Exercise 4.1, the reader is asked to approximate $\sqrt{3}$ in a similar manner, and to show that those approximations converge. In Section 3.10, it will be shown that these methods are special cases of a powerful method for finding zeros of functions.

Example 5. Consider the following sequence of numbers e_n, defined by

(4.3)
$$e_n = \left(1 + \frac{1}{n}\right)^n, \qquad n = 1, 2, \ldots .$$

The first 8 members of this sequence are, correct to 3 decimals,

$$e_1 = 2.0$$
$$e_2 = 2.250$$
$$e_3 = 2.370$$
$$e_4 = 2.441$$
$$e_5 = 2.488$$
$$e_6 = 2.521$$
$$e_7 = 2.546$$
$$e_8 = 2.565 .$$

Gazing at these numbers we can say only that they seem to be getting bigger, but it is premature to hazard a guess whether they converge to a limit. If only we could look at e_n for larger values of n, like $n = 1000$. Fortunately we can, with the aid of a hand calculator, provided we carry out the raising of numbers to a high power intelligently. We can raise any number to the fourth power by squaring it twice:

$$x^4 = (x^2)^2.$$

Similarly the eighth power can be evaluated by three squarings,

$$x^8 = ((x^2)^2)^2,$$

and in general, if $n = 2^k$, x^n can be evaluated by performing k squarings. So we evaluate, without excessive labor, the following additional members of the sequence, correct to 3 decimals:

$$e_{16} = 2.638$$
$$e_{64} = 2.698$$
$$e_{256} = 2.712$$
$$e_{1024} = 2.717 .$$

Looking at these numbers we hazard the guess that the sequence converges, and that the first 2 digits of the limit are

$$2.71 .$$

In Chapter 5 we shall give a rigorous proof of the convergence of the sequence (4.3) and explain its importance.

The trouble with the definition of convergence (p. 12) is that it requires knowledge of the limit a of the sequence $\{a_n\}$, whereas in all interesting cases, such as Examples 4 and 5, precisely this information is lacking. Fortunately, to tell whether a sequence converges or not, it suffices to compare members of the sequence with each other. We claim:

Convergence criterion. *A sequence a_n converges if, no matter how large we choose m,*

$$|a_n - a_k| < 10^{-m}$$

for n and k both $> N$. N depends on m.

To establish our claim, we have to show that a sequence satisfying the convergence criterion converges to a limit a. We shall construct this limit digit by digit.

For the sake of making the discussion simple, suppose that all members of the sequence lie between 0 and 1. If, from a certain point on, every member a_n of the sequence has the same first digit, this common digit is taken to be the first digit of a. In symbols: if, for $n > N_1$, all a_n begin with $.d_1$, i.e., if

$$a_{N_1+1} = .d_1 \ldots, a_{N_1+2} = .d_1 \ldots, \ldots, a_n = .d_1 \ldots, \quad \text{for all } n > N_1,$$

d_1 shall be the first digit of a. Similarly, if there is a point beyond which all a_n have the same second digit, i.e., if $a_j = .d_1 d_2 \ldots$ for all $j > N_2$, d_2 shall be the second digit of a. Subsequent digits are constructed analogously; and if this construction works for all digits, then the limit a, for any given m, has the same first m digits as all a_k with $k > N_m$. Hence $|a_k - a| < 10^{-m}$ for $k > N_m$, so the sequence converges to a by the definition of convergence.

Suppose now that the construction described above breaks down at some point, say at the p-th digit. This means that the members a_n of the sequence, from a certain point on, do *not* have the same p-th digit *no matter how far in the sequence we go*. In symbols: For every N, no matter how large, the a_n with $n > N$ do *not* all have the same p-th digit. This means that at least two of the numerals among the candidates $0, 1, 2, \ldots 8, 9$ must occur as the p-th digit in infinitely many of the a_n; for, if only one of them, say d_p, occurred infinitely often, then we could go so far out in the series that the finite number of members with numerals other than d_p in their p-th place have all been passed at some point, and we are then in the situation where, for $n > N_p$, all a_n have d_p as their p-th digit and our construction would work.

Now, among the numericals that occur infinitely often as p-th digit of the a_n, take the largest, and denote it by l. We show next that the convergence criterion dictates that the limit a of the sequence has l as its p-th digit, whereas from the $(p + 1)$-st digit on all entries are zero. For, given any m, there is an N such that $|a_n - a_k| < 10^{-m}$ for all $n, k > N$. On the other hand, we can choose an a_n with l as its p-th digit and an a_k with a smaller numeral as its p-th digit (since there are infinitely many of both kinds), so that

$$a_k \le .d_1 d_2 \ldots d_{p-1} l \le a_n;$$

hence

$$0 \le .d_1 d_2 \ldots d_{p-1} l - a_k \le a_n - a_k,$$

and

$$0 \le a_n - .d_1 d_2 \ldots d_{p-1} l \le a_n - a_k.$$

But since $|a_n - a_k| < 10^{-m}$, we may conclude that

$$|a_k - .d_1 d_2 \ldots d_{p-1} l| < 10^{-m} \quad \text{and} \quad |a_n - .d_1 d_2 \ldots d_{p-1} l| < 10^{-m},$$

for $n, k > N$. This, according to the definition of convergence, shows that the a_n tend to $a = .d_1 d_2 \ldots d_{p-1} l$. $\square$

We close this section by stating and proving the *rules of arithmetic for convergent sequences*:

Suppose that $\{a_n\}$ and $\{b_n\}$ are convergent sequences,

$$\lim a_n = a, \qquad \lim b_n = b.$$

Then the sequences $\{a_n + b_n\}$ and $\{a_n b_n\}$ also converge, and

(i) $\lim(a_n + b_n) = a + b$
(ii) $\lim a_n b_n = ab$.

Furthermore, if a is not zero, then all but a finite number of the a_n also differ from zero, and

(iii) $\lim \dfrac{1}{a_n} = \dfrac{1}{a}$.

We remark that these rules are nearly self-evident. They assert that, if the numbers a_n and b_n are close to the numbers a and b, then their sum, product, and reciprocal are close to the sum, product, and reciprocal of a and b themselves. If this weren't so, we would be unable to do any arithmetic at all; for we never actually operate with a and b, only with approximations to them.

We present a formal proof, not so much because we think that the reader needs convincing, but as an exercise in proving things about convergent sequences.

PROOF. Since a_n and b_n are both convergent, given any m,

(4.5) $\qquad |a_n - a| < 10^{-m}, \qquad |b_n - b| < 10^{-m} \qquad$ for $n > N$.

By the triangle inequality,

$$|a_n + b_n - a - b| \leq |a_n - a| + |b_n - b| \leq 10^{-m} + 10^{-m}$$
$$= 2 \cdot 10^{-m} < 10 \cdot 10^{-m} = 10^{-m+1}.$$

This proves (i).

The proof of (ii) is simplest in the case that a and b are less than 1 in absolute value. Then the same holds for a_n and b_n for n large enough.

(4.6) $$|a_n| < 1, \qquad |b_n| < 1, \qquad \text{for } n > N.$$

Then

$$a_n b_n - ab = a_n b_n - a_n b + a_n b - ab$$
$$= a_n(b_n - b) + (a_n - a)b.$$

By the triangle inequality

$$|a_n b_n - ab| \leq |a_n(b_n - b)| + |(a_n - a)b|,$$

Using (4.5) and (4.6) we get

$$|a_n b_n - ab| \leq 10^{-m} + 10^{-m} < 10^{-m+1};$$

this proves (ii) in the case $|a| < 1$, $|b| < 1$. The proof in the general case is not more difficult; it only takes a little more writing (see Exercise 4.7).

To prove (iii) we assume for convenience that $|a|$ is greater than 1; then so is $|a_n|$ for n large enough. Then

$$\frac{1}{a_n} - \frac{1}{a} = \frac{a - a_n}{a_n a},$$

so that

$$\left| \frac{1}{a_n} - \frac{1}{a} \right| = \frac{|a - a_n|}{|a_n||a|} < |a_n - a| < 10^{-m}.$$

This proves (iii) for $|a| > 1$. The proof for any nonzero a is no more difficult, see Exercise 4.8. $\square$

EXERCISES

4.1 Define a sequene u_n by

$$u_1 = 2,$$

$$u_{n+1} = \frac{1}{2}\left(u_n + \frac{3}{u_n}\right), \qquad \text{for } n = 1, 2, \ldots.$$

Using a hand calculator calculate the first six terms of this sequence, carrying 8 digits. Do the numbers indicate that the sequence converges? Is u_6 a good approximation to $\sqrt{3}$? Using the convergence proof for approximations to $\sqrt{2}$ in the text as a model, can you prove the convergence of the above approximations to $\sqrt{3}$?

4.2 Assume that the sequence u_n defined in Exercise 4.1 converges to the limit u which is $\neq 0$. Using the relation

$$u_{n+1} = \frac{1}{2}\left(u_n + \frac{3}{u_n}\right),$$

and the arithmetic of convergent series, prove that u satisfies

$$u = \frac{1}{2}\left(u + \frac{3}{u}\right),$$

which implies that

$$u^2 = 3.$$

4.3 Define the sequence v_n by

$$v_1 = -2,$$

$$v_{n+1} = \frac{1}{2}\left(v_n + \frac{3}{v_n}\right), \quad \text{for } n = 1, 2, \dots .$$

Calculate the first 6 terms of this sequence.
(a) Does the sequence seem to converge?
(b) Is v_6 a good approximation to $\sqrt{3}$?
(c) What seems to be the relation between v_n of Exercise 4.3 and u_n of Exercise 4.1? Can you prove your surmise?

4.4 (a) How would you calculate x^5 by performing only 3 multiplications?
(b) Can you calculate x^9 by performing only 4 multiplications?
(c) How many multiplications do you need to calculate x^{2^n}?

4.5 Prove that if $a_n \to a$, then $ka_n \to ka$, k any number.

4.6 Suppose $\lim_{n \to \infty} a_n = a$, and the corresponding digits of a_n and a disagree for all $n > N$. Show then that all but a finite number of digits of a are zero.

4.7 Prove statement (ii) of the rules of arithmetic for convergent sequences without the restriction $|a| < 1$, $|b| < 1$. [*Hint*: Use the fact that all but a finite number of a_n, b_n satisfy $|a_n| < C$, $|b_n| < C$, where C is some constant larger than a and b.]

4.8 Prove statement (iii) of the rules of arithmetic for convergent sequences without the restriction that $|a| > 1$. [*Hint*: Since $a \neq 0$, there is an integer k such that $\bar{a} = 10^k a > 1$; apply the proof on p. 18 to $\bar{a}_n = 10^k a_n$ and $\bar{a}$, but choose N so large that $|\bar{a}_n - \bar{a}| < 10^{-m-k}$ for $n > N$.]

4.9 The Fibonacci sequence $\{F_n\}$ is defined by

(4.7) $F_0 = 0, \quad F_1 = 1, \quad F_k = F_{k-2} + F_{k-1}, \quad k = 2, 3, \dots .$

(a) Write down the first six terms of this sequence.

(b) Denote by r_k the quotient

$$r_k = \frac{F_k}{F_{k-1}}, \qquad k = 2, 3, \dots .$$

Using (4.7) we see that the ratio r_{k+1} is related to r_k as follows:

(4.8)
$$r_{k+1} = \frac{F_{k+1}}{F_k} = \frac{F_k + F_{k-1}}{F_k} = 1 + \frac{1}{r_k}.$$

Write a computer program to evaluate these ratios, printing the value of r_p for $p = 10, 20, 30,$ and 40.

(c) Assume that r_n tends to a limit $r \neq 0$. Using (4.8) and the rules of arithmetic for convergent sequences, show that $r = 1 + 1/r$, and therefore

$$r = \frac{1}{2}(1 + \sqrt{5}).$$

Does the numerical evidence in (b) support this contention?

1.5* Infinite sums

We start with the best known and most beloved of sums, the *finite geometric series*

(5.1)
$$1 + x + x^2 + x^3 + \cdots + x^{n-1}.$$

We claim that the value of this sum of n terms is

$$\frac{1 - x^n}{1 - x}.$$

For, denote the sum (5.1) by a_n; the sum

$$xa_n = x + x^2 + \cdots + x^n$$

consists of the same terms as a_n except that the first term of a_n is missing, and there is the extra term, x^n; so

$$xa_n = a_n - 1 + x^n$$

from which we get, for $x \neq 1$, that

(5.2)
$$a_n = \frac{1 - x^n}{1 - x}.$$

We have shown, in Example 3 of Section 1.4, that if $|x| < \frac{1}{2}$ then x^n tends to zero. We shall show in Section 1.6 that x^n tends to zero whenever $x < 1$. It follows then that the sequence a_n, given by (5.2), tends to $1/(1 - x)$. We shall write this relation in the suggestive form of an *infinite* sum

(5.3)
$$1 + x + x^2 + \cdots = \frac{1}{1 - x} \qquad \text{for } |x| < 1.$$

* This section may be omitted at the first reading.

The meaning of any kind of infinite sum is defined analogously. Let $b_1, b_2, \ldots$ be an infinite sequence of numbers; the infinite sum, traditionally called *infinite series*,

(5.4)
$$b_1 + b_2 + \cdots$$

is called *convergent* if the sequence a_n of finite sums (also called *partial sums*)

$$a_n = b_1 + b_2 + \cdots + b_n$$

converges to a limit. That limit is called the *value of the infinite sum* (5.4). An infinite series which does not converge is called *divergent*.

Since we shall deal with quite a few infinite and finite series, it pays to introduce a compact notation. The *summation* sign, $\sum$, is defined as follows:

$$b_1 + b_2 + \cdots + b_n = \sum_{j=1}^{n} b_j.$$

The letter j is called the summation index; we shall omit the summation index under the $\sum$ if this causes no ambiguity; thus we shall write

$$\sum_{1}^{n} b_j = b_1 + b_2 + \cdots + b_n.$$

Moreover, when dealing with an infinite series, it is convenient to write

$$\sum_{1}^{\infty} b_j = b_1 + b_2 + \cdots.$$

If the series converges to a limit a, then

$$a = \lim_{n \to \infty} \sum_{1}^{n} b_j = \sum_{1}^{\infty} b_j.$$

In terms of this notation the infinite series (5.3) can be written as

(5.5)
$$\sum_{0}^{\infty} x^j \quad \text{or as} \quad \sum_{1}^{\infty} x^{j-1}.$$

We have proved the convergence of the infinite series (5.5) by explicitly evaluating the finite sums (5.1), and by explicitly determining the limit of the sequence of these sums. Clearly, this method works only for infinite series whose partial sums can be written in a form that exhibits the convergence of the sequence a_n. In our example, this form was (5.3). However, the most useful infinite series have partial sums that cannot be so expressed. Therefore it is very important to develop criteria which can be used to study the convergence of such series.

Convergence of the series

$$\sum_{1}^{\infty} b_j$$

21

was defined to mean convergence of the sequence a_n of partial sums

(5.6)
$$a_n = \sum_1^n b_j.$$

Convergence of the sequence a_n means that, for N large enough, all members of the sequence beyond the N-th are within 10^{-m} of the limit. By the convergence criterion for sequences, this can also be expressed thus: For any given m, there is an N so that

$$|a_n - a_k| < 10^{-m} \quad \text{for all } n, k \text{ greater than } N.$$

Using the expression (5.6) for a_n in terms of b_j, and assuming that $k < n$, we can express the difference of a_n and a_k as

$$a_n - a_k = \sum_{k+1}^n b_j.$$

The convergence of the infinite series can therefore be expressed directly in terms of the b_j's:

No matter how large m,

(5.7)
$$\left| \sum_{k+1}^n b_j \right| < 10^{-m}$$

for all k, n greater than N, N some number depending on m.

Example 1. Consider the infinite repeating decimal $0.621621621\ldots$ which may be written

$$\frac{621}{10^3} + \frac{621}{10^6} + \cdots = 621 \cdot 10^{-3}[1 + 10^{-3} + 10^{-6} + \cdots].$$

The factor in brackets, by (5.3), converges to

$$a = 1 + 10^{-3} + 10^{-6} + \cdots = \frac{1}{1 - 10^{-3}} = \frac{1000}{999},$$

so the value of the decimal is $621/999 = 23/37$. The partial sums $a_1 = 1$, $a_2 = 1 + 10^{-3}, \ldots$, can be written, by (5.2), as

$$a_n = \frac{1 - 10^{-3n}}{1 - 10^{-3}}.$$

The difference between the n-th and k-th partial sums is

$$a_n - a_k = \sum_{j=k+1}^n 10^{-3(j-1)} = 10^{-3k} \sum_{j=1}^{n-k} 10^{-3(j-1)} = 10^{-3k} a_{n-k},$$

where a_{n-k} is the $(n-k)$-th partial sum given by the above formula with n replaced by $n-k$. Thus, for our example,

$$\sum_{k+1}^{n} b_j = a_n - a_k = \frac{1}{999}\left[\left(\frac{1}{10^3}\right)^{k-1} - \left(\frac{1}{10^3}\right)^{n-1}\right].$$

Since $1/999 < 10^{-2}$, and since $(10^{-3})^{n-1} < 10^{-3k+3}$, we have

$$\left|\sum_{k+1}^{n} b_j\right| < 10^{-m},$$

provided that $3k - 2 > m$, i.e., $k > \frac{1}{3}(m+2)$. This will be the case for all n, k greater than N, where N is any integer exceeding $\frac{1}{3}(m+2)$.

We present now a very simple and immensely useful criterion for the convergence of infinite series. It is called the

Comparison criterion. *Suppose $\sum_1^\infty b_j$ is a convergent infinite series all of whose terms are positive:*

$$0 < b_j, \quad j = 1, 2, \dots .$$

Suppose that each of the numbers c_j has absolute value less than b_j:

(5.8) $$|c_j| < b_j, \quad j = 1, 2, \dots .$$

Then the infinite series

$$\sum_1^\infty c_j$$

also converges.

Whenever inequality (5.8) holds we say that the series $\sum b_j$ *dominates* $\sum c_j$. The comparison criterion can now be expressed more succinctly:

Every infinite series dominated by a convergent series is itself convergent.

PROOF. The proof follows immediately; we have to show that, for any m,

(5.9) $$\left|\sum_{k+1}^{n} c_j\right| < 10^{-m},$$

provided that k and n are large enough. Applying repeatedly the triangle inequality, we get

$$\left|\sum_{k+1}^{n} c_j\right| \le \sum_{k+1}^{n} |c_j| .$$

According to (5.8), each $|c_j|$ is less than b_j; so

$$\sum_{k+1}^{n} |c_j| < \sum_{k+1}^{n} b_j.$$

23

Since $b_j > 0$, and the infinite series $\sum_1^\infty b_j$ is assumed to converge, by (5.7)

$$\left|\sum_{k+1}^n b_j\right| = \sum_{k+1}^n b_j < 10^{-m}$$

holds for $k, n > N$. But then, since we have already shown that

$$\left|\sum_{k+1}^n c_j\right| < \sum_{k+1}^n b_j,$$

also $|\sum_{k+1}^n c_j|$ is less than 10^{-m}. □

Remark. The proof remains valid if inequality (5.8) holds for all but a finite number of j, because (5.9) does not involve the first k terms of the series, and N may be taken so large that (5.8) holds for all $j > N$.

It is worth pointing out that we have shown not only that the infinite series $\sum c_j$ converges, but also that it converges as fast or faster than the series which dominates it. Our proof shows that, if N is chosen so that the finite sum $\sum_1^N b_j$ differs by less than 10^{-m} from the infinite sum $\sum_1^\infty b_j$, then likewise $\sum_1^N c_j$ differs by less than 10^{-m} from the infinite sum $\sum_1^\infty c_j$.

It takes some experience and ingenuity to hit upon an appropriate convergent series $\sum b_j$ with positive terms which dominates a given series $\sum c_j$. We now describe a general method for constructing convergent series with positive terms. We start with an increasing, convergent sequence $\{a_n\}$:

$$a_0 < a_1 < a_2 < \cdots, \quad \text{and} \quad \lim a_n = a.$$

We set

(5.10) $$b_j = a_j - a_{j-1};$$

clearly, since $\{a_j\}$ is increasing, each b_j is positive. The partial sums

(5.11) $$\sum_1^n b_j = (a_n - a_{n-1}) + (a_{n-1} - a_{n-2}) + \cdots + (a_1 - a_0)$$

telescope into the single difference

$$a_n - a_0$$

which converges to $a - a_0$. The next few examples illustrate the power of this idea; further examples are in the exercises and real life.

Example 2. Set

$$a_n = 1 - \frac{1}{n+1};$$

clearly, this is an increasing sequence, tending to 1. Set

$$b_j = a_j - a_{j-1} = \frac{1}{j} - \frac{1}{j+1} = \frac{1}{j(j+1)};$$

we conclude that the series

$$\sum \frac{1}{j(j+1)}$$

converges.

Note that, if $\sum b_j$ converges to a limit b, then $\sum kb_j$ converges to kb, for any constant k.

Example 3. We claim that the infinite series

$$\sum_1^\infty \frac{1}{j^2}$$

converges. We shall verify that it is dominated by the convergent infinite series

$$\sum_1^\infty \frac{2}{j(j+1)}.$$

We have to show that for all but finitely many j,

$$\frac{1}{j^2} < \frac{2}{j(j+1)}.$$

To prove this inequality, we invert it and obtain the equivalent inequality

$$j^2 > \frac{j(j+1)}{2}$$

Multiplying by 2 and dividing by j, we get the equivalent inequality

$$2j > j + 1.$$

Subtracting j from both sides we get the equivalent inequality

$$j > 1,$$

which is certainly true for $j = 2, 3, \ldots$. This proves that $\sum_1^\infty 1/j^2$ is dominated and completes the proof of its convergence.

Example 4. We claim that the infinite series

$$1 - \frac{1}{2} + \frac{1}{3} - \frac{1}{4} + \cdots = \sum_1^\infty \frac{(-1)^{j+1}}{j}$$

converges. We shall prove this by using the comparison criterion; one's first impulse is to compare this series with

$$\sum_1^\infty \frac{1}{j}.$$

25

Alas, this doesn't work since, as we shall show in Example 6 of this section, the infinite series $\sum_1^\infty 1/j$ diverges. We have to be cleverer than that: we shall bracket two consecutive terms as follows

(5.12) $(1 - \frac{1}{2}) + (\frac{1}{3} - \frac{1}{4}) + (\frac{1}{5} - \frac{1}{6}) + \cdots = \sum_1^\infty \left(\frac{1}{2j-1} - \frac{1}{2j} \right).$

The j-th term of this series is

$$\frac{1}{2j-1} - \frac{1}{2j} = \frac{1}{(2j-1)2j}.$$

We claim that this series is dominated by the convergent series (see Example 2) $\sum 1/j(j+1)$, i.e., that for all but finitely many j

$$\frac{1}{(2j-1)2j} < \frac{1}{j(j+1)}.$$

To prove this inequality, we invert it and obtain the equivalent inequality

$$(2j-1)2j > j(j+1).$$

Dividing both sides by j, we obtain the equivalent inequality

$$(2j-1)2 > j+1.$$

Adding $2 - j$ to both sides, we get the equivalent inequality

$$3j > 3,$$

which is certainly true for $j > 1$. This proves that (5.12) is dominated by a convergent series, and thus, according to the comparison criterion, the bracketed series (5.12) converges. It follows easily that $\sum_1^\infty (-1)^{j+1}/j$ also converges.

Example 5. We recall the notation $n!$, read "n factorial," defined for positive integers n as the product of all integers from 1 to n:

$$n! = 1 \cdot 2 \cdot \cdots \cdot (n-1) \cdot n.$$

We agree to define 0! as 1. This is a mere convenience which turns out to be consistent in all the contexts where the symbol ! is used. For the first few values of n we have

$$1! = 1, \quad 2! = 2, \quad 3! = 6, \quad 4! = 24, \quad \text{and} \quad 5! = 120.$$

We shall investigate the convergence of the infinite series

(5.13) $$\sum_0^\infty \frac{1}{j!}.$$

With the aid of a hand calculator we can easily compute to seven decimals the finite sums $\sum_0^n 1/j! = e_n$ for n up to, say, 10:

$$e_0 = 1.0000000$$
$$e_1 = 2.0000000$$
$$e_2 = 2.5000000$$
$$e_3 = 2.6666666$$
$$e_4 = 2.7083333$$
$$e_5 = 2.7166666$$
$$e_6 = 2.7180555$$
$$e_7 = 2.7182540$$
$$e_8 = 2.7182787$$
$$e_9 = 2.7182790$$
$$e_{10} = 2.7182818 \ .$$

These numbers look like members of a convergent sequence, whose limit is a number with the first four digits

$$2.7182.$$

To prove this we shall use the comparison criterion. We shall compare our series with the convergent geometric series

(5.14)
$$\sum_1^\infty \frac{1}{2^j},$$

a promising choice, since the denominator of the j-th term of (5.14) has the j equal factors 2, while the denominator of (5.13) has j increasing factors $1, 2, 3, \ldots, j$. To show that the j-th term of the series (5.13) is less than the corresponding term of the comparison series (5.14), we invert the desired inequality

$$\frac{1}{j!} < \frac{1}{2^j}$$

and get the equivalent inequality

(5.15)
$$j! > 2^j.$$

This certainly holds for $j = 4$:

$$4! = 24 > 16 = 2^4.$$

We claim that it is true for all $j > 4$; for, when j is increased by 1, the left side of (5.15) is multiplied by $j + 1$ whereas the right side is merely doubled: this shows that since (5.15) is true for $j = 4$, it is true for $j = 5, 6$, etc.

The sum of the first ten terms differs from the sum of the infinite series by $\sum_{11}^\infty 1/j!$ which is

$$e - e_{10} = \sum_{11}^\infty \frac{1}{j!} < \sum_{11}^\infty \frac{1}{2^j} = \frac{1}{2^{11}} [1 + \tfrac{1}{2} + \cdots]$$

$$= \frac{1}{2^{11}} \cdot \frac{1}{1 - \tfrac{1}{2}} = \frac{1}{2^{10}} = \frac{1}{1024} < \frac{1}{10^3} \ .$$

27

This shows that the first four figures of the infinite sum (5.3) are indeed 2.718. This number looks very much like the number we encountered in Example 5 of Section 1.4 as the limit of the sequence $[1 + (1/n)]^n$. Later on, in Chapter 5, we shall prove that the two numbers are indeed exactly equal.

Example 6. We shall investigate the infinite series

(5.16)
$$\sum_1^\infty \frac{1}{j},$$

called the *harmonic series*. Its finite sums

$$h_n = \sum_1^n \frac{1}{j}$$

have the following approximate values up to $n = 10$:

$$h_1 = 1$$
$$h_2 = 1.50$$
$$h_3 = 1.83$$
$$h_4 = 2.08$$
$$h_5 = 2.28$$
$$h_6 = 2.43$$
$$h_7 = 2.57$$
$$h_8 = 2.69$$
$$h_9 = 2.80$$
$$h_{10} = 2.90 .$$

We shall prove now that the series (5.16) *diverges* by constructing a comparison series whose terms are *smaller* than those of (5.16), but which diverges. Denote the comparison series by $\sum_1^\infty b_j$, with

(5.17)
$$b_j < \frac{1}{j}.$$

It will then follow that the series (5.16) diverges, for its convergence would, according to the comparison criterion, imply the convergence of $\sum b_j$. We choose b_j as follows:

$$b_1 = 1, \qquad b_2 = \tfrac{1}{2}, \qquad b_3 = b_4 = \tfrac{1}{4}, \qquad b_5 = b_6 = b_7 = b_8 = \tfrac{1}{8}.$$

The general pattern is: For j such that $2^{k-1} < j \le 2^k$, k an integer, we choose

$$b_j = \frac{1}{2^k}.$$

Clearly inequality (5.17) holds:

$$b_j = \frac{1}{2^k} \le \frac{1}{j}.$$

There remains to show that $\sum b_j$ diverges:

$$(5.18) \qquad \sum_1^{2^k} b_j = b_1 + (b_2) + (b_3 + b_4) + (b_5 + \cdots + b_8) + \cdots$$

$$+ (b_{2^{k-1}} + \cdots + b_{2^k}).$$

There are k parentheses; the p-th contains 2^{p-1} terms, all equal to $1/2^p$. Therefore each bracketed sum is $\frac{1}{2}$, and the total sum (5.18) is $1 + (k/2)$. As k increases, $1 + (k/2)$ increases without bound, so this sequence has no limit; therefore the series (5.18) diverges.

EXERCISES

5.1 How large does n have to be in order for

$$a_n = 1 - \frac{1}{n+1}$$

to be within $1/10$ of $\lim a_n$? within $1/100$? within $1/1000$?

5.2 (a) How large does n have to be in order for

$$\sum_1^n \frac{1}{j^2}$$

to be within $1/10$ of the infinite sum? within $1/100$? within $1/1000$? Calculate the first, second and third digit after the decimal point of $\sum_1^\infty 1/j^2$.
(b) Answer the same questions about the series

$$1 - \tfrac{1}{2} + \tfrac{1}{3} - \tfrac{1}{4} + \cdots.$$

Use a hand calculator whenever you can, otherwise write a computer program.
Note. The sum of the infinite series in part (a) is $\pi^2/6$ and of that in part (b) is the natural logarithm of 2. Using tabulated values of π and of log 2, verify that you have determined correctly the first three digits of these infinite sums.

5.3 (a) Prove that the sequence

$$\left\{ 1 - \frac{1}{\sqrt{n+1}} \right\}$$

converges to the limit 1.
(b) Prove that the infinite series

$$\sum_1^\infty \left(\frac{1}{\sqrt{j}} - \frac{1}{\sqrt{j+1}} \right)$$

converges.
(c) Prove that the infinite series

$$\sum_1^\infty \frac{1}{j^{3/2}}$$

converges.

(d) How large does n have to be in order for

$$\sum_1^n \frac{1}{j^{3/2}}$$

to be within 1/10 of the infinite sum? within 1/100? within 1/1000?

5.4 A ball, dropped from a height h, rebounds to $(6/7)h$. It continues to bounce, each time reaching 6/7 of the height it reached in the preceding bounce. Compute the total distance traversed by the ball.

5.5 Although, by the comparison test, we have shown that the harmonic series diverges, what might you conclude about the sequence $\{c_n\}$ defined by the difference

$$c_n = \sum_{j=1}^n \frac{1}{j} - \log n?$$

Before you hazard a guess, complete the following table using a digital computer:

n	$\sum_1^n \frac{1}{j}$	$\log n$	c_n
1	1	0	1
10			
50			
100			
1000			
5000			
10000			

5.6 From the result obtained in Exercise 5.5, estimate for what value of n the sum

$$\sum_1^n \frac{1}{j}$$

is greater than 10^6. Note from your answer that if you interpret n as the number of seconds required to form this sum (one division and addition per second), the time required is greater than the known age of the universe!

5.7 Express the following repeating decimals as ratios of two integers, i.e., in the form $p/q, q \neq 0$:
(a) 8.013013...
(b) 3.01238888...
(c) 0.19999...
(d) 2.17891789... .

1.6 The least upper bound

We have remarked in Section 1.3 that the decimal representation is ideally suited for deciding which of two given numbers is larger; we shall exploit this systematically.

To decide which of the two numbers a and b is greater we start by comparing their integer parts. If one of these is larger than the other, the number with the larger integer part wins, i.e., is the larger of the two. If their *integer parts* are equal, both numbers are still in the running; and to make a decision, look at their first decimal digits. If these are equal, we go on to the next digit to the right. Since the two numbers are different, not all their digits agree, and we are sure to reach a decision after a finite number of steps.

We now describe an obvious modification of this process suitable for picking out the largest of a *finite* set S of numbers. We examine the integer parts of all numbers in this set. In this finite collection of integers there will be a largest, and clearly all numbers in the set S whose integer part is less than this are out of the competition. We call the remaining numbers *eligible* and examine their first digits. In this finite collection, there is a largest; we eliminate as ineligible all numbers whose first digit is less than this largest numeral and repeat this weeding-out process with the next digit to the right. The numbers which remain eligible after j steps have the same digits to the left of the j-th; if the numbers of S are distinct (no two equal), after a finite number of steps there is only one left eligible, the largest. If the numbers of S are not distinct, we may have several winners, all equal.

We turn now to an infinite set S of numbers and try to apply the above procedure for finding the largest number in it. Assume that the elements of S are positive. It could happen that among the integer parts of the numbers there is no largest. In this case the set S contains arbitrarily large numbers, and the search for the largest is clearly futile. We expressly rule out this possibility and assume:

The set S is bounded from above; i.e., all numbers in S are less than some number k. Such a number k is called an *upper bound* for S. We might as well assume that the upper bound k is 1 (if not, divide each number by k).

We examine the first digits of the infinitely many numbers in S and keep, as eligible, only those with largest first digit. We repeat the process on the remaining eligible numbers with each successive digit; after j steps all remaining eligible numbers have the same digits to the left of the j-th. However even if the numbers of the set S are distinct, there is no reason to expect that after finitely many steps the set of eligible numbers will be narrowed down to a single one.

Consider the following example: the set S consists of the numbers a_n, $n = 1, 2, \ldots$, defined by

$$a_n = \underbrace{.11 \ldots 1}_{n \ 1\text{'s}} 000 \ldots$$

i.e., the decimal expansion

$$a_n = 0.d_1 d_2 \ldots$$

has $d_i = 1$ for $i \le n$, and $d_i = 0$ for $i > n$. Clearly the eligible numbers remaining after j steps are all numbers a_n with $n \ge j$, and there are infinitely many of these. We have not reached a decision in this weeding out process.

31

We get around this difficulty as follows: define the number s by setting its j-th digit s_j equal to the j-th digit d_j of any number which remains eligible after j steps. By construction, s has the property that each of its digits is at least as great as the corresponding digit of any eligible number in S. Thus s is $\geq$ any number in S; if s belonged to S it would be the largest element of S. However there is no reason why s should belong to S; e.g., in the example given above, s has all ones in its decimal expansion and so clearly is not one of the numbers a_n. We claim that s is the *least upper bound* of S, i.e., *an upper bound which is smaller than any other upper bound*.

We have already shown that s is an upper bound. Let m be any number smaller than s. Denote by j the first position in which the digits of s and m differ, i.e., the largest integer j such that for $n < j$, $s_n = m_n$ and since m is smaller than s,

$$m_j < s_j.$$

On the other hand S was so constructed that there is a number x in the set S all whose digits d_i agree with the digits s_i of s for $i \leq j$. Clearly m is less than that number x, and so m is *not* an upper bound for S. This result is so important that it deserves restatement and a special name:

The least upper bound theorem. *Every set of numbers bounded from above has a least upper bound.*

There is a very intuitive geometrical interpretation of this result: Imagine pegs put in the number line at all points of the set S. Let k be an upper bound for S; this means that k is to the right of every point of S. Put the point of your pencil at k and move it as far to the left as the pegs will let it go. The point where the pencil gets stuck is the least upper bound of S; see Figure 1.8.

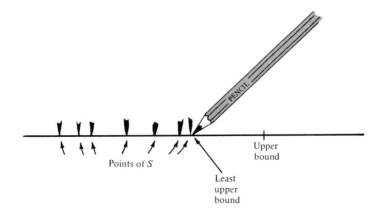

Points of S

Upper
bound

Least
upper
bound

Figure 1.8

We give now an application of this theorem. Let $a_1, a_2, \ldots$ be a *sequence which increases:*

$$a_n < a_{n+1}, \qquad n = 1, 2, \ldots .$$

Suppose also that *the sequence is bounded from above,* i.e., all a_n are less than some number k. We claim that *then the sequence converges,* and that its limit is the least upper bound of the set of numbers a_n. A glance at the position of the numbers a_n and k on the number line (see Figure 1.9) will convince most people of the truth of the statement. Here is a formal proof.

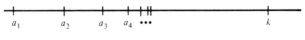

Figure 1.9

PROOF. Denote by s the least upper bound of the numbers a_n; this number s was so constructed that for any integer m there is a point a_n in the set such that the j-th digit of a_n and s are the same for all $j \leq m$. But then, since our sequence is increasing, for all $p > n$

$$a_n < a_p < s;$$

this shows that for $p > n$ all digits of a_p and s up to the m-th are the same; this proves that s is the limit of the sequence. ☐

Quite analogously one can deduce

The greatest lower bound theorem. *Every set of numbers bounded from below has a greatest lower bound.*

This can be applied to prove a convergence theorem about decreasing sequences; we summarize both results as the

Monotone convergence theorem. *An increasing (decreasing) sequence of real numbers which is bounded from above (below) is convergent.*

The adjective *monotone* means something that proceeds in the same direction; when applied to a sequence this means either increasing or decreasing.

We have shown earlier that if x is any number between 0 and $\frac{1}{2}$, then the sequence $\{x^n\}$ tends to zero. We will now show, with the aid of the monotone convergence theorem, that *the same conclusion holds for all x between 0 and 1.* For, clearly if $0 < x < 1$, the sequence $\{a_n\} = \{x^n\}$ decreases and is bounded from below by 0; therefore according to the monotone convergence theorem it converges to a limit a. We claim that this limit is zero. To see this consider the new sequence $\{b_n\}$ whose n-th term is

$$b_n = x a_n;$$

according to the rules of arithmetic for convergent sequences, $\{b_n\}$ converges and its limit b is x times the limit of $\{a_n\}$:

(6.1) $$b = xa.$$

On the other hand, we recognize that

$$b_n = a_{n+1},$$

i.e., that the sequence $\{b_n\}$ is the same as $\{a_n\}$ shifted over by one term. Therefore $\{b_n\}$ has the same limit as $\{a_n\}$, so relation (6.1) becomes

(6.2) $$a = xa.$$

Since $x < 1$, (6.2) implies that $a = 0$, just what we claimed.

Note that the convergence proof we have just given differs in one important respect from the demonstration of the convergence given earlier. In the earlier proofs we have stated exactly how large n has to be in order that a_n differ by less than 10^{-m} from the limit a. In contrast, our present proof gives no such information.

As an application of the monotone convergence theorem we prove the following: Let I_1, I_2, I_3, ... be a sequence of *closed* intervals which are "nested"; that is, each I_n contains the next one; see Figure 1.10.

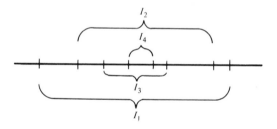

Figure 1.10

Assertion. *The intervals I_n have a point in common. This is called the nested interval property of real numbers.*

We base our proof on the monotone convergence theorem. Denote by a_n and b_n the left and right endpoints of I_n; the assumption of nesting means that the sequence $\{a_n\}$ increases, the sequence $\{b_n\}$ decreases, and that each b_n is greater than any a_n. So by the monotone convergence theorem, $\{a_n\}$ converges to some limit a, $\{b_n\}$ to some limit b. Clearly, each a_n is $\leq a$, and each $b_n \geq b$; and $a \leq b$. Since the intervals I_n are closed, this shows that the interval $[a, b]$ belongs to all the intervals I_n.

The distance of any two points in I_n is $\leq$ the length of I_n. Therefore, *if the length of the intervals I_n tends to zero as n tends to infinity, there is exactly one point common to all intervals I_n.* □

EXERCISES

6.1 Prove that $\sqrt{2}$ is the least upper bound of all rational numbers whose square is less than 2.

6.2 Take the unit square and, by connecting the midpoints of opposite sides, divide it into $2^2 = 4$ subsquares, each of side $\frac{1}{2}$. Repeat this division for each subsquare, obtaining $2^4 = 16$ squares whose sides have length $1/2^2$. Continue the process so that after n steps, there are 2^{2n} squares, each having sides of length 2^{-n}.

With the lower left corner as center and radius 1, inscribe a unit quarter-circle into the square (see Figure 1.11). Define by a_n the total area of those squares which, at the n-th step of the process, lie entirely inside the quarter-circle. For example

$$a_1 = 0, \qquad a_2 = \left(\frac{1}{2}\right)^2 = \frac{1}{4}, \qquad a_3 = 8\left(\frac{1}{2^2}\right)^2 = \frac{1}{2}, \ldots \ .$$

(a) Prove that $a_n \leq 1$ for $n = 1, 2, 3, \ldots$.
(b) Prove that, if $k \leq n$, then $a_k \leq a_n$. Denote by u the least upper bound of the numbers a_n defined above. [*Note*: $4u = \pi$.]

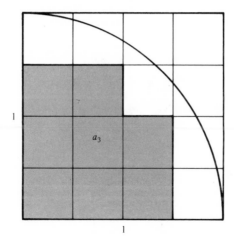

Figure 1.11

6.3 Let x be any number > 1.
(a) Show that the sequence $\{a_n\} = \{x^n\}$ is increasing.
(b) Prove that if the sequence $\{x^n\}$ were bounded from above, it would converge to some limit a.
(c) Prove that this limit satisfies the relation

$$a = xa.$$

(d) Conclude that the sequence $\{x^n\}$ is unbounded.

6.4 Deduce the greatest lower bound theorem from the least upper bound theorem.

Appendix 1.1 Irrationality of $\sqrt{2}$ and e

We shall show that there is no rational number r whose square is 2. That means that no number of the form p/q, p and q integers, satisfies

(A.1)
$$\left(\frac{p}{q}\right)^2 = 2.$$

We might as well assume that p/q is in lowest terms, i.e., that p and q have no common divisor; for, if they do, we divide both by their common divisor. Multiplying both sides of (A.1) by q^2, we get

(A.2)
$$p^2 = 2q^2.$$

The right side of this equation is clearly even; therefore p^2, the left side, must be even. We claim that is so only if p itself is even; for if p were *odd*, it would be of the form $2n + 1$, and its square would be

$$(2n + 1)^2 = 4n^2 + 4n + 1,$$

clearly odd.

Since p is even, it is of the form

$$p = 2k;$$

we substitute this into equation (A.2) and get

$$4k^2 = 2q^2.$$

Dividing both sides by 2 we get

$$2k^2 = q^2.$$

Now the left side is manifestly even; therefore q^2, the right side, must be even. As before we conclude that q is even. This makes both p and q even, contrary to our assumption at the beginning that p and q have no divisor in common. Thus we have arrived at a contradiction for which we have assumption (A.1) to blame; so (A.1) must be false, and this shows that the square of no rational number is 2.

EXERCISES

A.1 Show that $\sqrt{3}$ is irrational.

A.2 Show that the cube root of 2 is irrational.

Next we show that *the sum e of the infinite series*

(A.3)
$$\sum_{0}^{\infty} \frac{1}{n!} = e$$

is irrational.

36

We have shown in Section 1.5 that this infinite series converges.

Let q be any integer; we break the infinite sum (A.3) into two parts and write

$$e = \sum_{n=0}^{q} \frac{1}{n!} + \sum_{n=q+1}^{\infty} \frac{1}{n!}.$$

Multiply both sides by $q!$:

(A.4)
$$q!e = \sum_{n=0}^{q} \frac{q!}{n!} + \sum_{n=q+1}^{\infty} \frac{q!}{n!}.$$

Recall that $n!$ is the product of all whole numbers from 1 to n; clearly if n is less than q, $n!$ divides $q!$; this shows that each term in the first sum $(0 \leq n \leq q)$ in (A.4) is a whole number. We turn now to the second sum

(A.5)
$$\sum_{n=q+1}^{\infty} \frac{q!}{n!} = \frac{1}{q+1} + \frac{1}{(q+1)(q+2)} + \frac{1}{(q+1)(q+2)(q+3)} + \cdots.$$

Clearly, the m-th term of this series is 1 divided by the product of m numbers, each greater than q; thus the m-th term of this series is less than $1/q^m$ and so the sum of the infinite series (A.5) is less than the sum

$$\sum_{m=1}^{\infty} \frac{1}{q^m}.$$

This is an infinite geometric series whose sum is $1/(q-1)$, a number not greater than 1 for $q > 1$. This shows that for $q > 1$, the infinite series (A.5) is a number greater than 0 and less than 1. Since the first sum in (A.4) is a sum of integers, we conclude that *for $q > 1$, $q!e$ is not an integer.*

From this it is easy to deduce that e is not rational, i.e., not of the form p/q. For if it were, $q!e$ would be equal to $(q-1)!p$, an integer, contrary to what we have just demonstrated.

Appendix 1.2 Floating point representation

When we measure the accuracy of the approximation of a number x by x_{appr} in terms of the number of digits which are in agreement, we are measuring the accuracy in *absolute* terms. Often it is more important to know the *relative accuracy* of an approximation, i.e., instead of knowing the size of the *difference* $x - x_{appr}$, it is more important to know the size of the *ratio*

$$\left| \frac{x - x_{appr}}{x} \right|.$$

A convenient way of measuring this quantity is to write numbers in *normalized form*, i.e., in the form

(A.9)
$$x = y \, 10^n,$$

where n is a signed integer, and y is a number whose absolute value lies in the interval $[10^{-1}, 1)$. If x_{appr} is represented as

$$x_{appr} = y_{appr} \, 10^n,$$

n being the same integer, the relative size of the deviation between x and x_{appr} is measured by the number of digits of x and x_{appr} which agree. The digits of y_{appr} which agree with those of y are called the *significant digits* of x_{appr} in the representation (A.9).

In computing machines numbers are represented in the form (A.9), except that the base 10 is replaced by base 2 (base 8 in some machines). In that context (A.9) is called the *floating point* representation of the number x.

Example. The number -971.276 can be written as the normalized floating point number $-0.971276 \cdot 10^3$; the number 0.00295 can be written as $0.295 \cdot 10^{-2}$.

Functions 2

The idea of a function is the most important concept in mathematics; in this chapter we shall discuss functions of one variable, a simple but very important class.

2.1 The notion of a function

Rather than starting with a definition we shall first give a number of examples and then fit the definition to these. The first 5 examples, shown in Figures 2.1 to 2.5, describe functions graphically, the next 2 describe functions tabularly, and the last 4 give algebraic descriptions of functions.

Example 1

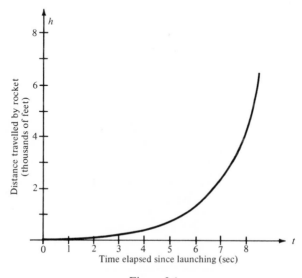

Figure 2.1

Example 2

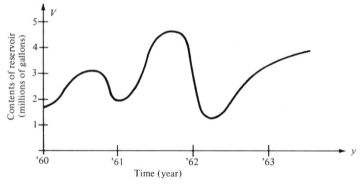

Figure 2.2

Example 3

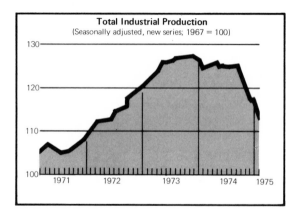

Figure 2.3

Example 4

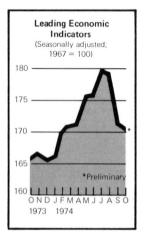

Figure 2.4

Example 5

Figure 2.5

Example 6

p	26	27	27	29	30	30
y	'60	'61	'62	'63	'64	'65

where p = price of milk in cents, and y = year.

Example 7

D	76	82	85	90	98	103	105
y	'55	'56	'57	'58	'59	'60	'61

where D = National debt in billions of dollars, and y = year.

Example 8

$$F = \tfrac{9}{5}C + 32,$$

where F = degrees in Fahrenheit, and C = degrees in centigrade.

Example 9

$$d = 16t^2.$$

where d = distance travelled by freely falling body, measured in feet, and t = time of fall, in seconds.

41

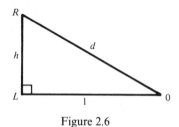

Figure 2.6

Example 10. Triangle RLO (see Figure 2.6) has a right angle at L, $LO = 1$, $LR = h$, so that

$$d = OR = \sqrt{1 + h^2}.$$

Example 11. The taxi fare in New York City is 65 cents for the first 1/6-th mile or portion of it, and 10 cents for each additional 1/6-th of a mile, or portion of it.

These examples contain information of the following kind: they enable you to determine

1. The distance travelled by a rocket at any time, up to 8 seconds after launch.
2. Content of a reservoir at any time between January '60 and June '62.
3. Total Industrial Production for the period March 1971 through February 1975.
4. Seasonally adjusted economic indicators for the period October 1973 through September 1974.
5. Daily sales of stocks on the New York Stock Exchange during a period in October–November 1974.
6. The price of milk for the years '60 to '65.
7. The national debt for the years '55 to '61.
8. The temperature in degrees Fahrenheit given the temperature in degrees centigrade.
9. How far a stone has dropped, if you know how long it has been falling.
10. The distance d from O to R, provided you know the distance from L to R.
11. The taxi fare, if you know the distance between your starting point and your destination.

Each of these examples enables one to determine the value of one quantity provided that the value of another is specified. Information of this kind is called a *function*; a more precise statement is contained in the following

Definition. A *function f* is a rule which assigns a definite number y to any number x lying in an interval or collection of intervals S. S is called the *domain* of the function f; y is called the *value* of the function f at the point x and is usually denoted in one of two ways:

$$f : x \to y \qquad \text{or} \qquad f(x) = y.$$

The set of all values which a function assumes is called its *range*. Thus the range of the function presented in Example 1 is all numbers between 0 and 40; in Example 6, the numbers 26, 27, 29, 30; in Example 11, all numbers which are of the form: 65 plus any whole multiple of 10.

The rules presented in our examples for determining the values of the functions are quite varied. In Examples 1–5 the rule is to read a graph; in Examples 6 and 7 the rule is to read a table. In Examples 8 and 9 we have to perform arithmetic operations; and in Example 10 we have to take a square root.

It is often distracting to look too closely at the sordid details that go into the workings of a function. To avoid cluttering up our minds we follow the practice of design engineers and regard a function as a *black box* (see Figure 2.7):

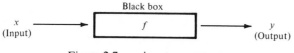

Figure 2.7 x: input; y: output.

Any number x from the domain of the function is a possible *input*; the *output* is the value of the function. All possible outputs constitute its range.

The exact domain of a function is important to keep in mind in some cases, not so important in others. For example, a function that describes the price of milk during the years 1945 to 1955 contains different information from the one that describes the price of milk during the years 1950 to 1960. On the other hand, take the function in Example 8, relating degrees in centigrade to Fahrenheit by the formula

$$F = \tfrac{9}{5}C + 32.$$

Since there is no temperature lower than absolute zero, the domain of this function is all values $C \geq -273$, yet there is no harm in regarding the function above as being defined for all C; in general, there is no harm in defining a function over a domain which is *larger* than the domain over which the function is eventually used. In particular, if a function is defined by a formula, we adopt the convention that, unless otherwise specified, its domain consists of all real numbers for which the formula is meaningful; we call this the *natural domain* of the function defined by the formula. For example, the natural domain of the function $f(x) = 1/(x - 1)$ is all numbers $x \neq 1$.

EXERCISES

1.1 Consider the function $f : h \to d$, defined by

$$d = \sqrt{1 + h^2}.$$

Suppose we choose the domain of f to be all h between 0 and 1; what is the range of f?

1.2 Define the function $f : x \rightarrow y$ over the domain $-1 \leq x \leq 1$ as follows:

$$f(x) = \begin{cases} -x & \text{for } -1 \leq x < 0 \\ x & \text{for } \quad 1 \leq x \leq 1. \end{cases}$$

(a) Draw the graph of this function (see Section 2.5).
(b) What is the range of this function?

2.2* Functions of several variables

As in Section 2.1, we give first a number of examples, then abstract the concept from these.

Example 1

$$d = 16t^2 + vt,$$

where d = distance in feet travelled by a stone thrown downward with initial velocity v, t = time of fall, in seconds and v = initial velocity, in feet per second.

Example 2

$$V = \tfrac{1}{3}\pi r^2 h,$$

where V = volume of right circular cone, r = radius of circular base, and h = height of cone.

Example 3

$$W = \left(1 + \frac{p}{100}\right)^n,$$

where W = worth of one dollar after n years, invested at p percent per annum.

Example 4

$$h = \sqrt{a^2 + b^2},$$

where h = length of hypotenuse of right triangle whose legs have lengths a and b.

Each of these examples enables one to calculate a certain quantity, provided that *two* others are specified. In Example 1 we have to know both, the initial velocity and the time of flight, before we can determine the distance travelled by the flung stone. In Example 4 we have to know the lengths of *both* legs before we can calculate the length of the hypotenuse, etc. Information of this kind is called a function of two variables.

---

* This section may be omitted at first reading.

Definition. *A function f of two variables* is a rule which assigns a definite number z to any pair of numbers x and y; x is called the first argument or independent variable of f, y the second. We use one of two notations

$$f : x, y \to z \quad \text{or} \quad f(x, y) = z.$$

The *domain* of f is all pairs x, y for which the rule is defined; it often consists of all x in one interval S_1 and all y in another S_2.

A function of three or more variables is defined quite analogously.

In all four examples presented above the rule which assigns the functional value was a formula, i.e., a recipe given in terms of arithmetic operations. The graphical description of a function of two variables is not practical, since it requires the construction of a three dimensional model. A tabulation of a function of two variables is cumbersome, but possible.

As before, it pays to think of a function of two variables as a black box; only in this case the input is a pair of numbers, in a definite order (see Figure 2.8).

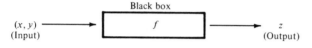

Figure 2.8 (x, y): input; z: output.

In a later chapter we shall study functions whose output is not a number, but a point in the plane.

EXERCISES

2.1 Define some function f of three variables x, y and z, for all real numbers x, y and z.

2.2 A function f of two variables is called *symmetric in x and y* if for all x, y in its domain

$$f(x, y) = f(y, x).$$

Which of the following functions, defined for all real values x, y, is symmetric?

(a) $f(x, y) = \sqrt{x^2 + y^2}$.

(b) $f(x, y) = 2x + 3y$.

(c) $f(x, y) = xy$.

(d) $f(x, y) = \begin{cases} 3x - y & \text{for } x \leq y \\ 3y - x & \text{for } y < x. \end{cases}$

2.3 Composite functions

In this section we describe a very natural way of making a new function out of two others. We start with a simple example (see Figure 2.9):

A rocket is launched vertically from point L; the distance $h(t)$ of the rocket R from the launching point at time t is described in Example 1, Section 2.1. An observation post O is located 1 mile from the launching site; we are asked to determine the distance d of the rocket from the observation post as a function of time.

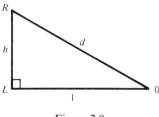

Figure 2.9

To solve this problem we recall that, in Example 10, Section 2.1, the distance d was described by the following function of h:

$$d = \sqrt{1 + h^2}.$$

Therefore the distane of R to O at time t is

$$d(t) = \sqrt{1 + h(t)^2}.$$

The process which builds the function $d(t)$ out of the two functions described in Examples 1 and 10 is called *composition*; the resulting function is called the *composite* of the two component functions.

Let us represent the two component functions by black boxes marked 1 and 2; the process of composition can be represented schematically as in Figure 2.10. That the output of one function serves as input for another is,

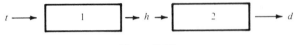

Figure 2.10

as we shall see, a typical situation in mathematics and all its applications. It reflects a chain-like interconnection between quantities. We now give a precise definition of composition:

Let f and g be two functions, and suppose that the range of g is included in the domain of f. Then the composite function, denoted by f ∘ g, is the function whose value, for any x in the domain of g is f(g(x)). The construction is well described by Figure 2.11.

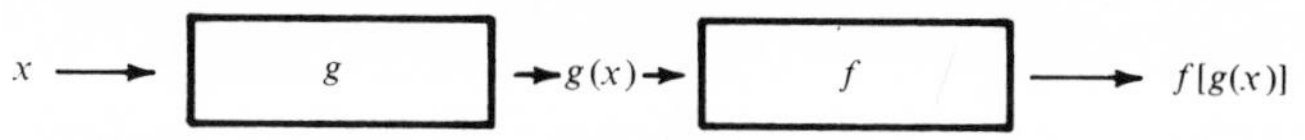

Figure 2.11

Example 1. If

$$f(g) = \frac{1}{g+1}, \qquad g(x) = x^2.$$

Then

(a) $f \circ g(x) = \dfrac{1}{x^2 + 1}$.

(b) $g \circ f(x) = \left(\dfrac{1}{x+1}\right)^2 = \dfrac{1}{x^2 + 2x + 1}$.

Notice that $f \circ g$ and $g \circ f$ are quite different functions. Thus *composition is not a commutative operation.* This is not surprising: using the output of g as input for f is quite different from using the output of f as input for g.

Example 2. If

$$f(x) = \frac{1}{x+1}, \qquad g(x) = \frac{1}{x-1}, \qquad h(x) = x^2 + 1.$$

Then

(a) $f \circ g(x) = \dfrac{1}{1/(x-1) + 1} = \dfrac{x-1}{x} = 1 - \dfrac{1}{x}$

(b) $g \circ h(x) = \dfrac{1}{x^2 + 1 - 1} = \dfrac{1}{x^2}$

(c) $(f \circ g) \circ h(x) = 1 - \dfrac{1}{x^2 + 1}$

(d) $f \circ (g \circ h)(x) = \dfrac{1}{(1/x^2) + 1} = \dfrac{x^2}{1 + x^2} = 1 - \dfrac{1}{1 + x^2}$.

Comparing (c) and (d) we notice that $(f \circ g) \circ h$ is the same as $f \circ (g \circ h)$; we shall now show that this is no accidental feature of our special choice of $f, g,$ and h but is true in general. Look at Figure 2.12 and observe that this *triple composite* function can be interpreted *either* as $(f \circ g) \circ h$ or as $f \circ (g \circ h)$.

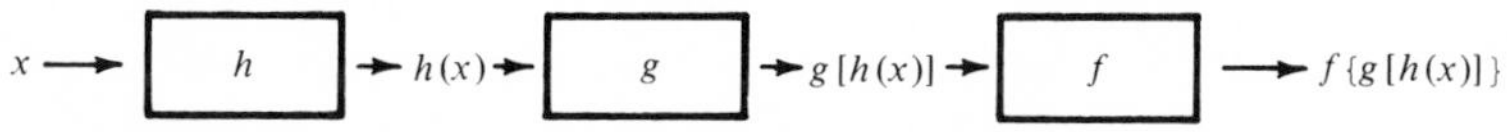

Figure 2.12

47

Therefore

$$(f \circ g) \circ h = f \circ (g \circ h)$$

is true for any three functions that can be composed. This property of composition is called *associativity*.

We look at some further examples:

Example 3. Let

$$f(x) = \frac{1}{x+1}, \qquad g(x) = \frac{1-x}{x}.$$

Then

(a) $f \circ g(x) = \dfrac{1}{(1-x)/x + 1} = \dfrac{x}{(1-x) + x} = x$

(b) $g \circ f(x) = \dfrac{1 - 1/(x+1)}{1/(x+1)} = \dfrac{(x+1) - 1}{1} = x$

Example 4. For

$$f(x) = 2x + 3, \qquad g(x) = \tfrac{1}{2}x - \tfrac{3}{2}$$

we see that

(a) $f \circ g(x) = 2(\tfrac{1}{2}x - \tfrac{3}{2}) + 3 = x$

(b) $g \circ f(x) = \tfrac{1}{2}(2x + 3) - \tfrac{3}{2} = x$

We ask the following question about a function g: if we know the output y, can we determine the input x? There are several good reasons for wanting to recoup the input of a function from its output, that is, for trying to "reverse" or "invert" the action of a function, so that it works backwards from output to input. For example, in electric circuit design, the input may be destroyed in the process of yielding a desired output. However, the same input is needed again to produce a new output via a different function; see Figure 2.13 for a schematic diagram. In many situations one wants to determine the input which would yield a desired output.

If a funtion g has the property that different inputs always lead to different outputs, i.e., if $x_1 \neq x_2$ implies $g(x_1) \neq g(x_2)$, then we can determine the input from the output. Such a function g is called *invertible*; its *inverse* f is defined by the diagram in Figure 2.14. In words: The domain of f is the range of g, and $f(y)$ is defined as that number x for which $g(x) = y$.

It is clear from this definition that if g is invertible and its inverse is f, then f too is invertible and its inverse is g. Furthermore the composite of a function g and its inverse f, in either order, is the *identity function*, i.e., one that carries x into x:

$$g \circ f(x) = f \circ g(x) = x$$

if f is the inverse of g.

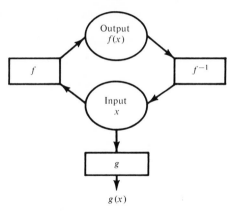

Figure 2.13 f^{-1} is the inverse of f.

The functions listed in Examples 3 and 4 are inverse to each other. Let us take a closer look at those of Example 3, where the formula defining $f(x) = 1/(x + 1)$ can be stated as follows:

1. Take any real number x (except -1) and add 1 to it, obtaining $x + 1$.
2. Take the reciprocal of the number so formed, getting $1/(x + 1)$.

Now, starting with the output $z = 1/(x + 1)$, how can we recapture the input x? By "undoing" each of the above steps, in the reverse order:

2^{-1}. Take the reciprocal of the output $z = 1/(x + 1)$, obtaining $1/z = x + 1$.

1^{-1}. From the number so formed, subtract 1, obtaining

$$\frac{1}{z} - 1 = x + 1 - 1 = x.$$

Note that we have broken up the function f into the component functions

$$f_1(x) = x + 1, \qquad f_2(y) = \frac{1}{y},$$

so that

$$f(x) = f_2 \circ f_1(x) = \frac{1}{x + 1};$$

then we found the inverse $g_2(z) = 1/z$ of f_2, then the inverse $g_1(y) = y - 1$ of f_1. Their composition

$$g_1 \circ g_2(z) = \frac{1}{z} - 1 = \frac{1 - z}{z}$$

yields the inverse g of f.

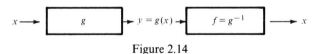

Figure 2.14

49

Many authors use the notation: inverse of $f = f^{-1}$. In this notation, what we have just shown is

(3.1) $$(f_2 \circ f_1)^{-1} = f_1^{-1} \circ f_2^{-1}.$$

Now suppose the output of f is obtained in a sequence of k steps: $f = f_k \circ \cdots \circ f_2 \circ f_1$; then

$$f^{-1} = (f_k \circ \cdots \circ f_2 \circ f_1)^{-1} = f_1^{-1} \circ f_2^{-1} \circ \cdots \circ f_k^{-1}$$

Indeed,

$$f \circ f^{-1} = (f_k \circ \cdots \circ f_2 \circ f_1) \circ (f_1^{-1} \circ f_2^{-1} \circ \cdots \circ f_k^{-1})$$
$$= f_k \circ \cdots \circ f_2 \circ (f_1 \circ f_1^{-1}) \circ f_2^{-1} \circ \cdots \circ f_k^{-1}$$
$$= f_k \circ \cdots \circ (f_2 \circ i \circ f_2^{-1}) \circ \cdots \circ f_k^{-1}$$
$$= \cdots \cdots \cdots \cdots \cdots \cdots \cdots \cdots \cdots \cdots \cdots \cdots$$
$$= i,$$

where i is the identity function, $i(x) = x$.

Here is another example: Define $g(x)$ to be x^2, and take the domain of g to consist of all positive numbers. Since the square of two different positive numbers is different, we find that g is invertible; its inverse is $f(x) = \sqrt{x}$.

Note that if we had defined $g(x) = x^2$ and taken its domain to be all real numbers, not just the positive ones, g would not be invertible. Thus invertibility depends crucially on what we take to be the domain of the function.

In later chapters we shall show how calculus can be used to decide if a given function is invertible, and how it can be used to devise practical methods for determining the inverse.

EXERCISES

3.1 Calculate both $f \circ g$ and $g \circ f$ for the following pairs of functions:

(a) $f(x) = \dfrac{x + 1}{x - 1}$, $\quad g(x) = x^2$.

(b) $f(x) = x + \dfrac{1}{x}$, $\quad g(x) = x - \dfrac{1}{x}$.

(c) $f(x) = |x|$, $\quad g(x) = x^3$.

(d) $f(x) = ax$, $\quad g(x) = bx$.

(e) $f(x) = x^a$, $\quad g(x) = x^b$.

(f) $f(x) = \log x$, $\quad g(x) = e^x$.

3.2 Calculate the triple composite $f \circ g \circ h$ for

$$f(x) = \log x, \quad g(x) = x^a, \quad h(x) = e^x.$$

3.3 Suppose that f and g are a pair of functions such that

$$f \circ g = g \circ f.$$

Let a be any invertible function, b its inverse. Define a new pair f_1, g_1 by

$$f_1 = a \circ f \circ b, \qquad g_1 = a \circ g \circ b.$$

Prove that

$$f_1 \circ g_1 = g_1 \circ f_1.$$

3.4 Show that the function

$$f(x) = (1 - x^n)^{1/n}, \qquad 0 \le x \le 1, \qquad n \text{ a positive integer},$$

is its own inverse by verifying that

$$f \circ f(x) = f[f(x)] = x.$$

Also show that each of the following functions of s:

$$s, \qquad -s, \qquad \frac{1}{s}, s \ne 0, \qquad \frac{s+1}{s-1}, s \ne 1$$

has the property that it is its own inverse. Are the functions

$$f(t) = 2 + \frac{1}{(t-2)}, t \ne 2$$

and

$$f(t) = \frac{at+b}{t-a}, t \ne a,$$

their own inverses?

3.5 Find the inverse f^{-1} and sketch both f and f^{-1} for

(a) $f(r) = r - 1$.

(b) $f(r) = (2r + 2)^{1/2}, r > -1$.

(c) $g(s) = \dfrac{s-3}{s+2}, s \ne -2$.

(d) $g(s) = s^2 + 4, s \ge 0$.

2.4 Sums, products, and quotients of functions

In this section we discuss three kinds of operations which make a new function out of two given functions f and g. Suppose that $f(t)$ and $g(t)$ denote the amounts of water contained at time t in two different reservoirs. Then the total amount of water reserve at time t is

$$f(t) + g(t).$$

The function whose value at time t is the sum of the values of the functions f and g is called *the sum* of f *and* g and is denoted by

$$f + g.$$

Suppose that the price of a certain item at time t is $f(t)$ dollars per pound and that $g(t)$ pounds of it are available. Then the total value in dollars of what is available is

$$f(t)g(t).$$

The function whose value at t is the product of the values of f and g at t is called the *product of* f *and* g, and is denoted by

$$fg.$$

Note that the sum and product of two functions is defined only at those points where both f and g are defined; in other words, the domain of $f + g$ and of fg is the *intersection* of the domains of f and of g (that is, the set of all points common to the domains of f and g).

Starting with the simplest functions, the *constant functions*

$$c(x) = c,$$

and the *identity function*

$$i(x) = x,$$

we can, by forming sums and products, build very complicated functions. For instance, forming the product of the identity function with itself repeatedly, we get the functions x^2, x^3, x^4, etc. Multiplying these functions by constants and adding them we get functions such as

$$1.8x^2 + 3.71x - 1.84$$

or

$$3x^3 - 4.27x^2 + 8.53x - 2.71,$$

etc. It is not too difficult to see that all functions p that can be obtained from the constant and identity functions by repeated additions and multiplications are of the form

(4.1)
$$p(x) = a_N x^N + a_{N-1} x^{N-1} + \cdots + a_0,$$

where the a's are constants, and N is a positive integer. Such a function is called a *polynomial*; the word refers to the fact that when written in form (4.1), p is the sum of many terms. The numbers $a_0, \ldots, a_N$ are called the *coefficients* of p. The highest power of x which occurs in p is called the *degree* of p.

We turn to the last of the arithmetic operations, *division*. Let f be any function; its *reciprocal*, denoted by $1/f$, is the function whose value at x is $1/f(x)$. The reciprocal of f is defined wherever f is defined and *not* equal to zero.

The following are examples of functions which can be built out of the constants and the identity function by adding, multiplying and forming reciprocals:

$$\frac{x+1}{x-1}, \qquad \frac{1}{x^2+1}, \qquad \frac{x^5+1+2x^2}{2+x^3}.$$

Multiplying a polynomial p with the reciprocal of another, q, yields the function

$$r = \frac{p}{q}.$$

Such a function, a ratio of two polynomials, is called a *rational function*. We claim that no other functions can be built out of the constants and the identity function by multiplication, division and addition; for, *the product, quotient and sum of two rational functions are rational*: The product

$$r_1 r_2 = \frac{p_1}{q_1} \frac{p_2}{q_2} = \frac{p_1 p_2}{q_1 q_2}$$

is rational; the quotient

$$\frac{r_1}{r_2} = \frac{p_1}{q_1} \bigg/ \frac{p_2}{q_2} = \frac{p_1 q_2}{q_1 p_2}$$

is rational. Finally, the sum

$$r_1 + r_2 = \frac{p_1}{q_1} + \frac{p_2}{q_2} = \frac{p_1 q_2 + p_2 q_1}{q_1 q_2}$$

is a rational function.

EXERCISES

4.1 Determine the product function fg where

$$f(x) = |x| - x, \qquad g(x) = |x| + x.$$

4.2 If f and g are polynomials of degree M and N respectively, what is the degree of their sum? of their product?

4.3 Define $f^K(x)$ as the product $f(x) \cdot f(x) \cdots f(x) = [f(x)]^K$ of the K equal factors $f(x)$ for any positive integer K. If f and g are polynomials of degree M and N, respectively, what is the degree of $f^K + g^L$ (L, K positive integers)? What is the degree of $f^K g^L$?

4.4 $f = -x^3 - 3x^2 - x + 3$, $\qquad g = 3x^3 - 3x^2 - 3x + 1$. Find the degree of $3f + g$.

4.5 Let $r(x) = \dfrac{x^2 + 5x + 6}{x + 2}$, $\quad p(x) = x + 3$.

(a) What is the domain of r?
(b) Find $r(-3)$, $p(-3)$; $r(-1)$, $p(-1)$; $r(2)$, $p(2)$.
(c) Are there any inputs x such that $p(x) \neq r(x)$?
(d) Find the composite $r \circ p(x) = r[p(x)]$. What is its domain?
(e) Let $s(x) = 1/p(x)$; form the composite $r \circ s(x) = r[s(x)]$.

2.5 Graphs of functions

The first five examples of functions presented in Section 2.1 were in graphical form; in this section we present a brief but systematic discussion of functions and their graphs.

First a brief review of the coordinate plane. Two perpendicular lines are chosen as the x and y *coordinate axes*; their point of intersection is called the *origin O*. It divides each axis into two rays. One ray of each axis (traditionally the right ray of the horizontal and upward ray of the vertical) is chosen as the *positive* half axis; see Figure 2.15. Given any point P, we construct through it

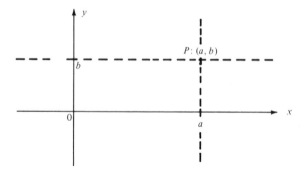

Figure 2.15

2 lines, one parallel to the y-axis, the other to the x-axis. The first line intersects the x-axis at some point whose *signed* distance from the origin we denote by a; the second line intersects the y-axis at some point whose signed distance from the origin is b. The numbers a and b are called the x *and* y *coordinates* of the point P, and the point P is denoted by (a, b). Given any two numbers a and b there exists exactly one point P with x-coordinate a and y-coordinate b; P is the intersection of two lines, one parallel to the y-axis, and the other parallel to the x-axis.

The *distance* between two points $P_1 = (a_1, b_1)$ and $P_2 = (a_2, b_2)$ is denoted by $d(P_1, P_2)$ and is given by the Pythagorean formula

$$d(P_1, P_2) = \sqrt{(a_1 - a_2)^2 + (b_1 - b_2)^2}.$$

The set of points of the form

$$P = (a, a)$$

form a straight line, sometimes called the *diagonal* in the x, y plane.

The points (a, b) and (b, a) are reflections of each other across the diagonal $x = y$.

The *graph* of a function defined in some interval S is the set of all points $(x, f(x))$, x in S. The equation

$$y = f(x)$$

is called the *equation of the graph*. We shall show next that various interesting properties of functions correspond to definite geometrical properties of their graphs. We present a "dictionary" of analytical properties of functions and equivalent geometric properties of their graphs:

1. f is an increasing function of x, i.e., $f(a) < f(b)$ whenever $a < b$.
2. f is a decreasing function of x, i.e., $f(a) > f(b)$ whenever $a < b$.
3. f is invertible.

4. f and g are inverses of each other.

5. f is even, i.e., $f(-x) = f(x)$.

6. f is odd, i.e., $f(-x) = -f(x)$.

1. The graph of f rises as x moves to the right.
2. The graph of f falls as x moves to the right.
3. Lines parallel to the x-axis intersect the graph in at most one point.
4. The graph of f is the reflection of the graph of g across the diagonal $x = y$.
5. The graph of f is symmetric with respect to the y-axis.
6. The graph of f is symmetric with respect to the origin.

We illustrate each of these geometrical properties in Figure 2.16a–f.

Functions that are either increasing or decreasing are called *monotonic*. If $f(x)$ is monotonic in its domain $a \le x \le b$, then the graph of its inverse, $f^{-1}(x)$, for all points x between $f(a)$ and $f(b)$, is simply the reflection of the graph of f across the diagonal $y = x$ (see Figure 2.17).

Notice that you can tell, by merely glancing at the graph of a function, whether that function is increasing, decreasing or neither. You can tell with equal ease whether a graph is symmetric with respect to the y axis, or the origin or neither. Thus the graphical representation of a function is not only a compact way of storing the information contained in a function, but it is a form of storage from which human observers can easily extract relevant information. However, for present-day digital computers the graphical form is *not* the most convenient for storing functions.

As we continue the study of calculus we shall come across further interesting properties of functions which correspond to easily discernible geometric

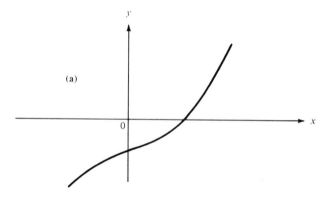

Figure 2.16   (a) Graph of an increasing function:
this function is invertible.

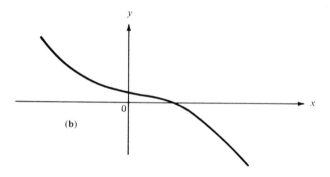

(b) Graph of a decreasing function: this function is invertible.

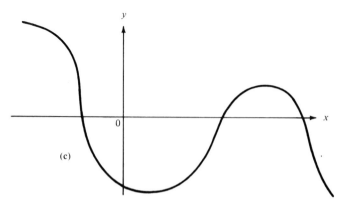

(c) Graph of a function which neither increases nor decreases:
this function is not invertible.

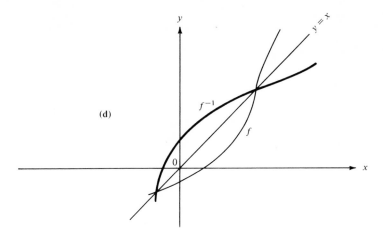

(d) Graph of a function f and its inverse f^{-1}.

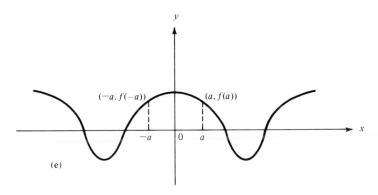

(e) Graph of an even function: $f(-x) = f(x)$.

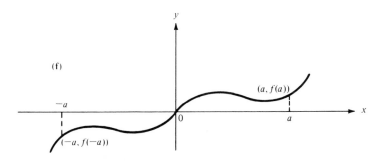

(f) Graph of an odd function: $f(-x) = -f(x)$.

57

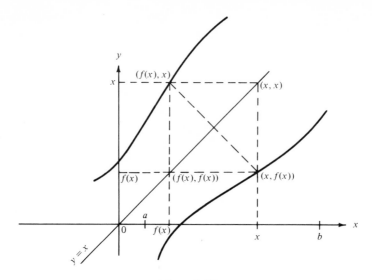

Figure 2.17

properties of their graphs. Such properties are continuity, discussed in Section 2.7, differentiability, discussed in Chapter 3, and convexity, discussed in Section 3.8.

EXERCISES

5.1 Given the two linear functions

$$l_1(x) = 3x + 4, \qquad l_2(x) = -x + 2.$$

(a) Find the point of intersection of the two linear functions.
(b) Graph the functions and check your answer obtained in (a).

5.2 Graph the second degree curves
(a) $y = x^2$.
(b) $x = y^2$.
(c) $y = x^2 - 2x$.
(d) $y = x^2 + 2x - 2$.
(e) $x = y^2 + 2x - 2$.

The graph of a function $y(x)$ which is quadratic in x is called a *parabola*.

5.3 Find the points of intersection by graphing
(a) $y = x^2$, $y = x$, and $y = -x$.
(b) $y = x^3$ and $y = x$.
(c) $y = x^3$ and $y = x^{1/2}$.
(d) $x = y^2 + 2$ and $y = x^2 - 3$.
(e) $y = x^2 - 12x + 2$ and $y = x - 1$.

5.4 Which of the functions in Figure 2.18 are increasing, decreasing, even, or odd?

5.5 Can an even function be decreasing or increasing in an interval containing the origin?

5.6 (a) Prove that if two functions are both increasing or both decreasing, their composite is increasing.

58

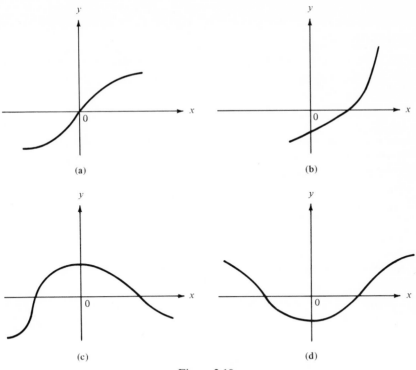

Figure 2.18

(b) Prove that if one of the functions is increasing and the other is decreasing, their composite is decreasing.

5.7 Prove than an odd function, defined in an interval containing the origin, passes through the origin.

5.8 We say that a function f consists of N monotonic pieces if its interval of definition can be divided into N adjacent subintervals on which the function f is alternately increasing and decreasing. For example, the function in Figure 2.19 consists of 5 monotonic pieces. Prove that if f consists of N monotonic pieces and g of M monotonic pieces, then their composite $f \circ g$ consists of at most MN monotonic pieces.

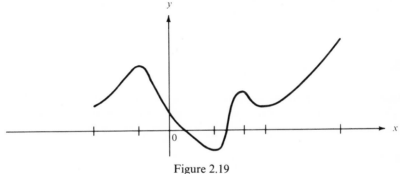

Figure 2.19

2.6 Linear functions

In this section we study the class of simplest functions, the so-called *linear* ones; these are polynomials of degree 1 as defined in Section 2.4. Every linear function l is of the form

(6.1) $$l(x) = mx + b,$$

where m and b are some given numbers. A linear function is certainly simple from the arithmetical point of view: to evaluate it we need to perform one multiplication and one addition.

The centigrade to Fahrenheit function, described in Example 8 in Section 2.1, is a fine specimen of a linear function:

$$F(C) = \tfrac{9}{5}C + 32;$$

here $m = 9/5$, $b = 32$.

It turns out that linear functions behave very simply under certain operations:

The sum of two linear functions is linear.

The composite of two linear functions is linear.

The proofs are straightforward. Let

$$l_1(x) = m_1 x + b_1, \qquad l_2(x) = m_2 x + b_2$$

be two linear functions. Then their sum

(6.2) $$(l_1 + l_2)(x) = (m_1 + m_2)x + b_1 + b_2,$$

and their composite

(6.3) $$(l_1 \circ l_2)(x) = m_1(m_2 x + b_2) + b_1$$
$$= m_1 m_2 x + m_1 b_2 + b_1$$

are both linear functions.

Which linear functions are invertible? If $m = 0$, then $l(x) = mx + b$ is independent of x and is therefore not invertible. If $m \neq 0$ we can solve

$$y = mx + b$$

for x as a linear function of y; we get

(6.4) $$x = \frac{1}{m} y - \frac{b}{m}.$$

Thus *every nonconstant linear function is invertible, and its inverse also is linear.*

On the other hand the product and quotient of two linear functions are not linear, except in very special cases. For example, the product of the linear functions

$$l_1(x) = 2x - 3, \qquad l_2(x) = x + 1$$

is $l_1 l_2(x) = 2x^2 - x - 3$, which is not linear; nor is their quotient

$$\frac{l_1(x)}{l_2(x)} = (2x - 3)\frac{1}{x + 1} = 2 - \frac{5}{x + 1}.$$

(See Exercise 6.10.)

Observe that the operations of addition, composition and inversion of linear functions do not lead out of the class of linear functions, whereas multiplication and division do; see Exercise 6.6.

What is the graph of a linear function like? We shall show that it is a *straight line*—hence the name linear.

The proof is based on the following property of straight lines: Let $P_1 = (x_1, y_1)$ and $P_2 = (x_2, y_2)$ be any pair of points on a line. The ratio

$$\frac{y_1 - y_2}{x_1 - x_2}$$

is *always* the same, no matter which pair of points on the line are chosen. The common value of these ratios is called the *slope* of the line.

To obtain a geometric interpretation of the slope we choose P_1 to be the point of intersection of the line with the x-axis and choose P_2 so that y_2 is positive; see Figure 2.20. Denote by θ the angle made by the ray $P_1 P_2$ with the ray starting at P_1 and going in the positive x-direction. According to elementary trigonometry

$$\tan \theta = \frac{y_2 - 0}{x_2 - x_1} = \text{slope}.$$

Given a line, denote its slope by m; let $P_1 = (x_1, y_1)$ be some point on the line, and denote by $P = (x, y)$ any other point on the line. According to the property of straight lines stated before,

(6.5)
$$\frac{y - y_1}{x - x_1} = m.$$

Multiplying by $x - x_1$ we get, after rearrangement, that

(6.6)
$$y = m(x - x_1) + y_1 = mx - mx_1 + y_1.$$

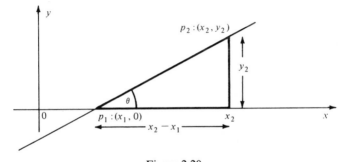

Figure 2.20

Note that Equation (6.6) is satisfied by $x = x_1$, $y = y_1$ as well, so that all points (x, y) on the line satisfy (6.6). Thus the originally given line is the graph of the *linear* function (6.6). We shall call (6.6) *the equation of the line through* $P_1 = (x_1, y_1)$, *with slope m*. In particular, the equation of the line through the point $(0, b)$ with slope m is

$$y = mx + b.$$

Since b and m are arbitrary, this shows that *the graph of any linear function is a straight line*.

Example. To find an equation of the straight line through the points $P_1 = (-1, 2)$, $P_2 = (2, 5)$, we calculate its slope:

$$m = \frac{2 - 5}{-1 - 2} = \frac{-3}{-3} = 1.$$

Using (6.6) with $m = 1$, $x_1 = -1$, $y_1 = 2$ we get

$$y = l(x) = x + 3.$$

Suppose two linear functions l_1 and l_2 have the same value at $x = s$ and at $x = t$, $s \neq t$:

$$l_1(s) = l_2(s) \quad \text{and} \quad l_1(t) = l_2(t).$$

Then the graph of both l_1 and l_2 passes through the two points

$$P_1 = (s, l_1(s)), \qquad P_2 = (t, l_2(t)).$$

Since there is only one line passing through these two distinct points, it follows that the graphs of l_1 and l_2, and so the functions l_1 and l_2 themselves, are the same. Thus we have shown, that *if two linear functions have the same value for two distinct inputs, then they have the same value for all inputs*.
Let's call the coefficient m occurring in the linear function

$$l(x) = mx + b$$

the *slope* of the linear function. Then Formulas (6.2), (6.3), and (6.4) can be expressed verbally as follows:

The slope of $l_1 + l_2$ is the *sum* of the slope of l_1 and the slope of l_2.
The slope of $l_1 \circ l_2$ is the *product* of the slope of l_1 and the slope of l_2.
The slope of the inverse of l is the *reciprocal* of the slope of l.

In conclusion we define a *linear function l of 2 variables x and y* as a function of the form

$$l(x, y) = mx + ny + b.$$

Obviously, the sum of two linear functions of 2 variables is again linear. A linear function of 3, 4, or any number of variables is defined similarly.

EXERCISES

6.1 Determine the linear functions whose graphs go through the following pairs of points:
(a) $P_1 = (1, 2)$, $P_2 = (3, 4)$.
(b) $P_1 = (-1.7, 0.4)$, $P_2 = (0.5, -2.8)$.
(c) $P_1 = (4.8, -1)$, $P_2 = (-2.3, 4.1)$.

6.2 Determine the linear functions whose graphs go through the point P and have slope m as given below:
(a) $P = (1, 2)$, $m = 1$.
(b) $P = (-0.8, 2.7)$, $m = 1.3$.
(c) $P = (0.5, -1.3)$, $m = -0.7$.

6.3 Determine the point of intersection of the pairs of lines which are the graphs of the following pairs of functions:
(a) $l_1(x) = 1.7x + 2$, $l_2(x) = x$.
(b) $l_1(x) = -0.4x + 1.5$, $l_2(x) = 1.3x - 1.2$.
(c) $l_1(x) = 2x + 3$, $l_2(x) = 3x + 4$.

6.4 Prove that if two lines are perpendicular, the product of their slopes is -1.

6.5 Find a point q on the graph of $l(x) = 2x + 3$ such that the line through q and the point $(1, 1)$ is perpendicular to the graph of $l(x) = 2x + 3$.

6.6 Show that every linear function has the property that $l(x + h) - l(x)$ does not depend on x. This property states, somewhat loosely, that: *equal increments in x produce equal increments in l.* Prove that if two functions have this property, so does their sum and composite.

6.7 Determine all pairs of linear functions l_1 and l_2 such that

$$l_1 \circ l_2 = l_2 \circ l_1.$$

6.8 x_f is called a fixed point of the function f if $f(x_f) = x_f$; that is, if f maps x_f into itself. Show that, if l_1 and l_2 are linear functions such that $l_1 \circ l_2 = l_2 \circ l_1$, a fixed point of l_1 is a fixed point of l_2.

6.9 Prove *algebraically* that if two linear functions l_1 and l_2 agree for two distinct values of x, then they agree for all x.

6.10 Show that the function $y = 1/(x + 1)$ is not linear by verifying that the quotients

$$\frac{y_1 - y_2}{x_1 - x_2} = \frac{1/(x_1 + 1) - 1/(x_2 + 1)}{x_1 - x_2} = \frac{-1}{(x_1 + 1)(x_2 + 1)}$$

are not constant.

2.7 Continuous functions

In this section we shall scrutinize the definition of function given in the first section of this chapter; according to that definition, a function f is a rule which assigns a definite value $f(x)$ to each number x in the domain where f is defined. Clearly, in order to find the value of $f(x)$, we have to know x.

But what does knowing x mean? It means, according to Chapter 1, being able to produce as close an approximation to x as is requested. This means that we never (or hardly ever) know x *exactly*; how then can we hope to determine $f(x)$? To find a way out of this dilemma we have to remember that determining $f(x)$ means being able to give as close an approximation to $f(x)$ as requested. So we can determine $f(x)$ if *approximate knowledge of x is sufficient for approximate determination of $f(x)$*. Approximate knowledge of x means that we know all digits of x up to the m-th; this is the same as saying that we know an interval of length 10^{-m} within the domain of f in which x lies. *If the values which f takes in this interval of length 10^{-m} lie in an interval of length 10^{-k}, this information about x suffices to determine all digits of $f(x)$ up to the k-th.* This property of the function f can be expressed as a

Continuity criterion. *In order for $f(x)$ and $f(y)$ to be so close that*

$$(7.1) \qquad\qquad |f(x) - f(y)| < 10^{-k},$$

it suffices for x and y to be so close that

$$(7.2) \qquad\qquad |x - y| < 10^{-m}.$$

The choice of m depends on k.

A function f which has this property for x in the domain of f is called *continuous on its interval of definition.*

Example 1. Any constant function, i.e., a function which has the same value for all x in its domain, is continuous. In fact, no function can be more continuous than that!

Example 2. The identity function $f(x) = x$ is continuous on any interval. Clearly it suffices to choose m equal to k.

Example 3. The function $f(x) = x^2$ is continuous on the interval $[-5, 5]$. For

$$f(x) - f(y) = x^2 - y^2 = (x + y)(x - y).$$

Since both x and y lie in the interval $[-5, 5]$ of length 10,

$$|x + y| \le 10,$$

and so

$$|f(x) - f(y)| = |x^2 - y^2| = |x + y||x - y| \le 10|x - y|.$$

If

$$|x - y| < 10^{-k-1},$$

then

$$|f(x) - f(y)| < 10^{-k}.$$

So in this case it suffices to choose $m = k + 1$.

Remark. $f(x) = x^2$ is continuous on every finite interval (see Exercise 7.3).

Example 4. The function

$$f(x) = \frac{1}{x}$$

is continuous on any closed interval not containing the origin. To see this we write

$$f(x) - f(y) = \frac{1}{x} - \frac{1}{y} = \frac{y - x}{xy}.$$

Let p be the endpoint of the interval closer to the origin. It follows that

$$|f(x) - f(y)| \le \frac{|y - x|}{p^2},$$

since, for any x in the domain of f, $|x| \le |p|$. If $|x - y| \le 10^{-m}$, then

$$|f(x) - f(y)| \le \frac{1}{p^2} 10^{-m} \le 10^{-m+2P},$$

where we choose P so large that

$$10^P > \frac{1}{p}.$$

Clearly choosing $m = k + 2P$ suffices to make $f(x)$ and $f(y)$ differ by less than 10^{-k}.

Example 5. The function $f(x) = \sqrt{x}$ is continuous in the interval $[0, 1]$. We shall determine $m(k)$ so that

$$|\sqrt{x} - \sqrt{y}| < 10^{-k} \quad \text{whenever} \quad |x - y| < 10^{-m}, \qquad x, y \text{ in } [0, 1].$$

To this end, we write

$$|\sqrt{x} - \sqrt{y}| = |\sqrt{x} - \sqrt{y}| \frac{\sqrt{x} + \sqrt{y}}{\sqrt{x} + \sqrt{y}} = \frac{|x - y|}{\sqrt{x} + \sqrt{y}}.$$

65

We claim that $\sqrt{x} + \sqrt{y} \geq \sqrt{|x - y|}$, and leave the proof of this inequality to the reader. We now increase the last member of the previous inequality by decreasing its denominator; thus

$$|\sqrt{x} - \sqrt{y}| = \frac{|x - y|}{\sqrt{x} + \sqrt{y}} \leq \frac{|x - y|}{\sqrt{|x - y|}} = \sqrt{|x - y|},$$

and

$$\sqrt{|x - y|} \leq 10^{-k} \quad \text{if} \quad |x - y| \leq 10^{-2k}.$$

The desired m, therefore, is $2k$. We shall see later in this section that the continuity of $\sqrt{x}$ in $[0, 1]$ follows, without any computation, from the inversion theorem.

Instead of presenting more special examples we shall prove a continuity theorem which will enable us to construct (or recognize) a host of continuous functions from functions known to be continuous. We recall that f is continuous on an interval S if a pair of outputs, $f(x)$ and $f(y)$, can be forced to differ by less than 10^{-k} by the restriction that the corresponding inputs, x and y, differ by less than a number 10^{-m} (which depends on the prescribed tolerance 10^{-k}).

Continuity theorem

(a) *The composite of two continuous functions is continuous.*
(b) *The sum of two continuous functions is continuous.*
(c) *The product of two bounded continuous functions is bounded and continuous.**
(d) *The reciprocal of a continuous function bounded away from zero is continuous.*

PROOF
(a) Let f and g be continuous functions on their intervals of definition; their composite $f \circ g$ is defined as

$$f \circ g(x) = f(g(x)).$$

Since f is continuous,

(7.3) $$|f(g_1) - f(g_2)| < 10^{-k}$$

provided that

(7.4) $$|g_1 - g_2| < 10^{-m},$$

* We show later that a continuous function defined on a closed finite interval is automatically bounded.

m depending on k. Since g is continuous,

(7.5) $$|g(x_1) - g(x_2)| < 10^{-m},$$

provided that

(7.6) $$|x_1 - x_2| < 10^{-n},$$

n depending on m. Now identify g_1 as $g(x_1)$ and g_2 as $g(x_2)$; then (7.5) and (7.4) are the same statements. The former is a consequence of (7.6), while the latter has (7.3) as consequence. All in all we have proved that (7.6) implies that

$$|f(g(x_1)) - f(g(x_2))| < 10^{-k};$$

this proves the continuity of the composite function $f \circ g$.

(b) Continuity of f and of g means that for x and y close enough,

$$|f(x) - f(y)| < 10^{-k} \qquad \text{and} \qquad |g(x) - g(y)| < 10^{-k}.$$

Clearly this implies that

$$|(f + g)(x) - (f + g)(y)| < 2 \cdot 10^{-k}.$$

We have introduced the modifying phrase "close enough" because the functions f and g may require different tolerances $|x - y|$ on their inputs, and so we choose the smaller of these distances.

(c) Boundedness means that there is a number b which is bigger than $|f(x)|$ and $|g(x)|$ for all x in the domains of definition of f and g:

(7.7) $$|f(x)| < b, \qquad |g(x)| < b.$$

Now write the difference of the two products as follows:

(7.8) $$f(x)g(x) - f(y)g(y) = f(x)(g(x) - g(y)) + (f(x) - f(y))g(y).$$

Since f and g are continuous, for x and y close enough

$$|f(x) - f(y)| < 10^{-n} \qquad \text{and} \qquad |g(x) - g(y)| < 10^{-n};$$

and, according to (7.7),

$$|f(x)| < b, \qquad |g(y)| < b.$$

Using these inequalities to estimate the right side of (7.8), we get

$$|f(x)f(x) - f(y)g(y)| < 2b \cdot 10^{-n}.$$

This implies the continuity of the product function fg.

(d) A function is said to be bounded away from zero if $|f(x)|$ is greater than some positive number p for every x in the domain of f. Denote by r the function

$$r(f) = \frac{1}{f}.$$

67

We have shown in Example 4 that this function of f is continuous on the domain $|f| > p$; therefore, according to part (a), also the composite function

$$r \circ f(x) = \frac{1}{f(x)}$$

is continuous. This completes the proof of the continuity theorem. ☐

We saw in Examples 1 and 2 that the constant function and the identity function are continuous. Since these functions are bounded on finite intervals, it follows from the continuity theorem that every function built out of these by addition and multiplication also is continuous. As we have observed in Section 2.4, every polynomial can be so built up, and so we conclude that *polynomials are continuous on finite intervals.*

A rational function, we recall from Section 2.4, is defined as the ratio of two polynomials p and q:

$$\frac{p}{q}.$$

Since according to the continuity theorem a quotient of continuous functions is continuous if the denominator is bounded away from zero, it follows from the result above about polynomials that *a rational function is continuous on any finite interval where the denominator is bounded away from zero.*

Now that we have recognized lots of functions as continuous we shall prove two theorems about them.

Intermediate value theorem. *Let f be a continuous function defined on some closed interval $I = [a, b]$. Let m be any number between the numbers $f(a)$ and $f(b)$; then there is a number c in the interval I where the value of f is m:*

$$f(c) = m.$$

The geometric evidence for the truth of this can be seen by glancing at the graph of f in Figure 2.21. Here we have chosen $f(a)$ to be $> f(b)$. In order for

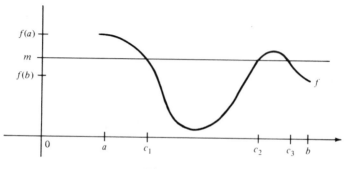

Figure 2.21

the graph to pass continuously from $f(a)$ to $f(b)$ as x goes from a to b, the graph must cross the line $y = m$ at least once. In Figure 2.21, there are in fact 3 crossings.

We proceed with an analytical proof.

PROOF. By hypothesis, $f - m$ has opposite signs at a and b, say positive at a and negative at b. Divide the interval I at its midpoint into two equal parts I_1 and I_2; see Figure 2.22. If at the midpoint $\frac{1}{2}(a + b)$ the value of f happens to be m, we have already attained our objective. If not, we select either the interval I_1 or I_2, depending on whether $f(\frac{1}{2}(a + b))$ is less than or greater than m. Note that this strategy always produces a subinterval at whose endpoints $f - m$ has opposite signs. Continuing in this fashion we obtain an infinite sequence of closed intervals I_n, each half as long as the previous, and each having the aforementioned property. According to the *nested interval property* proved in Section 1.6, the intervals I_n have exactly one point c in common. We claim that $f(c) = m$. For suppose not; then $f(c)$ is either greater, or less than m. Since f is assumed continuous, there is an interval J of length 10^{-k} around c such that the value of $f(x)$ in this whole interval is greater than m in the first case, less than m in the second case. On the other hand, the length of the interval I_n tends to zero as $n \to \infty$. As soon as the length of I_n is less than 10^{-k}, I_n is part of J; clearly the value of $f - m$ has the *same* sign at the endpoints of such an interval I_n, contrary to our construction of the intervals I_n. □

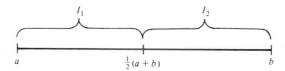

$$I_1 \qquad\qquad I_2$$

$$a \qquad\qquad \tfrac{1}{2}(a + b) \qquad\qquad b$$

Figure 2.22

The method just described for finding a number c where the value of f is m is called the *bisection* method. Let's see how well it works in practice. Suppose that the length $b - a$ of the originally given interval I is 1; then n steps of the procedure narrow the location of c to an interval of length $1/2^n$. For instance, 10 steps place c within an interval of length

$$\frac{1}{2^{10}} = \frac{1}{1024} < 10^{-3}.$$

Thus ten steps of the bisection method suffice to determine the first three digits of c after the decimal point. In Section 3.10, we shall describe a method for locating c that, under favorable circumstances, converges faster than the bisection method.

As Figure 2.21 shows, there may be several points c where f equals m; the bisection method finds only one of them. However, if f is *monotonic*, i.e., either increasing or decreasing, then there can be only one such value c.

This means a given output m is produced by only one input c. We recall that a function can be inverted if distinct inputs always yield different outputs. This leads us naturally to an

Inversion theorem. *Suppose that f is a continuous and monotonic function defined on an interval $[a, b]$. Then its inverse g is a continuous monotonic function defined on the interval $[f(a), f(b)]$.*

PROOF. As we have already noted, a monotonic function is invertible. According to the intermediate value theorem, for any m between $f(a)$ and $f(b)$

$$f(x) = m$$

has a solution c; on the other hand, the value $f(c)$ of a monotonic function at a point c between a and b lies between $f(a)$ and $f(b)$. This shows that the inverse g of f is defined precisely on the interval $[f(a), f(b)]$. There remains to show that the inverse g is continuous. Divide the interval $[a, b]$ into subintervals of length $\frac{1}{2} \cdot 10^{-k}$; denote by $a_1, \ldots, a_l$ the points of subdivision, i.e., $a_0 = a$, $a_l = b$, and

$$|a_{j+1} - a_j| = \tfrac{1}{2} 10^{-k}, \quad j = 0, 1, 2, \ldots, l - 1.$$

The values $f(a_j)$ of f at these points divide the range of f into an equal number of subintervals. Denote by d the length of the smallest of these; see Figure 2.23. Let y_1 and y_2 be any two points in the range of f:

$$f(x_1) = y_1, \qquad f(x_2) = y_2.$$

By definition of inverse,

$$g(y_1) = x_1, \qquad g(y_2) = x_2.$$

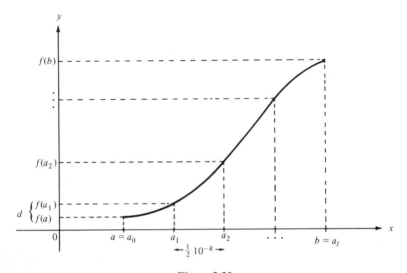

Figure 2.23

Suppose that

$$|y_1 - y_2| < d;$$

then y_1 and y_2 belong to the same or to adjacent intervals of the subdivision of the y-axis. Correspondingly, x_1 and x_2 belong to the same or to adjacent intervals of the subdivision of the x-axis. Since the intervals of subdivision of the x-axis have length $\frac{1}{2}10^{-k}$, two points x_1 and x_2 belonging to adjacent intervals have distance less than 10^{-k}:

$$|x_1 - x_2| < 10^{-k}.$$

Thus we have proved that $|y_1 - y_2| < d$ implies that

$$|g(y_1) - g(y_2)| < 10^{-k}.$$

This is precisely what continuity of g means.   □

As an application of the inversion theorem, take f as $f(x) = x^n$, n any integer. Clearly this is a continuous and increasing function on any interval $[0, b]$, and so it has an inverse g. The value of g at y is called the n-th *root* of y and is written with a fractional exponent:

$$g(y) = y^{1/n}.$$

Example 5 gave an independent proof for the case $n = 2$. In Figure 2.24 we graph several power functions and their inverses, the associated root functions.

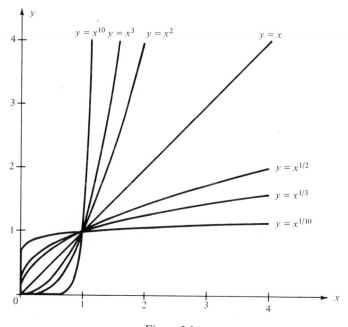

Figure 2.24

We shall see later that we can make important deductions about a function f from properties of its inverse, f^{-1}.

We now turn to another basic theorem concerning continuous functions.

Maximum value theorem. *A continuous function f defined on a closed finite interval $I = [a, b]$ takes on its maximum at some point c of I. In other words, the value of f at c is at least as great as the value of f at any other point of I: $f(c) \geq f(x)$ for all x in I.*

PROOF. First we show that such a function f is bounded. This means the value of f at any point in I is less than some number U:

$$f(x) < U.$$

For, since f is continuous, the difference between the value of f at x_1 and at x_2 is less than some constant, say d, provided that x_1 and x_2 don't differ too much:

$$|x_1 - x_2| < 10^{-k}.$$

Given any point x of the interval I, we can connect it through a chain of l points $a_1, a_2, \ldots, a_l = x$ to the left end point a so that two consecutive points have distances $< 10^{-k}$; see Figure 2.25. Since $f(a_i)$ differs from $f(a_{i-1})$ by less than d, $f(x)$ differs from $f(a)$ by less than $d \cdot l$. This construction can be carried out for all x in I and shows that the outputs $f(x)$ form a bounded set.

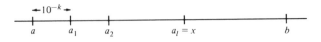

Figure 2.25

According to the least upper bound theorem (l.u.b.) of Section 1.6, every bounded set of numbers has a least (i.e., smallest) upper bound. We apply it to the *range* of the function $f(x)$ for all x in I.

Denote the least upper bound of the range of f by M; M is the *smallest* number such that

$$f(x) \leq M$$

for all x in I. We divide the interval I by its midpoint into two subintervals I_1 and I_2. Denote by M_1 and M_2 the least upper bounds of f over I_1 and I_2, respectively. Clearly, since M is an upper bound for f over I, M_1 and M_2 are both $\leq M$. On the other hand the larger of the two numbers M_1 and M_2 is an upper bound for f over both I_1 and I_2; since M is the smallest upper bound for f over I, it follows that either M_1 or M_2, or possibly both, are equal to M. We choose that interval I_1 or I_2 over which the l.u.b. of f equals M.

Proceeding in this fashion, we construct a sequence of intervals, each contained in the previous one, such that the l.u.b. of f over each equals M. The length of the n-th subinterval is $1/2^n$. According to the nested interval principle (see Section 1.6), these intervals have exactly one point c in common. We claim that $f(c) = M$. For, suppose on the contrary, that $f(c) < M$; then there would be a positive number p such that

$$f(c) = M - p.$$

Since f is continuous, $f(x)$ differs little from $f(c)$ if x lies sufficiently close to c:

$$f(x) < M - \tfrac{1}{2}p \quad \text{for} \quad |x - c| \leq 2^{-N}.$$

All points x of the N-th interval (of length 2^{-N}) satisfy that condition. Therefore it would follow that $M - \tfrac{1}{2}p$ is an upper bound for f there, contrary to the property that the smallest upper bound for f over the N-th subinterval is M. We were led to this contradiction by assuming that $f(c)$ was less than M. This leaves us with

$$f(c) = M$$

as the only tenable alternative. Thus the function f reaches its maximum at c.

$\square$

Quite analogously we can show that there is a point where the function f reaches its minimum.

The key argument in the construction of c was the proposition that if an interval is divided into two parts, the least upper bound of f over at least one of the parts is equal to the least upper bound of f over the whole interval. Since we have presented no effective method for deciding which of the two parts has that property, our proof yields no practical procedure for finding even approximately the point where f assumes its maximum.

In the second Fortran program (see p. 487) we present an effective method for locating the maximum of a function $f(x)$ defined in an interval $[x_1, x_2]$ which works for functions in a special class. The method compares enough function outputs to determine in which of two adjacent subintervals the function has a maximum.

EXERCISES

7.1 We have shown in Section 2.6 that a linear function l can be characterized by the condition that its graph has a constant slope, i.e., that

$$\frac{l(x_1) - l(x_2)}{x_1 - x_2}$$

has the same value for all distinct pairs of numbers x_1, x_2. Let f be any function; we shall say that *its slopes are bounded by m* if

$$\left| \frac{f(x_1) - f(x_2)}{x_1 - x_2} \right| \leq m$$

for all distinct pairs of numbers x_1, x_2 in its domain.

    (a) Prove that a function f whose slopes are bounded is continuous.

    (b) Prove that if the slopes of f are bounded by m, and those of g are bounded by n, then

        (i) the slopes of $f + g$ are bounded by $m + n$.

        (ii) the slopes of $f \circ g$ are bounded by mn.

    (c) Prove that if the slopes of f are bounded by m and the slopes of the inverse of f are bounded by n, then $nm \geq 1$.

7.2 Prove that the function $f(x) = |x|$ is continuous.

7.3 Prove that $f(x) = x^2$ is continuous on the interval $[-b, b]$.

7.4 Write a computer program which, given a function f having opposite sign at two points a and b, uses the bisection method to find an interval of length $< 10^{-m}$ between a and b containing a point where the function f vanishes. Use this program to

    (a) solve $x^3 - 5x^2 + 3x + 2 = 0$;

    (b) construct a 6 digit table of cube roots of integers < 20.

    (c) Compare your program with the one on page 485.

2.8 Convergent sequences of functions

In Chapter 1 we described a method for constructing a sequence of approximations to the square root of two, which is applicable to the square root of any positive number x. It goes like this: The first approximation, s_1, is arbitrary; for the sake of definiteness, let us choose

(8.1)
$$s_1 = x.$$

All subsequent approximations are constructed, one after another, according to the rule

(8.2)
$$s_{n+1} = \frac{1}{2}\left(s_n + \frac{x}{s_n}\right).$$

The first few of these are

$$s_1 = x$$

(8.3)
$$s_2 = \frac{1}{2}(x + 1)$$

$$s_3 = \frac{1}{2}\left[\frac{1}{2}(x + 1) + \frac{2x}{x + 1}\right] = \frac{1}{4}(x + 1) + \frac{x}{x + 1}.$$

The formulas for s_n become more and more complicated, but one fact stands out clearly: *Each s_n is a function of* x. In fact, each s_n is a rational function of x. The reason for this is that each $s_n(x)$ is obtained by addition, multiplication and division performed on the previous s_{n-1} and x; arithmetic operations performed on rational functions lead to rational functions.

In Section 1.4, we started with an arbitrary first approximation u_0 to $\sqrt{2}$; here we begin with the approximation x to $\sqrt{x}$. In a manner analogous to the argument given earlier, we find an estimate for how close $s_n(x)$ lies to $\sqrt{x}$:

$$(8.4) \qquad |s_n(x) - \sqrt{x}| \le K\left(\frac{1}{2}\right)^{n-1},$$

where K is $|s_1 - \sqrt{x}| = |x - \sqrt{x}|$.

Suppose we restrict our attention to the interval $0 \le x \le 1$. For x in this interval,

$$|x - \sqrt{x}| \le \frac{1}{4},$$

because $(\sqrt{x} - \frac{1}{2})^2 = x - \sqrt{x} + \frac{1}{4} \ge 0$, so

$$\frac{1}{4} \ge \sqrt{x} - x = \sqrt{x}(1 - \sqrt{x}) \ge 0.$$

Therefore, for such x we may replace the constant K in (8.4) by $1/4$:

$$(8.5) \qquad |s_n(x) - \sqrt{x}| \le \frac{1}{4}\left(\frac{1}{2}\right)^{n-1} = \left(\frac{1}{2}\right)^{n+1}.$$

This inequality shows that the larger n is, the closer the function $s_n(x)$ lies to $\sqrt{x}$. Specifically, since $(\frac{1}{2})^{10} = \frac{1}{1024} < 10^{-3}$, we see that for $n > 3m$ and for $0 \le x \le 1$,

$$|s_n(x) - \sqrt{x}| < 10^{-m}.$$

In Section 1.4, we showed that a sequence of numbers converges to $\sqrt{2}$. We have now shown that the sequence of functions $s_n(x)$ converges to $\sqrt{x}$ for $0 \le x \le 1$.

Generalizing this situation we say that *a sequence of functions g_n, defined on an interval S, converges to the function g, defined on S, if for n large enough, g_n differs little from g; that is, no matter how large an integer m we choose,*

$$|g_n(x) - g(x)| < 10^{-m}$$

for all x in S and all n greater than some integer N. How large we take N depends, of course, on m.

A symbolic way of writing "the functions g_n converge to g on S" is

$$\lim_{n \to \infty} g_n = g \text{ on } S, \qquad \text{or just} \qquad g_n \to g \text{ on } S.$$

It is important to indicate the interval S where the convergence takes place; for, a sequence of functions g_n could easily converge on a small interval and not on a larger one. The sequence

$$g_n(x) = x^n,$$

for example, converges to zero on any closed interval inside $(-1, 1)$, but does *not* converge on any interval which contains a point outside of $(-1, 1)$.

As we shall see, functions constructed by differentiation and functions obtained by solving differential equations are defined as limits of sequences. It is no exaggeration to say that almost all interesting functions are defined as limits of sequences of simpler functions. Therein lies the importance of the concept of a convergent sequence of functions.

Since most of the functions we shall study are continuous, we investigate next what convergence means for sequences of continuous functions. It turns out that continuity is preserved under the operation of taking limits.

Theorem. *Let f_n be a set of functions, each defined and continuous on an interval S. If the sequence $\{f_n\}$ converges to f on S, then f is continuous on S.*

PROOF. If $\{f_n\}$ converges, then for n large enough

$$(8.6) \qquad |f_n(x) - f(x)| < 10^{-m}$$

for all x on S. Since we have assumed f_n to be continuous, we know that for x and y in S close enough, say

$$|x - y| < 10^{-k},$$

$f_n(x)$ and $f_n(y)$ differ by less than 10^{-m}:

$$(8.7) \qquad |f_n(x) - f_n(y)| < 10^{-m}.$$

Now write

$$f(x) - f(y) = (f(x) - f_n(x)) + (f_n(x) - f_n(y)) + (f_n(y) - f(y)).$$

Referring to the right hand side we see that the absolute values of the first and third terms are less than 10^{-m} by (8.6), that of the middle term is less than 10^{-m} by (8.7). So using the triangle inequality we get

$$|f(x) - f(y)| < 3 \cdot 10^{-m}$$

for $|x - y| < 10^{-k}$. This proves the continuity of f on S. $\qquad\square$

The *rules of arithmetic* for convergent sequences of functions are the same as for convergent sequences of numbers. Suppose $f_n \to f$ and $g_n \to g$ on S. Then

(i) $f_n + g_n \to f + g$ on S.
(ii) If f and g are bounded, $f_n g_n \to fg$ on S.
(iii) If $1/f$ is bounded, then for n large enough f_n is not zero on S, and $1/f_n \to 1/f$ on S.

The proofs are the same as for sequences of numbers, see Section 1.4, except that we have to add the phrase "for all x in S" at the end.

Note that if the functions f_n, g_n are continuous and the interval S is closed and finite, then, as explained in Section 2.7, the condition in (ii) that f and

g be bounded is automatically fulfilled. Similarly, in (iii) it suffices to require that f be $\neq 0$ on S; this implies the boundedness of $1/f$. In fact, $|1/f|$ will be bounded by $1/m$, where m is the *minimum* of $|f|$ on S.

As we remarked earlier, almost every interesting function g is defined as the limit of some known sequence of functions g_n. In defining convergence of a sequence of functions g_n, we used the limit function g. But if we have no knowledge of g (except the approximations furnished by g_n), how do we determine whether or not the sequence g_n converges? The answer lies in the convergence criterion for sequences of numbers (see p. 16): A sequence of numbers a_n converges if, for n and k large enough, a_n and a_k differ little from each other, i.e., no matter how large an integer m we choose,

$$|a_n - a_k| < 10^{-m}$$

for n and k large enough. How large depends on m. We formulate this principle for functions:

Convergence criterion for functions. *A sequence of functions g_n defined on an interval S converges if, no matter how large an integer m we choose,*

$$(8.8) \qquad |g_n(x) - g_k(x)| < 10^{-m}$$

for all x in S and all n and k large enough—how large depends on m.

Since for a convergent sequence g_n and g_k differ little from the limit g if n and k are large enough, it follows that they differ little from each other. To show the converse we have to construct a function g which is the limit of the sequence. We accomplish this as follows:

At any point x of S, the sequence of *numbers*

$$a_n = g_n(x)$$

satisfies the convergence criterion for numbers. Therefore a_n tends to some limit a; this number a is taken to be the value of g at x:

$$a = \lim_{n \to \infty} a_n = \lim_{n \to \infty} g_n(x) = g(x).$$

This way g is defined at each point x of S; since (8.8) is valid for all x in S, $|g_n(x) - g(x)| < 10^{-m}$ for all x in S for n large enough. This proves that the function g is the limit of the sequence g_n.

EXERCISES

8.1 Prove that the sequence

$$g_n(x) = \frac{x}{n}$$

tends to zero on every finite interval.

8.2 Let g be a continuous function defined on an interval $[a, b]$. Prove that

$$g_n(x) = g\left(x + \frac{1}{n}\right)$$

tends to g on every interval $[a, c]$ with $c < b$.

8.3 Let f denote a function continuous on the whole real axis.
(a) Suppose $g_n \to g$ on S; prove that

$$f \circ g_n \to f \circ g \text{ on } S.$$

(b) Suppose that $g_n \to g$ on S, and $f_n \to f$ on the real axis. Prove that

$$f_n \circ g_n \to f \circ g \text{ on } S.$$

8.4 The members g_n of a convergent sequence of functions are, as n increases, better and better approximations on S to the limit g. Very often the improved approximation g_{n+1} is obtained from the previous approximation by adding a correction term:

(8.8) $$g_{n+1} = g_n + t_{n+1},$$

where t_n is a function defined on S.
(a) Show that with $g_0 = 0$, the n-th approximation can be expressed as the sum

(8.9) $$g_n(x) = \sum_{j=1}^{n} t_j(x).$$

(b) Denote by the number b_j an upper bound on S for the absolute value of t_j:

(8.10) $$|t_j(x)| \le b_j \text{ for } x \text{ in } S.$$

Suppose that the infinite sum

$$\sum_{1}^{\infty} b_j$$

is convergent. Prove that the sequence of functions g_n in (8.9) converges. [*Hint*: Use the comprarison theorem in Section 1.5.]
    The limit g of (8.9) as $n \to \infty$ is denoted as an infinite sum of functions:

$$g = \sum_{1}^{\infty} t_j.$$

If the limit exists, we say that the infinite sum converges.
(c) Prove that the infinite sum

$$\sum_{1}^{\infty} x^j$$

converges on any proper subinterval of $[-1, 1]$.
(d) Show that

$$\sum_{0}^{\infty} \frac{x^j}{j!}$$

converges on every finite interval S.

(e) Prove that

$$\sum_1^\infty \frac{x^j}{j^2}$$

converges on the interval $[-1, 1]$, but on no larger interval.

2.9 Algorithms

In Section 2.1 a function was defined as a rule which assigns an output y to any input x. We remarked that it is distracting to look at the sordid details that go into the workings of a function, and to avoid cluttering our minds with details it is useful to regard a function as a black box.

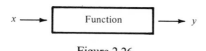

Figure 2.26

This section is about the science and art of black boxes, how to manufacture them to specifications at the least expense.

We have not yet said what an algorithm is but we have already said what an algorithm does: it computes a function. Here are some examples:

Example 1. Compute $f(x) = x^{16}$. The most obvious way of computing this function is to multiply x by x, multiply the product again by x, etc., fifteen times in all. Another way of proceeding is to square x obtaining x^2, then square the result, etc. four times in all, obtaining successively x^2, x^4, x^8 and x^{16}. Both procedures compute the same function, but they do it quite differently; the second algorithm employs only 4 multiplications in contrast to the first one, which employs 15 multiplications. So the second algorithm is faster and more efficient than the first.

Example 2. Compute $f(x) = x^n$, n any power of 2. The most obvious way of computing x^n is to perform multiplication by x, $n - 1$ times. A more sophisticated method is to square x repeatedly; this will involve fewer (in fact, $\log_2 n$) squarings. As will be shown in Section 5.2, $\log_2 n$ is much less than $n - 1$.

Example 3. Compute the value of a polynomial

$$p(x) = a_n x^n + a_{n-1} x^{n-1} + \cdots + a_1 x + a_0,$$

where $a_0, a_1, \ldots, a_n$ are arbitrary given numbers. The direct way to compute $p(x)$ is to compute $x^2, x^3, \ldots, x^n$, at the cost of $n - 1$ multiplications, then to compute $a_n x^n, a_{n-1} x^{n-1}, \ldots, a_1 x$ at the cost of n multiplications, then to add all these products to a_0 at the cost of n additions. Total cost: $2n - 1$ multiplications and n additions. A cleverer algorithm for calculating $p(x)$ is this: First calculate

$$a_n x + a_{n-1};$$

then calculate

$$(a_n x + a_{n-1})x + a_{n-2};$$

then calculate

$$((a_n x + a_{n-1})x + a_{n-2})x + a_{n-3},$$

and so on. After n steps we obtain $p(x)$; since each step needs 1 multiplication and 1 addition, the total cost is n multiplications and n additions; if addition time is negligible compared to multiplication, the saving in the second method over the first involves a factor of almost 2.

We have seen in the last 3 examples that the same function can be evaluated by different algorithms. We would like to remind the reader that this is the rule rather than the exception; take this example from elementary algebra:

Example 4. Bring the sum

(9.1) $$\frac{5}{x+3} + \frac{2}{x+2}$$

to a common denominator. The answer is

(9.2) $$\frac{7x + 16}{x^2 + 5 + 6}.$$

Formulas (9.1) and (9.2) represent the same function, but each leads to a different algorithm for evaluating that function. We need one addition and one division to evaluate each term in (9.1), and one more addition to get their sum. Total cost: 2 divisions and 3 additions. We need one multiplication and one addition to evaluate the numerator in (9.2); to evaluate the denominator, we need, using the fast method described in Example 3, 2 multiplications and 2 additions. One more division is needed to compute the quotient. Total cost: 3 multiplications, 1 division and 3 additions. If addition time is negligible compared to multiplication time, and if division time is approximately equal to multiplication time, then the evaluation of (9.2) takes almost twice as much time as that of (9.1).

Example 5. We now generalize Example 4 to any rational function

(9.3) $$r = \frac{p}{q}.$$

By division any rational function can be written as the sum of a polynomial plus a quotient of form (9.3), where the degree of p is less than that of q. We shall show in Appendix 2.1 that any such rational function whose denominator q has real and distinct roots $s_1, s_2, \ldots, s_n$ can be written in the form

(9.4)
$$\frac{p(x)}{q(x)} = \frac{c_1}{x - s_1} + \frac{c_2}{x - s_2} + \cdots + \frac{c_n}{x - s_n},$$

where $c_1, c_2, \ldots, c_n$ are constants. This is called the *partial fraction expansion* of p/q. Let us compare the number of operations needed for evaluating either side of (9.4). To evaluate each term on the right side takes one addition and one division; with the $n - 1$ additions needed to tote up the sum, the total cost is n divisions and $2n - 1$ additions. On the other hand, the cost of evaluating $q(x)$ is n multiplications and $n - 1$ additions. If p is of degree $n - 1$, the evaluation of $p(x)$ takes $n - 1$ multiplications and $n - 1$ additions. Assuming as before that addition time is negligible compared to multiplication time and division time is approximately equal to multiplication time, we conclude that using the partial fraction expansion saves computing time by a factor of 2.

In order to calulate the partial fraction expansion (9.4), we need to find all the roots of q, which is no mean numerical task and requires many multiplications, additions and divisions (how many depends on how accurately we need to know s_j). Doesn't this extra labor cancel out the saving resulting from the fewer operations needed to evaluate r in the new form? The answer to this question depends entirely on the number of different values of x for which the value of $r(x)$ is needed! If we need it for less than, say, 10 different values, then it doesn't pay to put r in the form (9.4). On the other hand if we expect to be called upon to evaluate $r(x)$ for *several thousand* different values of x, then the operations needed to find the values of s_j and c_j is a small price to pay for the greater speed gained in the evaluation of $r(x)$.

Many computers on the drawing boards today are designed to operate in parallel, i.e., to be able to perform several arithmetic operations simultaneously. Such a machine can perform n of the additions required in the denominators simultaneously, and the n divisions indicated in

$$\frac{c_1}{x - s_1} + \cdots + \frac{c_n}{x - s_n}$$

simultaneously. Now the resulting n terms are added, two at a time (by $n/2$ adders), to form an intermediate set of $n/2$ inputs which are again added, two at a time (by $n/4$ adders), and so on. This process requires $\log_2 n$ additions. The total operation count is 1 division and $1 + \log_2 n$ additions. It is worth pointing out that no such saving is possible in the evaluation of a polynomial q written in the form

$$(\cdots((a_n x + a_{n-1})x + a_{n-2})x + \cdots)x + a_0.$$

The multiplications, being nested, must be performed serially; thus there is no way to take advantage of the capabilities of a parallel computer.

So far we have considered only the question of operation count; this determines the operation time, hence the expense of function evaluation. Specifications for a black box also include accuracy requirements. The next example deals with a frequently occurring problem related to loss of accuracy when an algorithm computes the difference of two almost equal numbers.

Example 6. How to evaluate the function

$$(9.5) \qquad f(x) = \frac{x}{\sqrt{x^2 + 1} + x}.$$

Using the identity $(a + b)(a - b) = a^2 - b^2$ with $a = \sqrt{x^2 + 1}$, $b = x$, we can rationalize the denominator by multiplying both numerator and denominator by $(a - b)$, and write $f(x)$ in the form

$$(9.6) \qquad f(x) = x\sqrt{x^2 + 1} - x^2.$$

Algorithms for computing f based on either formulas (9.5) or (9.6) employ the same number of operations.

We now show that for large x, the algorithm based on (9.5) is much more accurate. Suppose our computer can accommodate 8 digits. Let's take $x = 10^4$; then $x^2 + 1 = 10^8 + 1$ is recorded as 10^8, and the algorithm based on (9.6) yields the function value $f(10^4) = 0$. On the other hand, on the same computer (9.5) yields $f(10^4) = 1/2$ which differs from the exact value by less than 10^{-8}!

A different example where accuracy is lost will crop up unexpectedly in Section 7.3. See also Exercise 9.1.

Most functions of interest are not given by formulas involving a finite number of operations, but, as in the case of the square root function, are defined as the limit of a converging sequence of functions. Constructing an algorithm for computing approximately the values of such a function involves

(a) choice of an approximating function which differs from the given function by less than the prescribed tolerance, and
(b) construction of an efficient algorithm which evaluates the approximate function with an error less than the prescribed tolerance.

If the function is to be evaluated at only a few points, one wants to know how long it takes to program the algorithm?

We are now ready to try to define what an algorithm is: It is a string of instructions for performing a sequence of arithmetic and logical operations on the input x in order to produce an output y. This definition is a little vague; we haven't defined the phrase "string of instructions" nor "logical operation," although most people have an idea what each means. The precise formulation of the concept toward which we are groping is this

Definition. An *algorithm* is a precisely formulated procedure wherein requested outputs are obtained by a definite, unambiguous finite chain of operations.

Being precisely defined, algorithms can be programmed for computers. In Section 5.3 we shall present a variety of algorithms for evaluating the exponential and logarithmic functions, and Section 7.3 deals with the evaluation of sine, cosine, and their inverses.

EXERCISES

9.1 Assume that the quadratic equation

$$y^2 + by + c = 0$$

has real and distinct roots. Write a computer program for finding the *smaller* of the roots. [*Hint*: Remember Example 6.]

9.2 In Section 2.8 we have shown that

$$|s_n(x) - \sqrt{x}| \leq \left(\frac{1}{2}\right)^{n+1}, \quad 0 \leq x \leq 1,$$

for the sequence $s_n(x)$ defined by

$$s_1 = x,$$

$$s_{n+1}(x) = \frac{1}{2}\left(s_n(x) + \frac{x}{s_n(x)}\right), \quad n = 1, 2, \ldots \ .$$

As $n \to \infty$, $s_n(x)$ tends to $\sqrt{x}$. Now we ask the reader to derive the following estimate for $s_n(x) - \sqrt{x}$:

(a) Define the quantity r_n by

$$r_n = \frac{s_n}{\sqrt{x}}.$$

The convergence of $s_n(x)$ to $\sqrt{x}$ is expressed by saying that r_n tends to 1. Show that

$$r_{n+1} - 1 = \frac{1}{2r_n}(r_n - 1)^2.$$

(b) Write

$$s_{n+1} - \sqrt{x} = \frac{1}{2}\left(s_n + \frac{x}{s_n}\right) - \sqrt{x}$$

in the form

$$s_{n+1} - \sqrt{x} = \frac{1}{2s_n}[s_n^2 - 2s_n\sqrt{x} + x] = \frac{1}{2s_n}(s_n - \sqrt{x})^2 \geq 0$$

to observe that $s_{n+1} \geq \sqrt{x}$. Use this result to show that

$$0 \leq r_{n+1} - 1 \leq \frac{1}{2}(r_n - 1)^2.$$

(c) Show that, when $1 \le x \le 4$, $r_0 \le 2$.

(d) Show, using (b) and (c), that for $1 \le x \le 4$

$$r_1 - 1 \le \frac{1}{2}$$

$$r_2 - 1 \le \left(\frac{1}{2}\right)^3$$

$$r_3 - 1 \le \left(\frac{1}{2}\right)^7.$$

(e) Show that for all n,

(9.7) $$r_n - 1 \le \left(\frac{1}{2}\right)^{2^n - 1}$$

for x in $1 \le x \le 4$. Compare the estimate (9.7) with actual numerical calculations.

9.3 Let n be any positive integer; n can be written as a sum of powers of 2:

(9.8) $$n = 2^{k_1} + 2^{k_2} + \cdots + 2^{k_j} = n_1 + n_2 + \cdots + n_j.$$

Use the law of exponents

(9.9) $$x^n = x^{n_1} x^{n_2} \ldots x^{n_j}$$

to compute x^n.

(a) How many multiplications are needed to compute x^n using (9.8), (9.9)?

(b) Take advantage of the binary representation of numbers in computers to write a program implementing this method for evaluating x^n.

Appendix 2.1 Partial fraction expansion

In Section 2.4 we defined "rational function" $r(x)$ to be a quotient

$$\frac{p(x)}{q(x)}$$

of polynomials. In this appendix we shall derive partial fraction expansions for rational functions whose denominators q are of arbitrary degree n, under the restriction that the equation $q(x) = 0$ has n distinct solutions.

It is enough to consider rational functions whose numerators are of lower degrees than their denominators; for, if not, one can divide numerator by denominator to obtain a polynomial and a remainder whose degree is less than that of the denominator:

$$p = q \cdot (\text{polynomial}) + \text{remainder}$$
$$p/q = \text{polynomial} + \text{remainder}/q.$$

This is similar to writing an improper fraction as an integer plus a proper fraction.

We use the following elementary results of algebra:

Remainder theorem. Given any polynomial q of degree n and any number s, q can be written as

$$q(x) = a(x)(x - s) + q(s)$$

for any x, where $a(x)$ is a polynomial of degree $n - 1$.

We deduce from this the

Factor theorem. If s is a root of q, i.e., if $q(s) = 0$, then

$$q(x) = a(x)(x - s).$$

Repeated application gives this result: If $q(x)$ is zero at the distinct points $s_1, \ldots, s_k$, then

$$q(x) = a_k(x)(x - s_1) \cdots (x - s_k),$$

where a_k is a polynomial of degree $n - k$. In case $n = k$ we get a complete factorization of q:

(A.1) $$q(x) = a_n(x - s_1) \cdots (x - s_n),$$

a_n some number. Clearly, unless $a_n = 0$, the right side above is not zero for any x different from the s_j, which shows that *a nonzero polynomial q of degree n has at most n roots.*

Our aim is to write r in the form

(A.2) $$r(x) = \frac{p(x)}{q(x)} = \frac{c_1}{x - s_1} + \frac{c_2}{x - s_2} + \cdots + \frac{c_n}{x - s_n}.$$

Suppose this is possible; multiplying both sides by $q(x)$, we get

(A.3) $$p(x) = c_1 \frac{q(x)}{x - s_1} + c_2 \frac{q(x)}{x - s_1} + \cdots + c_n \frac{q(x)}{x - s_n}.$$

Each term on the right is a polynomial of degree $n - 1$. Let

$$q_i(x) = \frac{q(x)}{x - s_i}.$$

Then

(A.4) $$q_i(x) = \prod_{j \neq i}(x - s_j)$$

[The symbol $\prod_{j=1}^{n} b_j$ denotes the product $b_1 \cdot b_2 \cdots b_n$. If the i-th factor is to be omitted from this product, we write $\prod_{j=1, j \neq i}^{n} b_j$, or simply $\prod_{j \neq i} b_j$.] Clearly, q_i is zero for $x = s_j, j \neq i$, and

$$q_i(s_i) = \prod_{j \neq i}(s_i - s_j) \neq 0.$$

We rewrite (A.3) as

(A.5) $$p(x) = c_1 q_1(x) + c_2 q_2(x) + \cdots + c_n q_n(x).$$

85

This has to hold for all x; therefore in particular it has to hold for $x = s_k$; using the fact that $q_i(s_j) = 0$ for $i \neq j$, we get

$$p(s_k) = c_k q_k(s_k).$$

This determines the numbers c_k as

(A.6)
$$c_k = \frac{p(s_k)}{q_k(s_k)}.$$

We claim that with this choice of c_k (A.5) holds for all values of x. To see this, form the difference of the two sides:

(A.7)
$$p(x) - \sum_1^n c_k q_k(x).$$

This function, being the difference of two polynomials of degree $<n$ is itself a polynomial of degree $<n$. The coefficients c_k were so chosen that (A.7) vanishes at the n points $s_1, \ldots, s_n$. Since we have shown above that a nonzero polynomial of degree $<n$ cannot have n roots, it follows that (A.7) is zero for all x, hence that A.5 holds for all x, and therefore that *the partial fraction expansion* (A.2) *holds for all $x \neq s_j$, provided that c_k is given by* (A.6).

EXERCISES

A.1 Write a computer program for finding the partial fraction expansion (A.2) for the case where the s_i are all real and distinct.

A.2 Write the partial fraction expansion for

(a) $\dfrac{3x^2 - 1}{(x^3 - x)}$

(b) $\dfrac{1}{(x + 1)(x - 2)(x - 3)}$

(c) $\dfrac{2.4x - 1.3}{x^2 - 3.8x + 2.4}$

(d) $\dfrac{3.85x - 1.34}{x^2 + 5x - 2.75}$.

Differentiation

3

Many interesting questions deal with the rate at which things change. Examples abound: What is the rate of inflation? How fast is inflation growing? How fast is the world's population increasing? In this chapter we define and discuss the concept of *rate of change* which, in mathematics, is called the *derivative*.

3.1 The derivative

Among the instruments on the dashboard of a car there are two that indicate quantitative measurements; the mileage meter and the speedometer. We shall investigate the relation between these two. To put the matter dramatically: Suppose your speedometer is broken; *is there any way of determining the speed of the car from the readings on the mileage meter* (so that, for example, you don't exceed the speed limit)? Suppose the mileage reading at 2 o'clock was 5268, and 15 minutes later it was 5280; then your *average* speed during that quarter hour interval was

$$\frac{\text{distance covered (miles)}}{\text{time interval (hours)}} = \frac{5280 - 5268}{0.25} = \frac{12}{0.25} = 48 \, \frac{\text{miles}}{\text{hour}}.$$

Denote by f the mileage reading as function of time, i.e., the mileage reading at any time t is $f(t)$. Then the *average speed* at time t, *averaged over a time interval of a quarter of an hour*, is

$$\frac{f(t + 0.25) - f(t)}{0.25}.$$

More generally, let h be any time interval; the *average speed over a time interval h* is

$$\frac{f(t + h) - f(t)}{h}.$$

This average speed is a function of time. We denote it by f'_h to indicate the dependence on h:

$$f'_h(t) = \frac{f(t + h) - f(t)}{h}.$$

Example 1. Suppose the mileage function $f(t)$ is described by the simple formula

$$f(t) = 5000 + 35t + 2.5t^2.$$

Then the average speed during the time interval h is

$$f'_h(t) = \frac{f(t + h) - f(t)}{h}$$

$$= \frac{5000 + 35(t + h) + 2.5(t + h)^2 - (5000 + 35t + 2.5t^2)}{h}$$

$$= \frac{35h + 5th + 2.5h^2}{h} = 35 + 5t + 2.5h.$$

Observe that as h gets smaller and smaller, $f'_h(t)$ tends to

$$35 + 5t.$$

This quantity, the limit of average speeds over shorter and shorter time intervals, is called the *instantaneous speed*; this is the quantity which is registered on the speedometer (with an unbroken cable).

We now formulate a general definition within the realm of functions, divorced from the concepts of speed and distance. Let f be a continuous function defined on some open interval S'. Let S be a closed interval contained in S'. We define the *average rate of change f'_h of f* over an interval h to be the function

(1.1) $$f'_h(t) = \frac{f(t + h) - f(t)}{h}.$$

Since f is defined on an interval larger than S, f'_h, called a *difference quotient*, is defined on S for h small enough; in other words, f'_h can be defined at the endpoints of S.

Suppose that, as h tends to zero, the functions f'_h converge to a limit on S; then the function f is called *differentiable* on S. The limit function is called the *derivative* of f, and is denoted by f':

$$\lim_{h \to 0} f'_h(t) = f'(t).$$

The value of f' at t, being the limit of the average rates of change over intervals $[t, t + h]$, is called *the instantaneous rate of change of f at t.*

Denote by Δf the difference

$$\Delta f = f(t + h) - f(t),$$

and by Δt the difference between $t + h$ and t, i.e.,

$$\Delta t = h.$$

Then the right side of (1.1) can be written as the quotient of differences

$$f'_h(t) = \frac{\Delta f}{\Delta t}.$$

The derivative, f', which is the limit of these difference quotients, is sometimes called the *differential quotient* and denoted by

$$f' = \frac{df}{dt}.$$

Another notation for the derivative is the D notation;

$$f' = Df.$$

This notation emphasizes the fact that the derivative is the result of certain operations performed on f. A process which produces a new function out of any given function f is called an *operator*.

Differentiation is an operator. Later in this text we shall see that, although basic, differentiation is only one of many operators of interest in calculus.

What does it mean that, as h tends to zero, the functions f'_h tend to f' on S? In Section 2.8, we defined convergence of a sequence $\{g_n\}$ of functions. This definition is easily generalized to families of functions, such as $\{f'_h\}$, which are defined for all sufficiently small values of $h \neq 0$.

We say that a family of functions $\{g_h\}$ defined on an interval S converges to g on S if, no matter how large an integer m we choose,

(1.2) $$|g_h(x) - g(x)| < 10^{-m}$$

for all x in S, provided that h is small enough; how small depends on m. It is easy to see that all the results in Section 2.8 carry over to such converging families of functions. In particular, the limit of a converging family of continuous functions is continuous.

We call a function $f(t)$ differentiable if its difference quotients $f'_h(t)$ tend to a limit $f'(t)$ in the sense of (1.2).

Since f was assumed continuous, its difference quotients (1.1) are continuous. It follows that their limit, f', is continuous. This shows that the derivative f' of a differentiable function f is a continuous function. There are other definitions of differentiability who do not require f' to be continuous; we shall ignore them.

The function presented before, $f(t) = 5000 + 35t + 2.5t^2$, is differentiable; for, in this case $f'_h(t) = 35 + 5t + 2.5h$ lies within 10^{-m} of

$$f'(t) = 35 + 5t,$$

provided that $h < 4 \cdot 10^{-m-1}$.

Example 2. Suppose $f(t)$ is the constant function k for all t. Then

$$\Delta f = f(t + h) - f(t) = k - k = 0 \quad \text{for all } h,$$

so

$$f'_h(t) = \frac{\Delta f}{h} = 0 \qquad \text{for all } h,$$

and

$$\lim_{h \to 0} f'_h(t) = f'(t) = 0.$$

Since a constant function does not change, this result

(1.3) $$k' = 0$$

is not surprising.

Example 3. Suppose $i(t)$ is the identity function, $i(t) = t$. Then

$$i'_h(t) = \frac{i(t + h) - i(t)}{h} = \frac{(t + h) - t}{h} = 1 \quad \text{for all } h,$$

so

(1.4) $$i'(t) = \lim_{h \to 0} i'_h(t) = 1.$$

We present now an example of a function which is *not* differentiable.

Example 4. $f(t) = |t|$; we claim that this function is not differentiable on any interval containing the point $t = 0$ in its interior. Recall the definition of absolute value,

$$f(t) = |t| = \begin{cases} t & \text{for } 0 \le t \\ -t & \text{for } t \le 0. \end{cases}$$

For h positive,

$$f'_h(t) = \frac{|t + h| - |t|}{h}$$

has the following values:

(1.5) $$f'_h(t) = \begin{cases} 1 & \text{for } 0 \le t \\ 1 + 2t/h & \text{for } -h \le t \le 0 \\ -1 & \text{for } t \le -h. \end{cases}$$

Figure 3.1 shows the graph of f'_h. It follows from (1.5) that:

For each $t > 0$, $f'_h(t)$ tends to 1 as h tends to 0.
For each $t < 0$, $f'_h(t)$ tends to -1 as h tends to 0.

This shows that, if f were differentiable, the values of its derivative would be

(1.6)
$$f'(t) = \begin{cases} 1 & \text{for } t > 0 \\ -1 & \text{for } t < 0. \end{cases}$$

But such an f' cannot be continuous on any interval containing $t = 0$. Since we have seen that the derivative of a differentiable function is continuous, we conclude that $f(t) = |t|$ is *not* differentiable on any interval containing the point $t = 0$. This example shows that *not all continuous functions are differentiable*; calculus deals with those which are. Many differentiable functions will be presented in the next section.

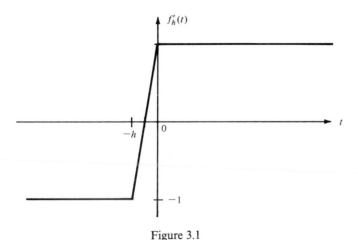

Figure 3.1

We close this section with a discussion of the importance of the derivative. One could argue persuasively that *changes* in magnitudes are often more important than magnitudes, and therefore the rate at which the value of a function changes from point to point or time to time is more relevant than its actual value. For example, it is often more useful to know if the outside temperature next day is going to be higher, lower, or the same as today than knowing tomorrow's temperature, but not today's.

Theoretical weather predictions are based on theories that relate the rate of change of meteorologically relevant quantities such as temperature, atmospheric pressure, humidity, etc., to factors which cause the change. The mathematical formulation of these theories involves equations relating the derivatives of these meteorological variables to each other. Almost all physical theories—mechanics, optics, theory of heat, sound, etc.—are formulated as equations involving derivatives; in Section 3.6 we shall describe some examples from mechanics. Chapter 8 contains more examples from mechanics and electric circuitry, and Chapter 9 deals with population dynamics and chemical kinetics.

We conclude with a brief *dictionary*:

speed ↔ rate of change of distance as function of time.
acceleration ↔ rate of change of speed as function of time.
angular velocity ↔ rate of change of angle as function of time.
density ↔ rate of change of mass as function of volume.
slope ↔ rate of change of height as function of horizontal distance.
curvature ↔ rate of change of angle between tangent to curve and horizontal
   as function of distance measured along the curve.
current ↔ rate of change of amount of electric charge as function of time.
marginal cost ↔ rate of change of production cost as function of number of
   items produced.

That so many words in common usage denote rates of change of other quantities is an eloquent testimony to the importance of the notion of derivative.

3.2 Rules of differentiation

We have explained in Chapter 2 that we often have occasion to form new functions out of given functions by addition, multiplication, division, composition, and inversion. In this section we shall show that the sums, products, quotients, composites, and inverses of differentiable functions are likewise differentiable, and we shall express their derivatives in terms of the component functions and their derivatives.

(i) *Derivative of sums.* Suppose f and g are differentiable functions. Then their sum is differentiable; its derivative is

(2.1)
$$\boxed{(f + g)' = f' + g'}$$

PROOF. Proof of this is based on the observation that the difference quotient of $f + g$ is the sum of the difference quotients of f and of g:

$$(f + g)'_h = \frac{f(x + h) + g(x + h)}{h} - \frac{f(x) + g(x)}{h}$$

$$= \frac{f(x + h) - f(x)}{h} + \frac{g(x + h) - g(x)}{h} = f'_h + g'_h.$$

Since f and g are differentiable, f'_h converges to f', g'_h to g' as h tends to 0; according to rule (i), Section 2.8, for the arithmetic of convergent sequences of functions, their sum $f'_h + g'_h$ converges to $f' + g'$. □

Quite analogously, the derivative of the sum of any finite number of differentiable functions is the sum of their derivatives.

(ii) *Derivative of products.* Suppose f and g are bounded, differentiable functions. Then their product is differentiable; its derivative is

(2.2)
$$\boxed{(fg)' = fg' + f'g}$$

We call this the *product formula*.

PROOF. To find the derivative of fg we form its difference quotient

$$(fg)'_h = \frac{f(x + h)g(x + h) - f(x)g(x)}{h}.$$

The right side can be written in the form

$$f(x + h)\frac{g(x + h) - g(x)}{h} + \frac{f(x + h) - f(x)}{h}g(x) = f(x + h)g'_h + f'_h g(x).$$

Since f and g are differentiable, $f'_h \to f'$ and $g'_h \to g$ as $h \to 0$; since f is continuous, $f(x + h) \to f(x)$. We now apply rule (ii), Section 2.8, according to which the product of two bounded convergent sequences of functions has, as its limit, the product of the limits of the individual sequences. Combined with rule (i) about sums of convergent sequences this yields that $(fg)'_h$ converges to $fg' + f'g$ as asserted. ☐

An important special case is when $f = k$ is a constant; then $f' = 0$ and (2.2) gives

(2.2)$_k$
$$\boxed{(kg)' = kg' \quad \text{for } k \text{ constant}}$$

These two rules enable us to calculate the derivative of any function which has been built by addition and multiplication out of functions whose derivatives we already know. Since we know the derivative of any constant function—it is zero—and of the identity function i—it is 1—we can differentiate every function built out of these by addition and multiplication.

Example 1. Consider the linear function $l(x) = mx + b$. It is a sum of two terms; the first is the product of the constant function $f(x) = m$ and the identity function $i(x) = x$, whereas the second term is the constant function $g(x) = b$. To find $l'(x)$, we use (2.1) and obtain

$$l'(x) = (mi(x))' + (b)'.$$

We apply (2.2) to the first term and get, by means of (1.4), $(mi(x))' = m$. By (1.3), $b' = 0$. Thus the derivative of a linear function is equal to its slope:

$$l'(x) = m.$$

93

There is nothing surprising about this result; according to Section 2.6 any difference quotient of a linear function is equal to the slope of the graph of the function.

We proceed to determine the derivative of polynomials.

Example 2. $p(x) = x^2$ is a product

$$p(x) = i(x)i(x),$$

where $i(x) = x$. Using the product formula with $f = i$, and $g = i$ and that $i' = 1$, we get

$$p' = i'i + ii' = 2i,$$

hence

$$(x^2)' = 2x.$$

Example 3. $p(x) = x^3$; we write this as a product

$$p = i^2 i.$$

Applying the product formula with $f = i^2$, $g = i$, we get, using Example 2,

$$(x^3)' = (x^2)'x + x^2(x)' = 2x \cdot x + x^2 \cdot 1 = 3x^2.$$

Example 4. $p(x) = x^4$. Writing $p = fg$, with $f(x) = x^3$, $g(x) = x$, we get, using the product formula and Example 3, that

$$(x^4)' = 4x^3.$$

The pattern is clear, and we are ready to show that *for any nonnegative integer n*

(2.3) $$(x^n)' = nx^{n-1}.$$

PROOF. We use mathematical induction,* i.e., we show that if the result holds for $n - 1$, then it holds for n. Since the result holds for $n = 1$, its validity will then follow for $n = 2, 3, \ldots$, for all n. We write

$$x^n = x^{n-1}x = f(x)g(x)$$

and apply the product formula; using the validity of $(x^{n-1})' = (n-1)x^{n-2}$, we get

$$(x^n)' = (x^{n-1})'x + x^{n-1}(x)' = (n-1)x^{n-2}x + x^{n-1} = nx^{n-1},$$

as asserted in (2.3). □

---

* Mathematical induction establishes a proposition for all positive integers by verifying its validity for 1, and showing that, if it is valid for all integers less than n, it is valid for n.

A polynomial, as described in Section 2.4, is of the form

$$p(x) = a_n x^n + a_{n-1} x^{n-1} + \cdots + a_0.$$

Using the rules for differentiating a sum and the formula already verified for the derivative of x^n, we obtain

$$p'(x) = n a_n x^{n-1} + (n-1) a_{n-1} x^{n-2} + \cdots + a_1.$$

Example 5. If

$$p(x) = x^2 - 3x + 2,$$

then

$$p'(x) = 2x - 3;$$

similarly, for

$$p(x) = x^3 + 5x^2 - 3x + 1,$$

we have

$$p'(x) = 3x^2 + 10x - 3.$$

We now turn to division.

(iii) *Derivative of reciprocals.* Let f be a differentiable function on an interval S, and suppose that $f \neq 0$ on S. Then $1/f$ is bounded on S. We shall show that $1/f$ is differentiable there, and that

(2.4)
$$\boxed{\left(\frac{1}{f}\right)' = -\frac{f'}{f^2}}.$$

PROOF. To prove differentiability we look at the difference quotient of $1/f$:

$$\left(\frac{1}{f}\right)'_h = \frac{[1/f(x+h)] - [1/f(x)]}{h} = \frac{f(x) - f(x+h)}{h f(x) f(x+h)}$$

$$= -f'_h(x) \frac{1}{f(x)} \frac{1}{f(x+h)}.$$

Since f is differentiable on S, the first factor, $-f'_h$, tends to $-f'$ as h tends to 0. The second factor is independent of h and the third tends to $1/f$. According to rule (ii), Section 2.8, the product of 3 convergent bounded sequences converges to the product of the three limits; this shows that $(1/f)'_h$ converges to $-f'(1/f)(1/f)$ on S, as asserted in (2.4). □

Example 6. Let $f(x) = x^k$, and denote the reciprocal function by

$$r(x) = \frac{1}{f(x)} = x^{-k}.$$

Using (2.4) and (2.3) we get

$$r' = \left(\frac{1}{f}\right)' = (x^{-k})' = \left(\frac{1}{x^k}\right)' = -\frac{kx^{k-1}}{(x^k)^2} = -kx^{-k-1}.$$

The above formula shows that *rule (2.3) for differentiating powers of x is valid also when the exponent n is a negative integer*, say $n = -k$.

Example 7. $f(x) = x^2 + 1$. Using (2.4) for the derivative of its reciprocal, we get

$$\left(\frac{1}{x^2 + 1}\right)' = -\frac{2x}{(x^2 + 1)^2}.$$

(iv) *Derivative of quotients.* Suppose f and g are differentiable functions on an interval S, $g \neq 0$ on S. Is their quotient f/g differentiable, and what is its derivative? We can answer this question by writing the quotient as a product

$$\frac{f}{g} = f \cdot \frac{1}{g}.$$

Using the product formula (2.2) and the reciprocal formula (2.4), we get

$$\left(\frac{f}{g}\right)' = f'\frac{1}{g} + f\left(-\frac{g'}{g^2}\right);$$

employing g^2 as common denominator we can write this as

(2.5)
$$\boxed{\left(\frac{f}{g}\right)' = \frac{f'g - fg'}{g^2}}.$$

A rational function is the quotient of two polynomials; since we know how to differentiate polynomials, we can, using the quotient formula (2.5), differentiate any rational function.

Example 8. For $f(x) = x$, and $g(x) = x^2 + 1$, the derivative of the quotient f/g is

$$\left(\frac{f}{g}\right)' = \left(\frac{x}{x^2 + 1}\right)' = \frac{1(x^2 + 1) - x(2x)}{(x^2 + 1)^2} = \frac{1 - x^2}{(x^2 + 1)^2}.$$

(v) *Derivative of composite functions.* We shall show that the composite $F(x) = f[g(x)] = f \circ g(x)$ of differentiable functions f and g is differentiable, and we shall determine the derivative F' of the composite function $f[g(x)]$. We recall the reasoning described in Section 2.7 to prove the continuity of the composite of continuous function; the gist of that argument was that, since g is continuous, a small change in its input x results in a small change in

g; since f is continuous, a small change in its input, g, in turn results in a small change in $f(g)$. Through this linkage, the net effect of a small change in the input x is a small change in the output $F(x) = f[g(x)]$. We shall now give a quantitative version of this argument employing the following notation: Let

$$k = g(x + h) - g(x),$$

so that

$$g(x + h) = g(x) + k.$$

Then the difference quotient of $f \circ g$ may be written as

$$F'_h = (f \circ g)'_h = \frac{f[g(x + h)] - f[g(x)]}{h} = \frac{f[g + k] - f(g)}{h}.$$

Since $[g(x + h) - g(x)]/k = 1$, we may rewrite this as

$$F'_h = (f \circ g)'_h = \frac{f(g + k) - f(g)}{k} \cdot \frac{g(x + h) - g(x)}{h}.$$

Now as h tends to zero, the second factor on the right converges to $g'(x)$; moreover, as $h \to 0$, also $k \to 0$ so that the first factor on the right tends to $f'(g)$. Since the limit of a product is the product of the limits, we conclude that *the derivative of the composite function is the product of the derivatives of the components*

(2.6)
$$\boxed{F' = (f \circ g)' = (f' \circ g)g'}.$$

This formula is sometimes called the *chain rule*. Using the notation

$$F'(x) = \frac{dF(x)}{dx},$$

we can restate the chain rule for the composite function $F(x) = f[g(x)]$ in the form

$$\frac{dF}{dx} = \frac{df}{dg} \frac{dg}{dx}.$$

Our derivation of the chain rule contains one flaw: we divided and then multiplied by k, which is impossible when $k = 0$. We show now how to remedy this flaw; we write, this time honestly,

$$F'_h(x) = (f \circ g)'_h = \begin{cases} \dfrac{f(g + k) - f(g)}{k} \dfrac{g(x + h) - g(x)}{h} & \text{when } k \neq 0 \\ 0 & \text{when } k = 0, \end{cases}$$

and see what happens when h tends to 0. We claim that in either case $F'_h(x)$ is very close to $(f' \circ g)g'(x)$ when h is small. We have already shown this when $k \neq 0$. When $k = 0$,

$$g'_h(x) = \frac{g(x + h) - g(x)}{h} = \frac{k}{h} = 0.$$

Since for h small, $g'_h(x)$ is very close to $g'(x)$, it follows that $g'(x)$ is very small. Then so is the product $(f' \circ g)g'(x)$. But $F'_h(x) = 0$ when $k = 0$; this proves our contention. □

Example 9. Let $f(y) = y^2$ and $y = g(x)$; then $f'(y) = 2y$. Using the chain rule for the composite $F(x) = f[g(x)] = [g(x)]^2$, we get

$$F'(x) = (f \circ g)' = (g^2(x))' = 2g'(x)g(x).$$

This is, of course, the same result we get if we regard g^2 as a product and use the product formula (2.2).

Example 10. Let $f(y) = 1/y$ and $y = g(x)$; then $f'(y) = -1/y^2$. Using (2.6) with $y = g(x)$, we get

$$(f \circ g)' = \left(\frac{1}{g}\right)' = -\frac{1}{g^2}g'.$$

Of course this is the same result as given by the reciprocal formula (2.4).

Example 11. Suppose $f(y) = y^n$ and $y = g(x)$; then $f'(y) = ny^{n-1}$, so $(g^n)' = ng^{n-1}g'$.

Example 12. Consider the case where $f(y) = y^n$, and $y = g(x) = x^k$. Using Formula (2.3) to find f' and g' we get

$$F' = (f \circ g)' = ((x^k)^n)' = ny^{n-1}kx^{k-1} = nk(x^k)^{n-1}x^{k-1}$$

$$= nkx^{k(n-1)+k-1} = nkx^{nk-1}.$$

Had we used Formula (2.3) to differentiate $(x^k)^n = x^{kn}$ directly, we would have obtained the same result.

Example 13. Let f be any differentiable function, and let g be the linear function $g(x) = mx + b$ with slope m. Then $g' = m$, so

$$(f \circ g)' = (f(mx + b))' = mf'(mx + b).$$

In the special case where the function f is also linear, say $f(y) = ny + c$, with $f' = n$, we find for $y = g(x)$ that $(f \circ g)' = mn$. This shows that the result derived in Section 3.6—the slope of the composite of two linear functions is the product of their slopes—is merely a very special instance of the chain rule.

(vi) *Derivative of the inverse.* Suppose f is a differentiable function whose derivative is not zero on some interval S. We shall show in Section 3.3 that then f is monotonic on S and has an inverse g. In Exercise 2.6 we shall lead the reader to prove that the inverse g of f is differentiable. Here we shall calculate the derivative g'.

To say that f and g are inverse to each other means

$$f[g(x)] = x,$$

that is, $f[g(x)] = i(x)$, see Section 2.3. We differentiate both sides and, using the chain rule, we obtain

$$f'(g)g' = 1.$$

Therefore the derivative of the inverse g of the function f is the reciprocal of the derivative of the function f:

(2.7)
$$g' = \frac{1}{f'(g)}.$$

Example 14. If $f(x) = mx + b$, then $f' = m$. Therefore by (2.7), the inverse g of f has derivative $g' = 1/m$.

Example 15. Let $f(y) = y^2$. This function is invertible on the interval $y > 0$; its inverse is

$$g(x) = \sqrt{x}.$$

$f'(y) = 2y$; so, according to (2.7),

$$g'(x) = \frac{1}{f'[g(x)]} = \frac{1}{2g(x)} = \frac{1}{2\sqrt{x}}.$$

So the derivative of $g(x) = x^{1/2}$ is $g'(x) = \frac{1}{2}x^{-1/2}$. This shows that rule (2.3) for differentiating x^n holds for $n = 1/2$.

Example 16. Let $f(y) = y^k$, k a positive integer. This function is invertible on $y > 0$; its inverse is

$$g(x) = x^{1/k}.$$

By (2.3), $f'(y) = ky^{k-1}$, so by (2.7),

$$g'(x) = \frac{1}{f'(g)} = \frac{1}{kg^{k-1}} = \frac{1}{k}\frac{1}{x^{(k-1)/k}} = \frac{1}{k}x^{(1/k)-1}.$$

This shows that rule (2.3) for differentiating x^n is valid for $n = 1/k, k$ an integer.

In Exercise 2.2, the reader will be asked to show that *rule (2.3) for differentiating x^n is valid for all rational n, positive and negative.* As soon as we define x^n for n irrational, which shall be done in Chapter 5, we shall show that rule (2.3) for differentiating x^n holds for all real exponents n.

Example 17. We want to differentiate the function

$$F(x) = \sqrt{x^2 + 1}.$$

We observe that it is a composite function $f \circ g$, where

$$g(x) = x^2 + 1, \qquad f(y) = y^{1/2}.$$

According to Example 15, $f'(y) = 1/2y^{1/2}$, while $g'(x) = 2x$; so by the chain rule (2.6)

$$F'(x) = (\sqrt{x^2 + 1})' = 2x\frac{1}{2g^{1/2}} = \frac{x}{\sqrt{x^2 + 1}}.$$

Example 18. $F(x) = (x^3 + 1)^{1/3}$ is a composite function $f \circ g$, with $g(x) = x^3 + 1$, $f(y) = y^{1/3}$. Using Example 16 and the chain rule, we get

$$F' = [(x^3 + 1)^{1/3}]' = 3x^2\tfrac{1}{3}(x^3 + 1)^{-2/3} = \frac{x^2}{(x^3 + 1)^{2/3}}.$$

Example 19. Let $f(y) = y^2 + 2y + 3$; then $f'(y) = 2y + 2$, which is positive for $y > -1$. We shall show in the next section that $f' > 0$ implies that f is increasing, hence, by the inversion theorem of Section 2.7, that f is invertible on the interval $y > -1$. Indeed, solving

$$y^2 + 2y + 3 = x$$

for y, we get

$$y = g(x) = -1 + \sqrt{x - 2};$$

here we have chosen the root y which is > -1. Differentiating g we get

$$g'(x) = \frac{1}{2\sqrt{x - 2}}.$$

Now let us determine g', using (2.7):

$$g'(x) = \frac{1}{f'} = \frac{1}{2y + 2}.$$

When we express y in terms of x, we get the same expression as above for the derivative $g'(x)$.

These rules enable anyone to perform orgies of differentiation. In doing the exercises provided below, the first step is to write the function to be differentiated as sum, product, or composite of simpler functions; the second step is to determine the derivatives of these simpler functions; the third step is to use one of the six rules given in this section to express the derivative of the complicated functions in terms of the already known derivatives of the simpler functions.

Warning. The component functions, although simpler than the original function, may still be pretty complicated; to find their derivatives we may have to decompose them further as sums, products, or composites of still simpler functions. We shall give some hints how to do this.

EXERCISES

2.1 Find the derivatives of the following functions:

(a) $f(x) = x^5 - 3x^4 + 0.5x^2 - 17$
(b) $f(x) = x^4 - 8x^3 + 3x^2 - 2$
(c) $f(x) = (x + 1)/(x - 1)$
(d) $f(x) = (x^2 - 1)/(x^2 - 2x + 1)$
(e) $f(x) = \sqrt{x}/(x + 1)$
(f) $F(x) = \sqrt{x^3 + 1}$ [*Hint*: Take $f = \sqrt{g}$, and $g(x) = x^3 + 1$; then $F = f \circ g$.]
(g) $f(x) = (x + 1/x)^3$
(h) $F(x) = \sqrt{1 + \sqrt{x}}$
(i) $f(x) = (\sqrt{x} + 1)(\sqrt{x} - 1)$.

2.2 Let p and q be integers; find the derivative of

$$F(x) = x^{p/q}.$$

[*Hint*: $f = g^p$, $g(x) = x^{1/q}$; $F = f \circ g$.]

2.3 (a) Show that the functions

$$f(x) = \sqrt{x^2 - 1}, \quad \text{defined on } x > 1,$$

and

$$g(y) = \sqrt{y^2 + 1}, \quad \text{defined on } y > 0$$

are inverse to each other.
(b) Calculate f' and g'.
(c) Verify that

$$f'(g(y))g'(y) = 1 \quad \text{and that} \quad g'(f(x))f'(x) = 1.$$

2.4 (a) The function

$$f(x) = x^2 + 3x + 1, \quad \text{defined for } x > 0,$$

is invertible; find its inverse $g(y)$.
(b) Verify that

$$f'(g(y))g'(y) = 1, \quad \text{and that} \quad g'(f(x))f'(x) = 1.$$

2.5 Let n be an integer; $(x + 1)^n$ is a polynomial of degree n. Denote the coefficient of x^k by $c(n, k)$, i.e.,

$$(x + 1)^n = c(n, n)x^n + c(n, n - 1)x^{n-1} + \cdots + c(n, 1)x + c(n, 0).$$

(a) Prove that the derivative of $(x + 1)^n$ is $n(x + 1)^{n-1}$.

(b) Deduce from (a) that

$$kc(n, k) = nc(n - 1, k - 1).$$

(c) Deduce from (b) that

$$c(n, k) = \frac{n(n - 1) \ldots (n - k + 1)}{1 \cdot 2 \ldots k}.$$

2.6 It will be shown in the next section that a function f whose derivative is positive is increasing. According to the inversion theorem of Section 2.7, such a function f is invertible. In this exercise the reader is asked to prove that the inverse of f is differentiable.

Suppose f is differentiable, $f' > 0$; denote the inverse of f by g. For any x, set

$$f(x) = y; \quad \text{then} \quad g(y) = x.$$

(a) Show that for $h \neq 0$,

$$f(x + h) - f(x) = k \neq 0.$$

With this notation

$$f(x + h) = y + k, \quad \text{so that} \quad g(y + k) = x + h.$$

(b) Show that

$$\frac{g(y + k) - g(y)}{k} = \frac{h}{k} = \frac{h}{y + k - y} = \frac{h}{f(x + h) - f(x)}$$

tends to $1/f'(x)$ as k tends to 0.

2.7 Let $f(x) = x^3 + 2x^2 + 3x + 1$; denote by g the inverse of f. Show that

$$f(1) = 7, \quad g(7) = 1,$$

and calculate $g'(7)$.

3.3 Increasing and decreasing functions

Suppose a function f, defined on some interval, is *increasing*. Then the difference quotient

$$f'_h(x) = \frac{f(x + h) - f(x)}{h}$$

of f, which is defined on some smaller interval S, is *positive*. Suppose that f is differentiable on S; that is, as h tends to zero, the difference quotients f'_h converge on S to the derivative f' of f. Since it is the limit of a sequence of positive functions, f' is *nonnegative* on S. Thus we have shown: *Increasing differentiable functions have nonnegative derivatives*. Similarly, *decreasing differentiable functions have nonpositive derivatives*.

We now shall derive the converses of these propositions; these furnish

Criteria for monotonicity. *Suppose that the derivative of a differentiable function f is positive on an interval S; then f is increasing on S. Similarly, suppose that the derivative of a differentiable function f is negative on an interval S; then f is decreasing on S.*

PROOF. Take any two points a and b in S such that $a < b$. Since f' is continuous on $[a, b]$, f' attains its minimum m somewhere on $[a, b]$, and $m > 0$ since f' is positive everywhere on S.

The derivative f' of f on $[a, b]$ is the limit of the difference quotients f'_h. This means that for h small enough f' and f'_h differ by little on $[a, b]$. One may make h so small that f'_h and f' differ on $[a, b]$ by less than $m/2$. Since f' is greater than m on the interval $[a, b]$, it follows that f'_h is greater than $m/2$ on $[a, b]$. This shows that f'_h is *positive* on $[a, b]$ for h small enough:

$$0 < \frac{f(x + h) - f(x)}{h},$$

and this is the same as saying that for h small enough and $h > 0$,

$$f(x) < f(x + h)$$

for all x in $[a, b]$.

Take any two points a and b in S. Break up the interval $[a, b]$ into small enough intervals; see Figure 3.2. Combine the chain of inequalities

$$\begin{aligned} f(a) &< f(a + h) \\ f(a + h) &< f(a + 2h) \\ &\vdots \\ f(b - h) &< f(b) \end{aligned}$$

to get that

$$f(a) < f(b).$$

Since a and b were any two points of S such that $a < b$, we have proved the increasing character of f. The decreasing character of a function with negative derivative can be proved by an analogous argument. □

Figure 3.2

We shall now illustrate how useful the *monotonicity criteria* just derived are for detecting the increasing or decreasing character of functions.

Example 1. $f(x) = mx + b$; the derivative of this function is the constant function m. We deduce, using the monotonicity criterion, that f is increasing if m is positive, decreasing if m is negative. (Recall that the graph of the linear function $f(x) = mx + b$ is a straight line with slope m; in light of this geometric interpretation the above result is hardly astonishing.) Here is a more astonishing case:

Example 2. The function

$$f(x) = 3x^2 - 11x$$

is increasing on the interval $2 \le x$. To prove this, we shall show that the derivative of f,

$$f'(x) = 6x - 11,$$

is positive for $x \ge 2$ and appeal to the criterion for monotonicity. f' is a linear function with positive slope 6, therefore as explained in Example 1, f' is increasing. Since its value at $x = 2$, $f'(2) = 6 \cdot 2 - 11 = 12 - 11 = 1$, is positive, its value at any $x > 2$ is also positive.

Example 3. The quadratic function

$$f(x) = 3x^2 - 11x$$

is decreasing for $x \le 1$. To prove this, we shall show that

$$f'(x) = 6x - 11$$

is negative for $x \le 1$ and appeal to the criterion for monotonicity. Since we know from Example 2 that f' is increasing, it suffices to show that f' is negative at the right end point, $x = 1$, of that interval. But

$$f'(1) = 6 \cdot 1 - 11 = -5$$

is indeed negative.

Example 4. We want to decide on which intervals of the real axis the quadratic function

$$f(x) = x^2 + bx + c$$

is increasing, and on which intervals it is decreasing. According to the monotonicity criterion it suffices to determine those intervals on which the derivative of f,

$$f'(x) = 2x + b,$$

is positive and those on which f' is negative. Since f' is a linear function with slope 2, it is, according to Example 1, an increasing function. Let's find that value z of x where f' is zero:

$$f'(z) = 2z + b = 0.$$

Clearly,

$$z = -\frac{b}{2}.$$

Since f' is increasing, we conclude that $f'(x)$ is negative for $x < -b/2$, positive for $x > -b/2$. This in turn implies that *the quadratic function*

$$f(x) = x^2 + bx + c$$

is decreasing for $x < -b/2$, increasing for $x > -b/2$.

Example 5. In which intervals is the function

$$f(x) = \frac{x}{x^2 + 2x + 3}$$

increasing, and in which intervals is it decreasing? According to the monotonicity criterion it suffices to determine the intervals of positivity and negativity of f'. Using the rule for differentiating a quotient, we get

$$f'(x) = \frac{3 - x^2}{(x^2 + 2x + 3)^2} .$$

The denominator of this fraction is a square of a nonzero number and therefore positive. Thus the sign of f' is the same as the sign of its numerator $3 - x^2$. Clearly, this function is positive on the interval $(-\sqrt{3}, \sqrt{3})$, negative for $x < -\sqrt{3}$ and for $\sqrt{3} < x$.

Example 5 illustrates the power of the monotonicity criterion; it is far simpler to show that $f'(x) < 0$ for all $x > \sqrt{3}$ than to verify the inequality

$$\frac{b}{b^2 + 2b + 3} < \frac{a}{a^2 + 2a + 3}$$

for all numbers a and b satisfying $\sqrt{3} < a < b$.

The following refinement of the monotonicity criterion is useful: *A differentiable function whose derivative f' is non-negative on an interval S is nondecreasing there.*

PROOF. This statement can be derived by a trick from our monotonicity criterion. For suppose that $f' \geq 0$ on S. Let m be any positive number and define a new function f_m by

$$f_m(x) = f(x) + mx.$$

Clearly, $f'_m = f' + m$ is positive on S, and so according to the monotonicity criterion f_m is increasing; that is,

$$\text{for } x < y, \qquad f_m(x) < f_m(y).$$

Substituting the definition of f_m into this relation, we get that for $x < y$ and any $m > 0$,

$$f(x) + mx < f(y) + my.$$

But then $f(x)$ cannot be greater than $f(y)$. If it were, the above inequality would be violated for the *positive* quantity

$$m = \frac{f(x) - f(y)}{y - x}.$$

Thus we have shown that $f'(x) \geq 0$ implies

$$f(x) \leq f(y) \quad \text{for } x < y.$$

This asserts the nondecreasing character of f. A similar argument shows that if $f' \leq 0$ on S, f is *nonincreasing* on S. □

Now suppose the derivative f' of f is *zero* on S. Then f' is both ≥ 0 and ≤ 0, and so, according to the latest version of the monotonicity criterion, f is both nondecreasing and nonincreasing on S. Such a function f must have the same value at all points of S. If f had different values at two distinct points, there would be either a decrease or an increase. Thus we have shown:

A differentiable function whose derivative is zero on a whole interval S has the same value at all points of S.

Here is a telegraphic formulation of this result:
If $f' = 0$ on S, then $f = constant$ on S.
And here is an epigrammatic formulation of the same result:
If the rate of change of f is zero at every point of S then f doesn't change at all across S.

This result is basic to calculus and to all its applications; it is part of the *fundamental theorem of calculus*. We shall encounter many applications of it throughout this text. Here is the first:

Example 6. Describe all functions having a constant derivative. Suppose $l(x)$ has a constant derivative, say

$$l'(x) = m \qquad \text{on } S,$$

m a constant. The function mx has derivative m, and so $l - mx$ has derivative zero. According to the theorem just proved, $l - mx$ is a constant; call that constant b: $l - mx = b$, therefore

$$l(x) = mx + b.$$

Thus we have shown: *every function whose derivative is constant is a linear function.*

Example 7. What are all functions whose derivative is linear? Let q be such a function, i.e.,

$$q'(x) = mx + b.$$

The function $\frac{1}{2}mx^2 + bx$ has derivative $mx + b$, therefore $q(x) - (\frac{1}{2}mx^2 + bx)$ has derivative 0. According to the theorem such a function is constant:

$$q(x) - (\tfrac{1}{2}mx^2 + bx) = c,$$

therefore

$$q(x) = \tfrac{1}{2}mx^2 + bx + c.$$

Thus we have shown: *every function whose derivative is linear is a quadratic function.*

EXERCISES

3.1 Determine the intervals on which the following functions are increasing, and those where the functions are decreasing:

(a) $f(x) = x^3 + 5x^2 - 2x + 1$ (b) $f(x) = (x + 1)/(x - 1)$

(c) $f(x) = \sqrt{x^2 + 2} - \sqrt{x^2 + 1}$ (d) $f(x) = (x - 3)/(x^2 - 2x - 1)$

(e) $f(x) = x - (x - 1)^{1/4}, \quad 1 < x$ (f) $f(x) = x - \sqrt{x}, \quad 0 < x$.

3.2 Show that every function whose derivative is a quadratic polynomial is a polynomial of third degree.

3.3 Prove that every function whose derivative is a polynomial of degree $n - 1$ is a polynomial of degree n.

3.4 Prove directly (i.e., without using the monotonicity criterion derived in this section) that

$$f(x) = x^2 + bx + c$$

is an increasing function for $x > -b/2$, decreasing for $x < -b/2$.

3.5 Assume that all functions occurring in this exercise are defined and differentiable on an interval $[-a, a]$. Recall from Chapter 2 that a function f is called *even* if

$$f(-x) = f(x),$$

and a function g is called *odd* if

$$g(-x) = -g(x).$$

(a) Show that the derivative of a differentiable even function is odd.

(b) Show that the derivative of a differentiable odd function is even.

(c) Show that, if the derivative f' of a function is odd, then the function f is even.

(d) Show that, if g' is an even function, then g is odd if and only if $g(0) = 0$.

3.4 The geometric meaning of the derivative

Let f be a differentiable function defined on some interval. Let a and $a + h$ be two points which belong to that interval, and consider the two points $(a, f(a))$ and $(a + h, f(a + h))$, located on the graph of f; see Figure 3.3.

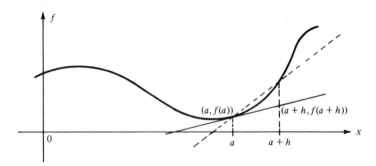

Figure 3.3

These two points determine a line called a *secant*. Its *slope*, as determined in Section 2.6, is

$$f'_h(a) = \frac{f(a + h) - f(a)}{h}.$$

As h tends to 0, this quantity tends to $f'(a)$, the derivative of f at a; and the secants through $(a, f(a))$ and $(a + h, f(a + h))$ tend to the line through the point $(a, f(a))$ with slope $f'(a)$. The limit of these secants is called the *tangent* to f at $(a, f(a))$.

The *slope* of the graph of f at this point is defined to be $f'(a)$.

Let us determine the linear function $l(x)$ whose graph is the tangent to f at $(a, f(a))$; according to Formula (6.6) in Chapter 2, that function is

(4.1) $$l(x) = f'(a)(x - a) + f(a).$$

We give below some examples of graphs of functions and lines tangent to them. In Figure 3.4a the tangent lies below the graph of the function, in Figure 3.4b it lies above the graph. Neither is the case in Figure 3.4c. The geometrical relation of the tangent to the graph is usually described in these words:

The tangent touches the graph.

We hope that the reader finds this descriptive phase appropriate, even for the graph of f in Figure 3.4c. The interesting and important questions "When does a tangent lie below the graph?," "When above it?," and "When does it cross the graph?" will be answered in Section 3.8.

Suppose f' is positive at every point of an interval S; then the graph of f has positive slope everywhere in S, which indicates that the graph of f rises from left to right, i.e., that f is an increasing function. This is the geometrical background of the monotonicity criterion. We now turn to some quantitative examples.

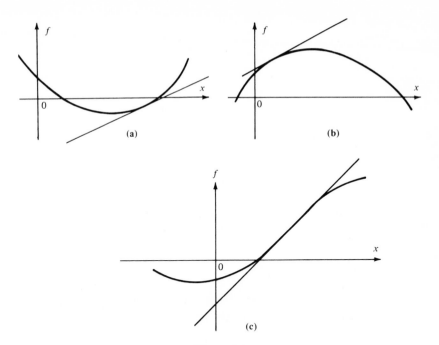

Figure 3.4

Example 1. The graph of the function $f(x) = \sqrt{1 - x^2}$, defined on the interval $-1 \le x \le 1$ is a semicircle with radius 1, centered at the origin; see Figure 3.5. The function f is differentiable on every closed subinterval which does not include the points $+1$ and -1, and

$$f'(a) = -\frac{a}{\sqrt{1 - a^2}}.$$

According to the foregoing discussion this is the slope of the tangent at the point $(a, \sqrt{1 - a^2})$. The slope of the line through the origin and the point $(a, \sqrt{1 - a^2})$ is

$$\frac{\sqrt{1 - a^2}}{a}.$$

The product of these two slopes is

$$-\frac{a}{\sqrt{1 - a^2}} \cdot \frac{\sqrt{1 - a^2}}{a} = -1.$$

According to Section 2.6, *two lines whose slopes have product* -1 *are perpendicular*. Thus we have given an analytical proof of the following well known fact of geometry: *The radius vector of a circle to a point is perpendicular to the tangent through that point.*

109

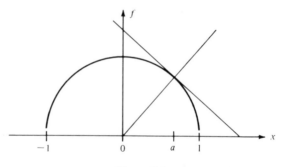

Figure 3.5

Our next example deals with the reflection of light from mirrors. We state the laws governing reflection; an explanation of these laws will be provided in the next section.

(a) In a uniform medium light travels in straight lines.
(b) When a ray of light impinges on a straight mirror, it is reflected; the angle i which the incident ray forms with the line *perpendicular* to the mirror equals the angle r which the reflected ray forms with the perpendicular; see Figure 3.6. The angle i is called the *angle of incidence, r the angle of reflection*.

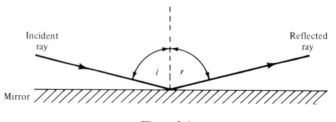

Figure 3.6

(c) The same rule governs the reflection of light from a curved mirror:

$$\text{angle of incidence} = \text{angle of reflection};$$

in this case the line perpendicular to the mirror is defined as the *line* (dashed in Figure 3.7) *perpendicular to the tangent to the mirror at the point of incidence.*

Example 2. We shall investigate reflections from a mirror located in the x, y-plane whose parabolic shape is described by the equation

(4.2) $$y = x^2.$$

We wish to calculate the path of rays which are reflections of incident rays *parallel to the y axis*; see Figure 3.7. In particular we wish to calculate the location of the point F where such a reflected ray intersects the y-axis. Denote

110

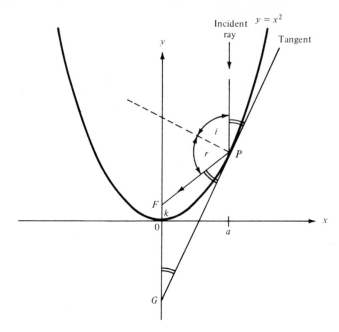

Figure 3.7

by P the point of incidence and by G the point where the tangent at P intersects the y-axis. The following geometric facts can be read off the above figure:

1. The angle FPG is complementary to the angle of reflection r.
2. The angle FGP is equal to an angle which is complementary to the angle of incidence i.

Since, according to the law of reflection, r equals i, we conclude that the triangle FGP is isosceles, the angles at P and G being equal. From this we conclude that the sides opposite are equal:

(4.3) $$PF = FG;$$

here PF denotes the distance from P to F. We shall calculate now the length of these sides. We denote the x coordinates of P by a; then according to (4.2) the y coordinate of P is a^2. We denote the y coordinate of the point F by k; according to the Pythagorean theorem

(4.4) $$PF^2 = a^2 + (a^2 - k)^2.$$

Next we calculate the y coordinate of the point G; since G is the intersection of the tangent with the y-axis, the y coordinate of G is the value at $x = 0$ of the linear function which, according to (4.1), is

$$l(x) = f'(a)(x - a) + f(a).$$

111

At $x = 0$,

$$l(0) = -f'(a)a + f(a).$$

In our case $f(x) = x^2$, so $f'(x) = 2x$, and

$$-f'(a)a + f(a) = -2a^2 + a^2 = -a^2.$$

The length FG is the difference of the y coordinates of F and G:

(4.5) $$FG = k - (-a^2) = k + a^2.$$

Using (4.4) and (4.5) to express the right and left sides of (4.3), we get

$$a^2 + (a^2 - k)^2 = (k + a^2)^2.$$

Now carry out the squarings, cancel common terms, add $2a^2k$ to both sides, and divide by a^2. The result is

$$4k = 1, \qquad \text{so } k = \tfrac{1}{4}.$$

This gives the surprising result that the location of the point F is *the same for all points* P, i.e., *the reflections of all rays parallel to the y axis pass through the point* $(0, \tfrac{1}{4})$. This point is called the *focus* of the curve (4.2); the curve itself is called a *parabola*.

Rays coming from a very distant object such as one of the stars are very nearly parallel. Therefore if a parabolic mirror is pointed so that its axis points in the direction of a star, all the rays will be reflected toward the focus; this principle is exploited in the construction of telescopes.

The rays from the sun are nearly parallel and therefore they can be focused pretty accurately by a parabolic mirror. This principle is exploited in the construction of solar furnaces, and, on a more modest scale, in the design of mirror cigarette lighters for outdoor use.

EXERCISES

4.1 Find a line which is tangent to the graphs of *both* functions,

$$f(x) = x^2 \quad \text{and} \quad g(x) = x^2 - 2x.$$

4.2 Find all tangents to the graph of

$$f(x) = x^2 - x$$

which go through the point (2, 1). Show that no tangent goes through the point (2, 3); can you find a geometrical explanation for this?

4.3 Find the equation of the line tangent to the parabola $y = 3x^2 - 4$ at the point $(1, -1)$.

4.4 Find the point of intersection of the tangents to the graphs of $f_1(x) = x^2 - 2x$ and $f_2(x) = -x^2 + 1$ at the points (2, 0) and (1, 0), respectively.

3.5 Maxima and minima

We have shown in Section 2.7 that every continuous function f on a closed interval S takes on its maximum at some point. In this section we shall show how to use calculus to locate that point (or those points in case there are several) when the function f is differentiable. The method rests on this simple observation: if the maximum of a differentiable function f occurs at some *interior* point c of S, i.e., a point that is not an endpoint, then $f'(c) = 0$. For, suppose that $f'(c) \neq 0$ say $f'(c) > 0$; since $f'(c)$ is the limit of $f'_h(c)$ as h tends to 0, it follows that, for h small enough,

$$f'_h(c) = \frac{f(c + h) - f(c)}{h}$$

is also positive. This implies that for all h small enough and *positive*

$$f(c + h) > f(c).$$

But since c is an interior point of S, for h small enough $c + h$ belongs to S, so that the above inequality violates the assumption that the maximum of f on S is assumed at c.

Similarly, $f'(c) < 0$ implies that for all h small enough and *negative*,

$$f(c + h) > f(c)$$

which, for an interior point c, contradicts the characterization of c as the point where f assumes its maximum. So only $f'(c) = 0$ is consistent with f achieving its maximum on S at the interior point c. The points in S where $f'(x)$ vanishes are often called *critical points* of f. We have proved the

Maximum theorem. *Let f be a differentiable function on a closed interval S. Then f achieves its maximum either at an endpoint of S, or at an interior point c where $f'(c) = 0$.*

Our analysis yields furthermore: *If the maximum of f is reached at the left endpoint a of S, then $f'(a) \leq 0$; if the maximum is reached at the right endpoint b, $f'(b) \geq 0$.*

A similar argument yields the

Minimum theorem. *The minimum of f on S is achieved either at one of the endpoints of S or at an interior point c where $f'(c) = 0$. If the minimum is achieved at the left endpoint a, $f'(a) \geq 0$; if the minimum is achieved at the right endpoint b, $f'(b) \leq 0$.*

Suppose $f'(x)$ is positive on S; then it is positive on any subinterval $[d, e]$, and according to the maximum theorem the only point where f can reach its maximum is the right endpoint, so

$$f(d) < f(e).$$

Since d and e can be any pair of points of S such that d is less than e, the above inequality shows that f is *increasing*. Thus the monotonicity criterion derived in Section 3.3 is a consequence of the maximum (and minimum) theorems presented here.

We shall now give a number of examples where we determine the maximum or minimum value of a function f on an interval S by examining the values of f at the endpoints of S and at those interior points of S where f' is zero.

Example 1. Consider the quadratic function

$$f(x) = x^2 + bx + c.$$

We can rewrite this by "completing the square":

$$f(x) = \left(x + \frac{b}{2}\right)^2 - \frac{b^2}{4} + c.$$

We see at a glance that the minimum of $f(x)$ is achieved at $x = -b/2$. We show now how to derive this result by calculus. To this end we look for critical points of f:

$$f'(x) = 2x + b.$$

Clearly

$$f'(x) \begin{cases} <0 & \text{for } x < -b/2 \\ =0 & \text{for } x = -b/2 \\ >0 & \text{for } x > -b/2; \end{cases}$$

it follows that f increases in any interval whose left endpoint is $-b/2$ and decreases in any interval whose right endpoint is $-b/2$. This proves that f achieves its minimum at $x = -b/2$.

Example 2. Find the largest and smallest values of the function

$$f(x) = x^3 + 5x^2 - x - 5$$

in the interval $I: -8 \le x \le 2$. At the endpoints, we have

$$f(-8) = -189, \qquad f(2) = 21.$$

Are there critical points in the interior of the interval I?

$$f'(x) = 3x^2 + 10x - 1 = 0$$

when

$$x = x_1 = \frac{-5 + 2\sqrt{7}}{3} \approx 0.097, \quad \text{and} \quad x = x_2 = \frac{-5 - 2\sqrt{7}}{3} \approx -3.43,$$

so x_1, x_2 both lie in $[-8, 2]$. Since

$$f(x_1) \approx -5.049 \qquad \text{and} \qquad f(x_2) \approx 16.9,$$

the left endpoint yields the smallest, the right endpoint the largest value of f in I.

Example 3. Find the largest and smallest values of the function given in Example 2 in the interval J: $-4 \le x \le 1$. At the endpoints of J, we have

$$f(-4) = 15, \qquad f(1) = 0,$$

so f attains its smallest value at $x = x_1$, its largest at $x = x_2$ (see Example 2).

Although the minimum problem of Example 1 can be solved by completing a square, no such algebraic method works for the cubic function in Examples 2 and 3. The method of calculus, however, is quite general.

Example 4. Determine the shape of the closed cylinder which has the largest volume among all cylinders of given surface area A.

Let r be the radius, h the height of the cylinder. Then its area is

(5.1) $$A = 2\pi r^2 + 2\pi rh = 2\pi r(r + h),$$

and its volume is

(5.2) $$V = \pi r^2 h.$$

We have expressed V as a function of two variables, r and h, but can eliminate one by means of the *constraint* (5.1):

$$h = \frac{A}{2\pi r} - r,$$

so that

$$V = f(r) = \pi r^2 \left[\frac{A}{2\pi r} - r \right] = \frac{Ar}{2} - \pi r^3, \qquad r > 0.$$

The derivative $f'(r) = \frac{1}{2}A - 3\pi r^2$ vanishes when $r = \sqrt{A/6\pi}$, so

$$h = 2\sqrt{\frac{A}{6\pi}} = 2r,$$

that is, the diameter of the maximal cylinder equals its altitude.

The surface area of the cylinder is proportional to the material needed to manufacture a cylindrical container. The above shape is *optimal* in the sense that it encloses the largest volume for a given amount of material. Examine the cans in the supermarket and determine which brands use the optimal shape.

Example 5. Find the largest and smallest values of the function
$$f(x) = x^4 + 2x^3 + 12x^2 - 17x + 1$$
in the interval $0 \le x \le 1$. At the endpoints, the values of f are
$$f(0) = 1, \qquad f(1) = -1.$$
Are there critical points in the interval $(0, 1)$? We compute f':
$$f'(x) = 4x^3 + 6x^2 + 24x - 17.$$
Note that
$$f'(0) = -17, \qquad f'(1) = 17,$$
i.e., that f' is negative at the left endpoint, positive at the right endpoint. Therefore by the intermediate value theorem (see Section 2.7) f' must vanish somewhere in the interval $(0, 1)$. We claim a little more: f' vanishes at exactly one point in $(0, 1)$. The reason for this is that f' is an *increasing* function. To see that f' is increasing, we verify that its derivative is *positive*. The derivative of f', called the *second derivative* of f and denoted by f'', is easy to calculate:
$$f''(x) = (f')' = 12x^2 + 12x + 24.$$
According to Example 1, the minimum is achieved at $x = -\frac{12}{24} = -0.5$, and the value of f'' there is
$$f''(-0.5) = \frac{12}{4} - \frac{12}{2} + 24 = 21 > 0.$$
But then it follows that $f''(x)$ is positive everywhere.

We have succeeded in using the methods of calculus to show that f' has exactly one zero z in the interval $(0, 1)$. But how do we find z? There is an algebraic formula due to Cardano for the roots of a cubic polynomial, but it is so cumbersome that we reject it in favor of the *method of bisection* described in Section 2.7. Using the program listed at the end of this text, we find that
$$z = 0.58801 \ldots .$$
As a check on the numerical calculation we calculate $f'(z)$:
$$f'(z) = 5.94 \cdot 10^{-6}$$
is gratifyingly small, indicating that z is indeed very close to the true zero of f'. Finally we calculate the value of f at z:
$$f(0.58801) = -4.320938.$$
This is less than both, $f(0)$ and $f(1)$, so -4.320938 is near the minimum of f.

Example 5 differs from Examples 1–4 inasmuch as we relied on a numerical method, rather than an algebraic one, for finding the zeros of f'. The typical maximum-minimum problems of real life fall far more frequently into the category of Example 5 than into the elementary category of the first four examples. Therefore the finding of zeros of functions is a frequently performed numerical task. This immediately raises the question if there are more efficient methods than bisection for finding zeros. The anser is yes; one such method, invented by Newton, will be described in Section 3.10.

Inequalities are important in mathematics, specially in calculus. One of the useful inequalities connects the *arithmetic mean*

$$\frac{a + b}{2}$$

and the *geometric mean*

$$\sqrt{ab}$$

of two positive numbers a and b. The inequality says that the latter never exceeds the former:

(5.3)
$$\sqrt{ab} \le \frac{a + b}{2}.$$

This is easy enough to prove algebraically; square both sides and multiply them by 4 to obtain the equivalent inequality

$$4ab \le a^2 + 2ab + b^2.$$

This is equivalent to

$$0 \le a^2 - 2ab + b^2,$$

which is obviously true, since the expression on the right is a perfect square.

In the next example, we shall show how to use the method of calculus, rather than algebra, to prove inequalities.

Example 6. The arithmetic mean of 3 numbers a, b, and c is defined to be

$$\frac{a + b + c}{3}.$$

Their geometric mean, defined when all 3 are positive, is

$$(abc)^{1/3},$$

We shall prove by calculus that the arithmetic mean of 3 positive numbers always exceeds their geometric mean, except when the 3 numbers are equal:

(5.4)
$$(abc)^{1/3} \le \frac{a + b + c}{3}.$$

To prove this, we begin by raising both sides to the third power and dividing by b. This yields the equivalent inequality

$$ac \leq \frac{1}{b} \left(\frac{a+b+c}{3} \right)^3,$$

equality holding only for $a = b = c$. This inequality can be phrased so: For any given positive numbers a and c, the function

(5.5)
$$f(b) = \frac{1}{b} \left(\frac{a+b+c}{3} \right)^3$$

is $\geq ac$ for all positive b. This statement is equivalent to the following: The minimum of f on the interval $b > 0$ is $\geq ac$.

We shall prove this last statement by finding the minimum value of f and verifying directly that it is not less than ac. We note first that for b very small, and likewise for b very large, $f(b)$ is very large; for, when b is close to 0, the first factor in (5.5) is very large while the second is at least $[(a + c)/3]^3$, and when b is very large, the second factor is bigger than $b^3/27$, so that $f(b)$ is bigger than $b^2/27$. It follows that f assumes its minimum at some interior point; according to the minimum theorem, $f'(b) = 0$ at this point. Differentiating (5.4) and setting the derivative equal to 0 gives

$$f'(b) = -\frac{1}{b^2} \left(\frac{a+b+c}{3} \right)^3 + \frac{1}{b} \left(\frac{a+b+c}{3} \right)^2 = 0.$$

Multiplying by b^2 and dividing by $[(a + b + c)/3]^2$ gives the equivalent relation

$$-\frac{a+b+c}{3} + b = 0,$$

valid only if

$$b = \frac{a+c}{2}.$$

Thus the minimum of f occurs at $b = (a + c)/2$; the value of f at this point is

$$f\left(\frac{a+c}{2} \right) = \frac{1}{(a+c)/2} \left(\frac{a+c+(a+c)/2}{3} \right)^3 = \left(\frac{a+c}{2} \right)^2.$$

At any other point b, the value of $f(b)$ is greater than $[(a + c)/2]^2$. Since according to (5.3)

$$\left(\frac{a+c}{2} \right)^2 \geq ac,$$

we conclude that

$$f(b) \geq ac;$$

equality can hold only if $b = (a + c)/2$, and if a and c are equal, i.e., when $a = b = c$. This completes the proof of inequality (5.4).

We remark that it is possible to give a purely algebraic proof of the inequality discussed in Example 6. It requires ingenuity. So does the calculus proof given in Example 6, but the calculus method provides a well defined starting point and works in cases where elementary methods fail.

Example 7. We wish to calculate the path of a ray of light which is reflected from a flat mirror. The path of a reflected ray going from point P to point Q is pictured in Figure 3.8. The ray consists of two straight line segments, one, the incident ray, leading from P to the point of reflection R, the other, the reflected ray, leading from R to Q. To determine the location of the point of reflection we shall use the following law of optics, called *Fermat's principle*:

Among all possible paths PRQ connecting two points P and Q via a mirror, light travels along the one which takes the least time to traverse.

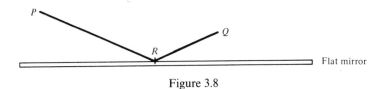

Figure 3.8

In a uniform medium like air, light travels with constant speed, so the time needed to traverse the path PQR equals its length divided by the speed of light. So the path taken by the light ray will be the *shortest* path PRQ. We choose the mirror to be the axis of a Cartesian coordinate system; see Figure 3.9. The coordinates of the point R are $(x, 0)$; denote the coordinates of the point P by (a, b), those of Q by (c, d). According to the Pythagorean theorem, the distances PR and RQ are

$$l_1(x) = PR = \sqrt{(x - a)^2 + b^2}, \qquad l_2(x) = RQ = \sqrt{(c - x)^2 + d^2}.$$

The total length l of the path is

(5.6) $$l(x) = l_1(x) + l_2(x) = \sqrt{(x - a)^2 + b^2} + \sqrt{(c - x)^2 + d^2}.$$

Since $l(x) \geq |x - a| + |c - x|$, for x large positive or negative $l(x)$ is very large, so that the function l assumes its minimum at a point where $l'(x) = 0$.

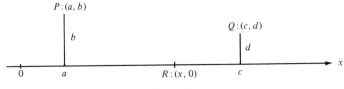

Figure 3.9

119

Differentiating l as given by (5.6), we get

$$l'(x) = \frac{x - a}{l_1(x)} - \frac{c - x}{l_2(x)}.$$

$l'(x) = 0$ means that

(5.7)
$$\frac{x - a}{l_1} = \frac{c - x}{l_2}.$$

The geometric meaning of this relation is illustrated in Figure 3.10, where the dashed line is perpendicular to the mirror at the point of reflection. The left side of (5.7) is the sine of the *angle of incidence,* defined as the angle i formed

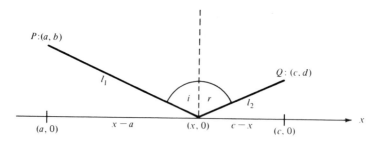

Figure 3.10

by the incident ray and the perpendicular to the mirror; the right side of (5.7) is the sine of the angle of reflection, defined as the angle r formed by the reflected ray and the perpendicular to the mirror. Therefore (5.7) asserts that

$$\sin i = \sin r;$$

and, since these angles are acute, this relation can be expressed by saying that *the angle of incidence equals the angle of reflection.* This is the celebrated *law of reflection.*

We now give a simple geometric derivation of the law of reflection; see Figure 3.11. We introduce the *mirror image P'* of the point P; i.e., the mirror is the *perpendicular bisector* of the interval PP'. Then any point R of the mirror is equidistant from P and P':

$$PR = P'R,$$

so that

$$l(x) = P'R + RQ.$$

The right side is the sum of two sides of the triangle $P'RQ$ and is therefore at least as great as the third side:

$$l(x) \geq P'Q.$$

120

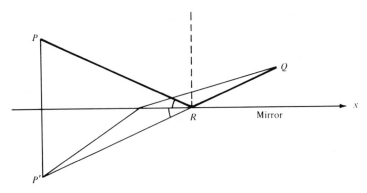

Figure 3.11

Equality holds only in the special case when the vertex R lies on $P'Q$: The two smaller angles formed by the mirror and the line through P' and Q are equal; since one is complementary to the angle of incidence, the other to the angle of reflection, the law of reflection follows.

Example 8. We now turn to light reflection from a curved mirror. This case can no longer be handled by elementary geometry; calculus, on the other hand, still gives the answer, as we shall now demonstrate. Again according to Fermat's principle of least time, the point of reflection R can be characterized as that point on the mirror which minimizes the total length

$$l = PR + RQ.$$

We introduce Cartesian coordinates with R as the origin, and the x axis tangent to the mirror at R. In terms of these coordinates the mirror is described by an equation

$$y = f(x), \quad \text{such that } f(0) = 0, \ f'(0) = 0;$$

see Figure 3.12. Denoting as before the coordinates of P by (a, b), those of Q by (c, d), l can be expressed as $l(x) = l_1(x) + l_2(x)$ or

(5.8) $l(x) = \sqrt{(x - a)^2 + (f(x) - b)^2} + \sqrt{(c - x)^2 + (f(x) - d)^2}.$

The minimum value of l is achieved at $x = 0$; therefore $l'(0) = 0$. Differentiating (5.8), we get

$$l'(x) = \frac{(x - a) + f'(x)(f(x) - b)}{l_1(x)} - \frac{c - x + f'(x)(f(x) - d)}{l_2(x)}.$$

Since $f'(0) = 0$ at R, the value of $l'(x)$ at $x = 0$ simplifies to

$$l'(0) = -\frac{a}{l_1} - \frac{c}{l_2} = 0.$$

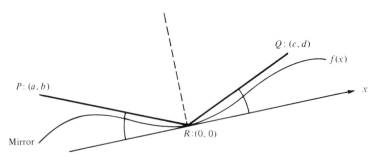

Figure 3.12

This equation agrees with (5.7) for the case of a straight mirror; we conclude as before that

angle of incidence = angle of reflction,

except that in this case these angles are defined as the ones formed by the rays with the line *perpendicular* to the *tangent* of the mirror at the point of reflection.

If we had not chosen the x-axis to be tangent to the mirror at the point of reflection, we would have had to use a fair amount of trigonometry to deduce the law of reflection from the relation $l'(x) = 0$. This shows that in calculus, as well as in analytic geometry, life can be made simpler by a wise choice of coordinate axes.

One difference between reflections from a straight and curved mirror is that, for a curved mirror, there may well be several points R which furnish a minimum of $PR + QR$. An observer located at P sees the object located at Q when he looks toward any of these points R; this can be observed in some funhouse mirrors.

Example 9. We shall study the *refraction* of light rays, that is, their passage from one medium into another, in cases where the propagation speed of light in the two media is different. A common example is the refraction at an air and water interface; see Figure 3.13. We rely as before on Fermat's optical

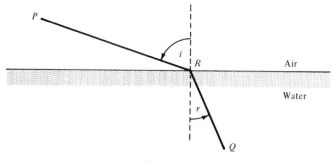

Figure 3.13

principle: among all possible paths PRQ, light travels along the one which takes the least time to traverse.

Denote by c_a and c_w the speed of light in air and water, respectively. The time it takes light to travel from P to R is PR/c_a, and from R to Q it takes RQ/c_w. The total time t is then

$$t = \frac{PR}{c_a} + \frac{RQ}{c_w}.$$

We introduce the line separating air and water as the x-axis. Denote as before the coordinates of P and of Q by (a, b) and (c, d) respectively, and the x-coordinate of R by x; then

$$PR = l_1(x) = \sqrt{(x - a)^2 + b^2}, \qquad RQ = l_2(x) = \sqrt{(c - x)^2 + d^2},$$

and so

$$t(x) = \frac{l_1(x)}{c_a} + \frac{l_2(x)}{c_w}.$$

As before we notice that $t(x) > |x - a|/c_a + |c - x|/c_w$; therefore, for x large positive or negative, $t(x)$ is large. It follows from this that $t(x)$ achieves its minimum at a point where $t'(x)$ is zero. The derivative $t'(x)$ is

$$t'(x) = \frac{l_1'(x)}{c_a} + \frac{l_2'(x)}{c_w} = \frac{x - a}{c_a l_1(x)} - \frac{c - x}{c_w l_2(x)}.$$

From the relation $t'(x) = 0$, we deduce that

$$\frac{c_w}{c_a} \frac{x - a}{l_1(x)} = \frac{c - x}{l_2(x)}.$$

As in Equation (5.7) of Example 7, the ratio $(x - a)/l_1$ and $(c - x)/l_2$ can be interpreted geometrically (see Figure 3.14) as the sines of the angle of incidence

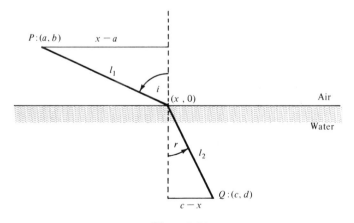

Figure 3.14

i and of the angle of refraction *r*, respectively:

(5.9)
$$\frac{c_w}{c_a} \sin i = \sin r.$$

This is the *law of refraction*, due to the physicist *Snell* and is often stated as follows: *When a light ray travels from a medium* 1 *into a medium* 2 *where the propagation speeds are* c_1 *and* c_2, *respectively, it is refracted so that the ratio of the sines of the angle of incidence to the angle of reflection is equal to the ratio* c_1/c_2 *of the propagation speeds.* The ratio c_1/c_2 is called the *index of refraction I*. Since the sine function does not exceed 1, it follows from the law of refraction that sin *r* does not exceed the index of refraction *I*; i.e., from (5.9)

(5.10)
$$\sin r = I \sin i \leq I.$$

The speed of light in water is less than in air; $I = c_w/c_a < 1$. It follows from inequality (5.10) that *r* cannot exceed a critical angle r_{crit} defined by $\sin r_{crit} = I$. An under water observer located at *Q* (see Figure 3.15) who looks in a direction which makes an angle θ with the perpendicular *greater* than r_{crit} *cannot see any point P above the water*, since such a refracted ray would violate the law of refraction. He sees instead reflections of under-water objects. This phenomenon, well known to snorklers, is called *total reflection*.

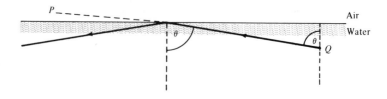

Figure 3.15

EXERCISES

5.1 Find the maximum and minimum values of the function
$$f(x) = 2x^3 - 3x^2 - 12x + 8$$
on each of the following intervals *S*:
(a) $[-2.5, 4]$ (b) $[-2, 3]$ (c) $[-2.25, 3.75]$.

5.2 The arithmetic mean of *n* numbers $a_1, \ldots, a_n$ is
$$\frac{a_1 + \cdots + a_n}{n};$$
the geometric mean, in case they are positive, is defined as
$$(a_1 a_2 \cdots a_n)^{1/n}.$$

Prove that for all sets of positive a_j

$$(a_1 \ldots a_n)^{1/n} \le \frac{a_1 + \cdots + a_n}{n},$$

equality holding only if all the a_j are equal. [*Hint*: Use the method described in Example 6 for the case $n = 3$.]

5.3 A manufacturer has outlets at 3 locations A, B and C; see Figure 3.16a. He wants to locate his plant at a point X so that his total shipping cost to all his outlets is as small as possible. Assuming that his shipments to all 3 outlets are equal, and that the cost of transportation is proportional to the distance, he will accomplish his aim if he chooses X so that the sum $s(X)$ of the distances from X to A, B and C is as small as possible.

In this exercise the reader is asked to locate point X using the simplifying assumption that points A and B are symmetric with respect to C, in the following sense: the line connecting C to the midpoint M of segment AB is perpendicular to that segment. (See Figure 3.16b.)

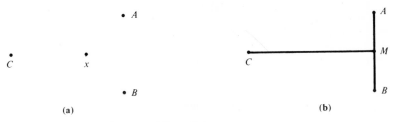

Figure 3.16

(a) Show that the minimizing point X lies on the segment CM.
(b) Introduce rectangular coordinates so that C is the origin and the ray CM is the positive x-axis. Denote the coordinates of A by (a, b); by symmetry, those of B are $(a, -b)$. The coordinates of X are $(x, 0)$, $0 \le x \le a$. Show that $s(X) = XA + XB + XC$ is given by the formula

$$s(x) = x + 2\sqrt{(a - x)^2 + b^2}.$$

(c) Show that if $a \le b/\sqrt{3}$, $s(x)$ achieves its minimum at $x = 0$, and if $a > b/\sqrt{3}$, $s(x)$ achieves its minimum at $x = a - b/\sqrt{3}$.

5.4 Consider an open cardboard box whose bottom is a square of edge length x, and whose altitude is y.
(a) The volume V and surface area S of the box are given by

$$V = x^2 y, \quad S = x^2 + 4xy.$$

Among all boxes with given volume find the one with smallest surface area. Show that this box is squat, i.e., $y < x$.

5.5 Consider a particle of unit mass moving on a line whose position at any time t is

$$x(t) = 3t - t^2.$$

Find the time when the particle has
(a) maximum displacement x;   (b) maximum velocity v.

5.6 Find the point on the graph of $y = x^2/2$ closest to the point $(6, 0)$.

5.7 (a) Find the positive number which exceeds its cube by the largest amount.

(b) Explain why you would expect the solution to be in the interval $[0, 1]$.

(c) What is the largest amount by which a positive number can exceed its cube?

5.8 (a) Find the positive number x such that the sum of x and its reciprocal is as small as possible.

(b) Show that there is no positive number x such that the sum of x and its reciprocal is a maximum.

(c) Use the arithmetic-mean–geometric-mean inequality to answer part (a).

5.9 Consider the function f defined by

$$f(x) = \frac{x}{x^2 + 1}.$$

(a) Find $f'(x)$.

(b) In which interval(s) does $f(x)$ increase?

(c) In which interval(s) does $f(x)$ decrease?

(d) Find the smallest value of $f(x)$ in $[-10, 10]$.

(e) Find the largest value of $f(x)$ in $[-10, 10]$.

5.10 Find the smallest value of the function

$$f(x) = 2x^5 - x^3 + 80x^2 + 10x + 2$$

in the interval $-1 \le x \le 2$ by writing a computer program, using the method of bisection.

5.11 Write a computer program which finds, approximately, the largest value taken on by a function f in an interval S, by performing the following steps:

(a) Evaluate $f(x)$ at the endpoints of S and at N equidistant points x_j between the endpoints of S; N to be specified by the user.

(b) If $f(x_j)$ is greater than both $f(x_{j-1})$ and $f(x_{j+1})$, the program finds a zero z of $f'(x)$ in the interval (x_{j-1}, x_{j+1}) by using Newton's method described in Section 3.10.

(c) The program evaluates $f(z_j)$ and determines the largest of these values. This largest value is then compared to the values of f at the endpoints.

3.6 One-dimensional mechanics

Mechanics is the study of the motion of collections of rigid bodies under given forces. In this section we shall show how calculus is used to formulate the basic laws of motion for particles. We shall show, in a very simple situation, how to use these basic laws to determine the motion.

A *particle* is an idealization in physics. A particle has no extent so that its position is a *point*; it has a *mass*, usually denoted by the letter m. In this section we shall investigate the motion of particles along straight lines. In this simple case the position of a particle is completely described by its distance x from an arbitrarily chosen point (the origin) on the line; x is taken to be positive if x lies to one side (chosen arbitrarily) of the origin and negative if the particle is located on the other side of the origin.

Since the particle moves, x is a function of the time t. The *derivative* $x'(t)$ of this function is the *velocity* of the particle, a quantity usually denoted by v:

(6.1) $$v = x' = \frac{dx}{dt}.$$

Note that v is positive if the x coordinate of the particle increases during the motion.

The velocity of a particle changes in general as time changes; the rate at which it changes is called the *acceleration*, and is usually denoted by the letter a:

(6.2) $$a = v' = \frac{dv}{dt}.$$

Newton's laws of motion relate the acceleration of a particle to its mass and the force acting on it. His first law asserts that, if no force acts on the particle, its velocity does not change; this can be expressed by saying that the acceleration of a particle on which no force acts is zero.

The second law deals with particles on which a force is acting. It is not possible to give a philosophical definition of force, but everybody has an intuitive understanding of the commonest kind of forces: the force exerted by a taut rope attached to an object, the force of a coiled spring, the force of gravity, the force of friction acting on a moving body, magnetic force acting on a piece of iron, and electrostatic force acting on an electrically charged object such as a piece of paper or synthetic cloth coming out of a clothes dryer.

Everybody understands intuitively that force has a *direction* and that forces can be compared with each other in strength. A stronger force acting on a particle causes a greater acceleration than a weaker force would acting on the same particle. We shall now describe a natural way of measuring the strength of any force. The consistency of this kind of measurement rests on two experimental facts:

1. *The acceleration caused by two forces in the same direction acting jointly on a particle is the sum of the accelerations that would be caused by each of the forces acting separately on the same particle.*
2. *The acceleration caused by a force acting on a particle is inversely proportional to the mass of that particle.*

In other words, a force acting on a particle causes an acceleration twice as large as that force would impart to another particle twice as massive.

127

Suppose a force, acting in the direction of the x-axis, causes a particle of mass m to accelerate at the rate a; *we define the magnitude f of that force to be*

(6.3)
$$f = ma.$$

In light of fact 2, f is consistently defined.

According to definition (6.3) a force acting along the directed x-axis is reckoned as negative if it imparts a negative acceleration (deceleration) to a particle travelling along the x-axis. There is nothing mysterious about this negative sign. It merely means that the force is acting in the negative direction along the x-axis.

We illustrate how Newton's law (6.3) can be used to describe motions in the specific situation when the force is that of *gravity*. According to a law that is again Newton's, the magnitude f_g of the force of gravity exerted on a particle of mass m is proportional to its mass:

(6.4)
$$f_g = gm.$$

The constant of proportionality g varies from point to point in space; at a point near the surface of the earth the direction of the force is toward the center of the earth, and the value of g is approximately

(6.5)
$$g_e = 32.17,$$

with the units feet per second per second, provided that distance is measured in feet, time in seconds. Near the surface of the moon the value of g is approximately ℺

(6.6)
$$g_m = 5.1.$$

Denote by x_e the distance from the surface of the earth of a falling body; denote by v_e the vertical velocity of this falling body. Since x was chosen to increase upward, and the force of gravity is downward, the force of gravity must be reckoned as $-g_e m$. Substituting this into Newton's law (6.3), we see that

$$-g_e m = ma_e,$$

where a_e is the acceleration of the falling body. Divide by m:

$$-g_e = a_e.$$

Recalling the definitions (6.1), (6.2) of velocity and acceleration, we can write this as

(6.7)$_e$
$$-g_e = \frac{dv_e}{dt},$$

where

(6.8)$_e$
$$v_e = \frac{dx_e}{dt}.$$

128

These are the equations governing "earth motion," i.e., vertical motion of a particle under the influence of gravity, and gravity alone, near the surface of the earth. The equations governing "moon motion" are entirely similar, except that the number $g_e = 32.17$ is to be replaced by the number $g_m = 5.1$. Since the equations of earth motion and moon motion are related so simply, one surmises that the motions themselves also might be simply related. This is indeed so as seen from the

Earth–moon theorem. *Suppose that $x_e(t)$ is earth motion; then*

$$(6.9) \qquad x_m(t) = x_e(kt), \qquad \text{with } k = \sqrt{\frac{g_m}{g_e}},$$

is moon motion.

PROOF. Differentiating (6.9) we get, using the chain rule, that

$$v_m(t) = \frac{dx_m}{dt} = \frac{d}{dt} x_e(kt) = kv_e(kt).$$

Differentiating this relation and using $(6.7)_e$ we get

$$a_m = \frac{dv_m(t)}{dt} = k\frac{d}{dt} v_e(kt)$$

$$= -k^2 g_e = -\frac{g_m}{g_e} g_e = -g_m.$$

Thus x_m satisfies the equation of moon motion; this proves the theorem. $\square$

Formula (6.9) says that the value of x_m at time t equals the value of x_e at time kt; this has a very interesting intuitive interpretation: The approximate value of k is slightly less than 0.4; thus earth motion is changed into moon motion by slowing down time by a factor of 0.4. More dramatically, if we project a movie picture taken of earth motion at 0.4 times the normal speed, we see motion on the screen which would simulate moon motion.

Television viewers of the landing on the moon may remember that during arid moments, while waiting for a moon walk to begin, the audience was treated to simulated moon walks, performed by men in astronaut gear on a simulated moonscape. Our analysis shows that if these simulated moon walks, filmed on earth, had been slowed down by a factor of 0.4, the resulting film would have looked much more like real moon walks! (Perhaps so real that most of the drama of moon walking would have vanished.)

We complete this discussion by deriving from Equations $(6.7)_e$ and $(6.8)_e$ explicit formulas for velocity and displacement in any earth motion as function of time. According to $(6.7)_e$ the derivative of v_e is $-g_e$. According to Example 6 of Section 3.3, v_e is of the form

$$(6.10)_e \qquad v_e(t) = -g_e t + b,$$

b some constant.

According to $(6.8)_e$, the derivative of x_e with respect to t is the linear function v_e given by $(6.10)_e$. According to Example 7 of Section 3.3 such a function x_e itself is quadratic:

$(6.11)_e$
$$x_e(t) = -\tfrac{1}{2}g_e\, t^2 + bt + c.$$

This expresses x_e as a function of t. There is a similar formula for x_m, with g_e replaced by g_m:

$(6.11)_m$
$$x_m(t) = -\tfrac{1}{2}g_m t^2 + bt + c.$$

The significance of the constants b and c entering these formulas is easily discerned. Set $t = 0$ in $(6.10)_e$ and $(6.11)_e$; we get

$$v_e(0) = b \qquad \text{and} \qquad x_e(0) = c.$$

In other words, b and c are the initial velocity and initial position, respectively, of the particle motion. Since the function $x_e(t)$ is determined once b and c are prescribed, we conclude that if two particles moving under the influence of the gravitational field of the earth near the surface of the earth have the same position and same velocity at time $t = 0$, then their positions are the same at any other time.

Any function of the form $(6.11)_e$ satisfies the equations of earth motion; therefore *the initial position and velocity of any earth motion can be prescribed arbitrarily at time $t = 0$; thereafter the motion is completely determined*. These two properties of earth motions can be summarized telegraphically as follows: *Initial position and velocity can be prescribed arbitrarily, and they determine the motion uniquely.*

Clearly, the same is true for moon motions. It is the *fundamental theorem of mechanics* that the same is true for motions under any kind of a force or combination of forces; it is true for motions which are not restricted to a line but take place in 3-dimensional space. We shall state precisely and prove this fundamental theorem of mechanics in the chapter on differential equations in Volume II.

If a number of different forces act on a particle, as they do in most realistic situations, according to fact 1 the effective force acting on the particle is the *sum* of the separate forces. For example, a body might be subject to the force of gravity f_g, the force of air resistance f_a, an electric force f_e and a magnetic force f_m; the effective force f is then

(6.12)
$$f = f_g + f_a + f_e + f_m,$$

and the equation governing motions under this combination of forces is

(6.13)
$$x' = v, \qquad mv' = f.$$

There is a tremendous advantage in being able to synthesize the force acting on a particle from various simpler forces, each of which can be analyzed separately.

Although we can write down the equation governing the motion of particles subject to any combination of forces, we cannot, except in simple situations such as earth and moon motions, write down simple formulas for the particle position as a function of time. Other interesting examples will be presented in Chapters 7 and 8. We shall explain briefly in Chapter 8 and in detail in Volume II how to apply, in complicated situations, numerical methods to calculate the position of a particle as function of time to an astonishingly high degree of accuracy in a matter of milliseconds, with the aid of electronic calculators.

EXERCISES

6.1 Suppose that the initial position and velocity of a particle are $x(0) = 0$, $v(0) = 10$; calculate position and velocity at time $t = 1$ and $t = 2$ when
(a) x describes earth motion;   (b) x describes moon motion.

6.2 What are the largest values of $x(t)$ during the motions described in Exercise 6.1?

6.3 (a) A person can jump 4 feet high on earth. Assume that his muscles can produce the same initial velocity if he jumps on the moon. How high can he jump on the moon?
(b) If a person can jump h feet on earth, how high can he jump on the moon?
(c) What initial velocity does he need in order to jump h feet on the moon?

3.7 Higher derivatives

All the differentiable functions f we have presented in examples so far had the property that their derivatives f' also turned out to be differentiable. Such functions are called *twice differentiable*, and the derivative of f' is called the *second derivative* of f, denoted variously by

$$(7.1)_2 \qquad\qquad f'' \quad \text{or} \quad \frac{d^2 f}{dx^2}.$$

We have met the second derivative of a function in Example 5 of Section 3.5, and again in the discussion of mechanics, Section 3.6, where the acceleration was defined by $a(t) = v'(t) = x''(t)$. Let us look at some additional examples:

Example 1. $f(x) = x^2 + 1$, $\quad f'(x) = 2x$, $\quad f''(x) = 2$.

Example 2. $f(x) = r - \sqrt{r^2 - x^2}$, r some positive number.

$$f'(x) = \frac{x}{\sqrt{r^2 - x^2}},$$

$$f''(x) = \frac{1}{\sqrt{r^2 - x^2}} + \frac{x^2}{(r^2 - x^2)^{3/2}} = \frac{r^2}{(r^2 - x^2)^{3/2}}.$$

131

Similarly we define a 3-times differentiable function f as one whose second derivative is differentiable. The derivative of the second derivative is called the third derivative of f; it is denoted by

(7.1)₃
$$f''' \quad \text{or} \quad \frac{d^3 f}{dx^3}$$

Example 3. The third derivative of f given in Example 1 is $f''' = 0$.

Generally, a function f is called *n-times differentiable* if its $(n - 1)$-st derivative is differentiable. The resulting function is called the *n-th derivative* of f and is denoted by

(7.1)ₙ
$$f^{(n)} \quad \text{or} \quad \frac{d^n f}{dx^n}.$$

We recall the operator notation Df for the derivative, introduced in Section 3.1. We denote by D^2 the operator consisting of performing the operation D twice. Then

$$D^2 f = f'',$$

and the *n*-th derivative is

(7.2)
$$D^n f = f^{(n)}.$$

The interpretations of the second derivative are interesting and important. If $x(t)$ denotes displacement at time t, dx/dt is the velocity v of the displacement. The derivative of velocity is called *acceleration*. Thus

$$\text{acceleration} = \frac{dv}{dt} = \frac{d^2 x}{dt^2};$$

in words, *acceleration is the second derivative of displacement*. In this language Newton's second law (6.3) can be stated as

(7.3)
$$f = m \frac{d^2 x}{dt^2}.$$

The geometric interpretation of the second derivative is no less interesting than the physical interpretation. We note that a linear function

$$f(x) = mx + b$$

has second derivative zero. Therefore a function with nonzero second derivative is not linear. Since a linear function can be characterized as one whose graph is a straight line, it follows that if $f'' \neq 0$, then the graph of f is not a straight line. This fact suggests that the size of $f''(x)$ measures in some sense the deviation of the graph of f from a straight line at the point x. The graph of the function

$$f(x) = r - \sqrt{r^2 - x^2}, \qquad -r < x < r,$$

of Example 2 is a semicircle of radius r; see Figure 3.17. The larger the value of r, the closer this semicircle lies to the x axis. The value of f'' at $x = 0$ is

$$f''(0) = \frac{r^2}{(r^2)^{3/2}} = \frac{1}{r}.$$

The larger r is, the smaller is the value of $f''(0)$; so in this case the smallness of $f''(0)$ does indicate that the graph of f is close to a straight line. In Volume II we shall systematically study this connection.

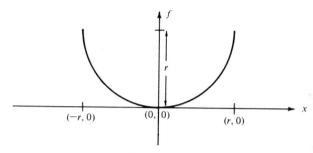

Figure 3.17

The definition of the second derivative involves a double limiting process, the first one used to calculate f', the second to calculate $(f')' = f''$. We raise here the question whether these two limiting processes can be combined; specifically, f' is defined by

$$f' = \lim_{h \to 0} f'_h ,$$

where

(7.4) $$f'_h(x) = \frac{f(x + h) - f(x)}{h} .$$

Similarly,

(7.5) $$f'' = \lim_{h \to 0} f''_h ,$$

where

(7.6) $$f''_h(x) = \frac{f'(x + h) - f'(x)}{h} .$$

The question is: can we, in (7.6), replace f' by f'_h; i.e., is it true that

(7.7) $$f'' = \lim_{h \to 0} f''_{hh} ,$$

where

(7.8) $$f''_{hh} = \frac{f'_h(x + h) - f'_h(x)}{h} = \frac{f(x + 2h) - 2f(x + h) + f(x)}{h^2} ?$$

133

f''_{hh} is called the *second difference quotient* of f; we want to know if it tends to the second derivative of f, as h tends to 0. Let us try some examples:

Example 4. $f(x) = x^2$; in this case $f'' = 2$. Using formula (7.8), we get

$$f''_{hh}(x) = \frac{(x + 2h)^2 - 2(x + h)^2 + x^2}{h^2}$$

$$= \frac{x^2 + 4xh + 4h^2 - 2(x^2 + 2xh + h^2) + x^2}{h^2} = \frac{2h^2}{h^2} = 2.$$

So in this case f''_{hh} is the same as f''.

Example 5. $f(x) = x^3$; in this case $f''(x) = 6x$. Using (7.8) we get, using the binomial theorem, that

$$f''_{hh}(x) = \frac{(x + 2h)^3 - 2(x + h)^3 + x^3}{h^2}$$

$$= \frac{x^3 + 6x^2h + 12xh^2 + 8h^3 - 2(x^3 + 3x^2h + 3xh^2 + h^3) + x^3}{h^2}$$

$$= \frac{6xh^2 + 6h^3}{h^2} = 6x + 6h.$$

In this case f'' and f''_{hh} differ by $6h$; clearly, as h tends to zero, this difference diminishes to zero; so in this case too (7.7) holds.

It turns out that (7.7) holds for all twice differentiable functions; a proof of this will be given in Section 3.8.

EXERCISES

7.1 Calculate, by repeated differentiation, the second derivatives of the functions
   (a) $f(x) = x^6 + 5x^4 - 3x^3 + x^2 - 1$
   (b) $f(x) = (x + 1)/(x - 1)$ (c) $f(x) = \sqrt{x}$
   (d) $f(x) = 1/\sqrt{x}$ (e) $f(x) = \sqrt{1 - x^2}$.

7.2 Use formula (7.7) to calculate the second derivatives of the functions
   (a) $f(x) = ax + b$ (b) $f(x) = 1/x$
   (c) $f(x) = x^4$ (d) $f(x) = x^n$.
   Verify the results by differentiating f twice.

3.8 Mean value theorems

In this section we shall use the monotonicity criterion to obtain quantitative information about functions. Specifically we shall investigate this question: if we have an estimate about the derivative f' of a function f, what can we deduce about f itself?

Suppose the estimate for f' is of the form

(8.1) $$f'(x) \le g'(x) \qquad \text{for all } x \text{ in } [a, b],$$

where g is some known function. We deduce immediately that

$$0 \le g' - f' = (g - f)',$$

i.e., that the derivative of $g - f$ is non-negative on the interval $[a,b]$. It follows then from the monotonicity criterion that $g - f$ is increasing on $[a, b]$:

$$g(a) - f(a) \le g(x) - f(x) \qquad \text{on } [a, b].$$

This can be rewritten as

(8.2) $$f(x) - f(a) \le g(x) - g(a) \qquad \text{on } [a, b].$$

We summarize our deduction:

Theorem 8.1. *If the derivatives of a pair of functions f and g satisfy inequality (8.1) on an interval, the functions themselves satisfy inequality (8.2) on that interval.*

A very special case of (8.1) occurs for $g'(x)$ constant:

(8.3) $$f'(x) \le M,$$

where M is a constant; set $g(x) = Mx$, so the conclusion (8.2) reads

(8.4) $$f(x) - f(a) \le M(x - a).$$

Another special case of (8.1) occurs when $g'(x)$ is bounded from below by a constant m:

(8.5) $$m \le g'(x).$$

In this case, set $f(x) = mx$, so that conclusion (8.2) reads

(8.6) $$m(x - a) \le g(x) - g(a).$$

Let us write f in place of g in (8.5), (8.6); then we can combine the two inequalities as follows:

If the derivative of a function f satisfies

(8.7) $$m \le f'(x) \le M \qquad \text{on } [a, b],$$

then

(8.8) $$f(a) + m(x - a) \le f(x) \le f(a) + M(x - a).$$

The function f', being a derivative, is a continuous function and so achieves its maximum and its minimum on $[a, b]$. We may choose M as the maximum of f', m as its minimum. We deduce from (8.8) that, if we express $f(x)$ as

$$f(x) = f(a) + H(x - a),$$

then H lies between m and M. According to the intermediate value theorem of Section 2.7, a continuous function defined on an interval $[a, b]$ takes on all values between its minimum and maximum; it follows that f' takes on the value H at some point c on $[a, b]$. Setting $x = b$ and $H = f'(c)$ in the above equation, we obtain the

Linear approximation theorem. *Let f be a differentiable function in $[a, b]$; then there is a number c in $[a, b]$ such that*

$$(8.9) \qquad\qquad f(b) = f(a) + f'(c)(b - a).$$

A useful equivalent form of this result is the

Mean value theorem. *Suppose f is differentiable on the interval $[a, b]$; then there is a number c in the interval $[a, b]$ where*

$$f'(c) = \frac{f(b) - f(a)}{b - a}.$$

The mean value theorem has an interesting geometric interpretation; see Figure 3.18. The difference quotient of f is the slope of the *secant* through the points $(a, f(a))$ and $(b, f(b))$ on the graph of f; the derivative $f'(c)$ is the

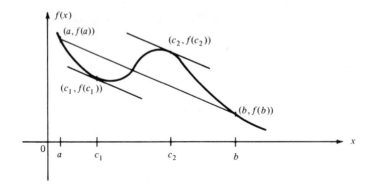

Figure 3.18

slope of the *tangent* to the graph at $(c, f(c))$. The mean value theorem asserts that *there is at least one point on the graph of f between a and b where the tangent is parallel to the secant* (in Figure 3.18 there are two such points, c_1 and c_2).

Example 1. $f(x) = x^2$; then $f' = 2x$, and the theorem asserts that there is a number c in $[a, b]$ such that

$$2c = \frac{b^2 - a^2}{b - a} = b + a.$$

This number c is

$$c = \frac{b + a}{2},$$

the *arithmetic mean* of a and b.

Example 2. $f(x) = 1/x$, $0 < a < b$. In this case $f(x) = -1/x^2$; the mean value theorem asserts the existence of a number c between a and b such that

$$-\frac{1}{c^2} = \frac{(1/b) - (1/a)}{b - a} = -\frac{1}{ab}.$$

This number is $c = \sqrt{ab}$, the *geometric mean* of a and b, clearly contained in $[a, b]$.

Emboldened with our success we pose the following question: *If we have some estimate about the second derivative p'' of a function p, what can we deduce about p itself?* Suppose the estimate for p'' is

(8.10) $\qquad\qquad p''(x) \leq M \qquad$ for all x in $[a, b]$,

M some known constant. Since p'' is the first derivative of p' and M the first derivative of Mx, we can apply Theorem 8.1 with $f = p'$ and $g = Mx$ and deduce from (8.2) that

(8.11) $\qquad p'(x) - p'(a) \leq Mx - Ma = M(x - a) \qquad$ on $[a, b]$.

Note that the left side is the derivative of

$$f(x) = p(x) - p'(a)x,$$

while the right side is the derivative of

$$g(x) = \tfrac{1}{2}M(x - a)^2;$$

thus (8.11) is of the form (8.1), and therefore (8.2) follows, with f and g as defined above:

$$p(x) - p'(a)x - p(a) + p'(a)a \leq \tfrac{1}{2}M(x - a)^2.$$

We regroup terms on both sides to get the inequality into a more compact form:

(8.12) $\qquad\qquad p(x) \leq p(a) + p'(a)(x - a) + \frac{M}{2}(x - a)^2.$

By an analogous argument we can deduce that, if p'' is bounded from below on $[a, b]$ by

$$m \leq p''(x), \qquad m \text{ some constant,}$$

then p itself is bounded from below on $[a, b]$ as follows:

$$p(a) + p'(a)(x - a) + \frac{m}{2}(x - a)^2 \leq p(x).$$

We combine both inequalities into one statement: *Let p be a twice differ-entiable function whose second derivative satisfies*

(8.13) $m \leq p''(x) \leq M$ on $[a, b]$;

then

$$p(a) + p'(a)(x - a) + \frac{m}{2}(x - a)^2 \leq p(x)$$

(8.14)

$$\leq p(a) + p'(a)(x - a) + \frac{M}{2}(x - a)^2.$$

The upper and lower bounds in (8.14) differ inasmuch as one contains the constant M, the other m. It follows that there is a number H between m and M, such that

(8.15) $p(x) = p(a) + p'(x - a) + \dfrac{H}{2}(x - a)^2.$

We may take for M the maximum of p'', p'' over $[a, b]$, for m the minimum. Appealing once more to the intermediate value theorem, we see that H equals the value of p'' at some point c on $[a, b]$. Taking $x = b$ in (8.15) we obtain the following generalization of the mean value theorem:

The quadratic approximation theorem. *Let p be twice differentiable in $[a, b]$; then there is a point c in the interval $[a, b]$ such that*

(8.16) $p(b) = p(a) + p'(a)(b - a) + \frac{1}{2}p''(c)(b - a)^2.$

Suppose b is close to a. Then c is even closer to a, and, since p'' is continuous, $p''(c)$ is close to $p''(a)$. We express this by writing

$$p''(c) = p''(a) + s,$$

where s denotes a quantity which is small when b is close to a. Substituting this into (8.16) we get

(8.16)' $p(b) = p(a) + p'(a)(b - a) + \frac{1}{2}p''(a)(b - a)^2 + \frac{1}{2}s(b - a)^2.$

This formula shows that, for b close to a, the first 3 terms on the right are a very good approximation to $p(b)$. Since the sum of the first 3 terms is a quad-ratic polynomial in $b - a$, we have shown that *over a short interval every twice differentiable function can be exceedingly well approximated by a quadratic polynomial.* In the derivation of this theorem we have taken b to be greater than a; nevertheless, as we shall show in Exercise 9.1, the theorem as stated is true also for $b < a$. We give some applications:

Local minimum theorem. *Let p be a twice differentiable function defined near* $x = a$, *and suppose that* $p'(a) = 0$, $p''(a) > 0$. *Then p has a local minimum at a, i.e.*

$$p(a) < p(b)$$

for all nearby points $b \neq a$.

PROOF. $p''(a) > 0$; by continuity of p'', $p''(x) > 0$ for all x close enough to a. Choose b so close to a that $p''(c) > 0$ for all c on the interval $[a, b]$. Using formula (8.16) for $p(b)$ and the facts that $p'(a) = 0$ and that $p''(c) > 0$, we get

$$p(b) = p(a) + \frac{p''(c)}{2}(b - a)^2 > p(a),$$

as asserted. $\qquad\square$

We leave it to the reader to prove the analogous

Local maximum theorem. *Let p be a twice differentiable function defined near* $x = a$, *and suppose that* $p'(a) = 0$, $p''(a) < 0$. *Then p has a local maximum at a, i.e.*

$$p(a) > p(b)$$

for all nearby points $b \neq a$.

Suppose p'' is non-negative in an interval containing a and b; then the last term on the right in (8.16) is non-negative, so that omitting it yields this inequality:

(8.17) $$p(b) \geq p(a) + p'(a)(b - a).$$

This inequality has a striking geometrical interpretation. We notice that the quantity on the right in (8.17) is the value at b of the linear function

$$p(a) + p'(a)(x - a).$$

The graph of this linear function is the line tangent to the graph of p at $(a, p(a))$, see Figure 3.19. So inequality (8.17) asserts that *the graph of p lies above its tangent lines.*

A function whose graph lies above its tangents is called *convex*; in this language inequality (8.17) says: *Every function whose second derivative is* ≥ 0 *is convex.*

Convex functions have another interesting property:

Convexity theorem. *Let p a twice differentiable function on* $[a, b]$, *and suppose that* $p'' > 0$ *there. Then for every x satisfying* $a < x < b$,

(8.18) $$p(x) < p(b)\frac{x - a}{b - a} + p(a)\frac{b - x}{b - a}.$$

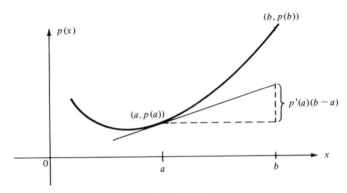

Figure 3.19

This theorem has an illuminating geometric interpretation. Denote by l the function standing on the right in (8.18):

$$l(x) = p(b) \frac{x - a}{b - a} + p(a) \frac{b - x}{b - a}.$$

l is a linear function whose value at $x = a$ and at $x = b$ agrees with the values of p at these points. Thus the graph of l is the *secant* of p on $[a, b]$; see Figure 3.20. Therefore inequality (8.18) says: *The graph of a convex function p on an interval $[a, b]$ lies below the secant on $[a, b]$.*

PROOF. We wish to show that $p - l \leq 0$ on $[a, b]$. According to the maximum value theorem of Section 2.7, $p - l$ reaches its maximum somewhere on $[a, b]$; denote that point by c. We claim that c is one of the endpoints; for, if c were an interior point, then according to the maximum theorem of Section 3.5, $(p - l)' = 0$ at c. Since l is linear, $l''(c) = 0$; and so

$$(p - l)'' = p'',$$

which is assumed positive. Applying the local minimum theorem to $p - l$, we conclude that at all points x near c, the value $p(x) - l(x)$ is greater than the value at c:

$$p(x) - l(x) > p(c) - l(c).$$

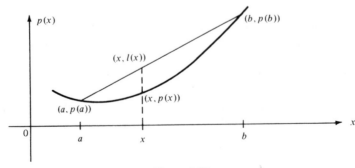

Figure 3.20

But this contradicts the characterisation of c as the point where $p - l$ is maximum. The only alternative remaining is that c is one of the endpoints. At an endpoint, p and l have the same value. This shows that the maximum of $p - l$ is 0, and that at all points x of $[a, b]$ other than the endpoints

$$p(x) - l(x) < 0.$$

This completes the proof of the convexity theorem.   □

A function whose graph lies *below* its tangent is called *concave*. The following analogues of results on convex functions hold:

Every function whose second derivative is negative is concave.
   The graph of a concave function p on an interval $[a, b]$ lies above the secant on $[a, b]$; see Figure 3.21.

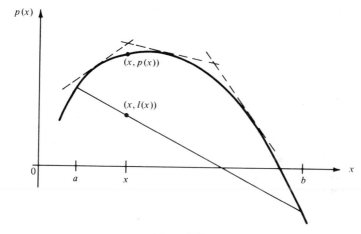

Figure 3.21

Next we apply the quadratic approximation theorem to answer a question raised at the end of Section 3.7: can the second derivative of a function be computed as the limit of second difference quotients given by Formula (7.6)? We now show that, as you may already have guessed, the answer is yes! According to (7.8) the second difference quotient is

$$f''_{hh} = \frac{f(x + 2h) - 2f(x + h) + f(x)}{h^2}.$$

By the quadratic approximation theorem with $a = x$, $b = x + h$, for a twice differentiable function f,

(8.19)$_+$ $f(x + h) = f(x) + f'(x)h + \tfrac{1}{2}f''(c)h^2,$

where c lies between x and $x + h$. Similarly, with $a = x$, $b = x + 2h$, we get

(8.20) $f(x + 2h) = f(x) + f'(x)2h + \tfrac{1}{2}f''(d)4h^2,$

141

where d lies between x and $x + 2h$. Subtracting twice $(8.19)_+$ from (8.20), adding $f(x)$ and dividing by h^2, we get this formula for the second difference quotient:

$$f''_{hh}(x) = 2f''(d) - f''(c).$$

As h tends to 0, c and d, being trapped in $[x, x + h]$ and $[x, x + 2h]$ respectively, tend to x. Since the derivative of a twice differentiable function is continuous, it follows that $f''(c)$ and $f''(d)$ both tend to $f''(x)$. So we have shown:

The second difference quotients f''_{hh} of a twice differentiable function f tend to f'' as h tends to 0.

Next we use the quadratic approximation theorem to answer the following question: how good an approximation is the difference quotient

(8.21)
$$\frac{f(x + h) - f(x)}{h}$$

to the derivative $f'(x)$? According to $(8.19)_+$

$$f(x + h) = f(x) + f'(x)h + \tfrac{1}{2}f''(c)h^2,$$

where c is some point between x and $x + h$. Subtracting $f(x)$ from both sides and dividing by h, we get

$$\frac{f(x + h) - f(x)}{h} = f'(x) + \tfrac{1}{2}f''(c)h.$$

For small h the point c lies close to x; and therefore, since f'' is continuous, it follows that for small h, $f''(c)$ is very close to $f''(x)$. So we conclude that for small h the derivative of f at x differs from the difference quotient (8.21) approximately by $f''(x)h$.

The difference quotient (8.21) is asymmetric in the sense that it favors one side of the point x. We turn now to the *symmetric* difference quotient

$$\frac{f(x + h) - f(x - h)}{2h}.$$

By the quadratic approximation theorem

(8.19)_-
$$f(x - h) = f(x) - f'(x)h + \tfrac{1}{2}f''(e)h^2,$$

where e lies between x and $(x - h)$. Subtracting this from $(8.19)_+$ and dividing by $2h$, we get the following expression for the symmetric difference quotient:

(8.22)
$$\frac{f(x + h) - f(x - h)}{2h} = f'(x) + \tfrac{1}{2}[f''(c) - f''(e)]h.$$

Both c and e differ by less than h from x; since f'' is continuous, it follows that for h small both $f''(c)$ and $f''(e)$ differ little from $f''(x)$. Thus we deduce from

(8.22) that the symmetric difference quotient differs from $f(x)$ by an amount sh, where

$$s = \tfrac{1}{2}[f''(c) - f''(e)]$$

is small when h is small.

Since the one-sided difference quotient differs from $f'(x)$ approximately by $f''(x)h$, we conclude that for small h *the symmetric difference quotient* (8.22) *is a better approximation to the derivative at x than the one-sided difference quotient* (8.21). We shall make good use of this observation in Sections 5.3 and 5.6.

EXERCISES

8.1 Show that the following functions are convex on the indicated intervals:

(a) $f(x) = x^2$ on the real axis

(b) $f(x) = -\sqrt{1 - x^2}$ on $[-1, 1]$

(c) $f(x) = 1 - \sqrt{1 - x^2}$ on $[-1, 1]$

(d) $f(x) = \dfrac{1}{x}$ on $0 < x$.

8.2 Show, using the quadratic approximation theorem, that for twice differentiable f, $(f(a) + f(b))/2$ differs from $f((a + b)/2)$ by less than $M(b - a)^2/8$, where M is an upper bound for f'' on $[a, b]$.

8.3 Show that

$$\frac{f(b) - f(a)}{b - a} = \frac{f'(b) + f'(a)}{2} + s(b - a),$$

where s is small when b is close to a.

8.4 (a) Show that if $f''(x) \neq 0$,

$$\frac{f(x + h) - f(x - h)}{2h} = f'(c),$$

where $c = x + sh$, s being a quantity which tends to 0 as h tends to 0.

(b) Show that if f is a quadratic polynomial, $c = x$.

(c) Determine what kind of function c is of h for $f(x) = x^3$, at $x = 0$.

8.5 (a) Show that, for twice differentiable f, the symmetric second difference quotients

(8.23)
$$\frac{f(x + h) - 2f(x) + f(x - h)}{h^2}$$

tend to f'' as h tends to zero.

(b) For the function $f(x) = x^3$, show that for small h the symmetric second difference quotient (8.23) is a better approximation to f'' than the asymmetric second difference quotient defined by (7.6).

(c) Do you think the same is true for any other twice differentiable f? Guess!

3.9* Taylor's theorem

We are ready to tackle the general problem: given upper and lower bounds for the n-th derivative, $f^{(n)}$, n any integer, of a function f in an interval, find upper and lower bounds for f over that interval. Generalizing (8.13), (8.14), we surmise that the following result holds:

Suppose that f is an n times differentiable function on an interval $[a, b]$, and denote by m and M the minimum and maximum, respectively, of $f^{(n)}$ over $[a, b]$; that is,

(9.1) $$m \leq f^{(n)}(x) \leq M, \qquad x \text{ in } [a, b].$$

Define the n-th degree polynomial $p_n = p_n(x, H)$ to be

(9.2) $$p_n(x, H) = f(a) + f'(a)(x - a) + \frac{f''(a)}{2!}(x - a)^2$$

$$+ \frac{f'''(a)}{3!}(x - a)^3 + \cdots + \frac{H}{n!}(x - a)^n.$$

[We remind the reader that $n!$ is an abbreviation for the product $1 \cdot 2 \cdot 3 \cdots (n - 1)n$.] Then

Taylor's inequality

(9.3) $$p_n(x, m) \leq f(x) \leq p_n(x, M)$$

holds for all x in $[a, b]$.

PROOF. We have already shown that (9.3) holds for $n = 1$ and 2 in (8.8) and (8.14) respectively. We shall prove it for all n *inductively*, i.e., we shall show that if the result is true for any particular number n, it is true for $n + 1$; thus we shall see that the result holds for $n = 3, 4$, etc.

Suppose f is an $(n + 1)$ times differentiable function, and

(9.4) $$m \leq f^{(n+1)}(x) \leq M.$$

We assume Taylor's inequality holds for functions whose n-th derivative is bounded, and we apply it to the function f', whose n-th derivative is $f^{(n+1)}$, and satisfies (9.4). Thus, we conclude from the induction hypothesis that f' satisfies

(9.5) $$q_n(x, m) \leq f'(x) \leq q_n(x, M),$$

where

(9.6) $$q_n(x, H) = f'(a) + f''(a)(x - a) + \cdots + \frac{H}{n!}(x - a)^n.$$

---

* This section may be omitted on the first reading.

The right side of inequality (9.5) is of the form (8.1), where g is a function whose derivative is $q_n(x, M)$; as may be easily checked by differentiation, $g = p_{n+1}(x, M)$ is such a function. Therefore, we conclude that (8.2) holds in $[a, b]$:

$$f(x) - f(a) \leq p_{n+1}(x, M) - p_{n+1}(a, M).$$

Since $p_{n+1}(a, M) = f(a)$, it follows that

$$f(x) \leq p_{n+1}(x, M) \qquad x \text{ in } [a, b],$$

which is the right half of inequality (9.3). The left half follows from the left half of inequality (9.5) and from (8.2) with the roles of f and g reversed. Thus, if (9.3) holds for n, it holds for $n + 1$. This concludes the proof of Taylor's inequality. $\qquad\square$

Since $p_n(x, H)$ is a continuous function of H, it follows from (9.3) that for each x in $[a, b]$ there is a value H between m and M such that

(9.7) $$f(x) = p_{n+1}(x, H).$$

According to the intermediate value theorem, every number H between the minimum m and maximum M of a continuous function $f^{(n+1)}$ over an interval is taken on at some point c of that interval. Therefore (9.7) can be expressed as

Taylor's formula with remainder. *Let f be an n times differentiable function in an interval $[a, b]$. Then*

(9.8) $$f(b) = f(a) + f'(a)(b - a) + \cdots$$

$$+ f^{(n-1)}(a) \frac{(b - a)^{n-1}}{(n - 1)!} + f^{(n)}(c) \frac{(b - a)^n}{n!},$$

where c lies between a and b.

In the derivation of this theorem we have exploited the fact that $a < b$; however, it is not hard to show that the theorem remains true if $a > b$; see Exercise 9.1.

Here are some applications of Taylor's formula.

Example 1. $f(x) = x^n$, n a positive integer. Then $f^{(k)}(x) = n(n - 1) \cdots (n - k + 1)x^{n-k}$; in particular, $f^{(n)}(x) = n!$ for all x. Therefore, according to Taylor's formula (9.8) with $b = 1 + y$, and $a = 1$,

$$(1 + y)^n = 1 + ny + \frac{n(n - 1)}{2!} y^2 + \cdots + y^n.$$

This is nothing but the binomial expansion, revealed here as a special case of Taylor's formula.

For other than polynomial functions f, Taylor's formula can be used to yield only an approximation to f over an interval. Inequality (9.3) is such an approximation; we derive now a useful variant of that inequality.

Denote by C_n the *oscillation* of $f^{(n)}$ on the interval $[a, b]$, i.e.

$$C_n = M_n - m_n,$$

where M_n is the maximum, m_n the minimum of $f^{(n)}$ over $[a, b]$. Denote by $t_n(x)$ the polynomial

(9.9) $\qquad t_n(x) = f(a) + f'(a)(x - a) + \dfrac{1}{2!} f'(a)(x - a)^2 + \cdots$

$$+ \frac{1}{n!} f^{(n)}(a)(x - a)^n;$$

we call t_n the n-th *degree Taylor polynomial of f at a*. The right side of (9.8) differs from $t_n(x)$ in that the last term has $f^{(n)}$ evaluated at c rather than a. Since $f^{(n)}(c)$ and $f^{(n)}(a)$ differ by at most the oscillation C_n of $f^{(n)}$, we see that

(9.10) $\qquad |f(x) - t_n(x)| \leq \dfrac{C_n}{n!}(x - a)^n \qquad$ for all x on $[a, b]$.

In particular,

(9.10)' $\qquad |f(x) - t_n(x)| \leq \dfrac{C_n}{n!}(b - a)^n \qquad$ for all x on $[a, b]$.

Suppose the function f is *infinitely many times differentiable*, i.e., has derivatives of all orders. Suppose further that

(9.11) $\qquad\qquad \lim_{n \to \infty} \dfrac{C_n}{n!}(b - a)^n = 0.$

Then inequality (9.10)' shows that, as n gets larger and larger, $t_n(x)$ gets closer and closer to $f(x)$. We can state this result in the following spectacular form:

Taylor's theorem. *Let f be an infinitely differentiable function on an interval $[a, b]$. Denote by C_n the oscillation of $f^{(n)}$ on the interval $[a, b]$; suppose that these numbers satisfy the limit relation (9.11). Then f can be represented at every point of the interval $[a, b]$ by the Taylor series*

(9.12) $\qquad\qquad f(x) = \sum_{k=0}^{\infty} \dfrac{1}{k!} f^{(k)}(a)(x - a)^k.$

PROOF. The meaning of the infinite sum on the right (see Exercise 8.4, Chapter 2) is this: form the finite sum with k ranging from 0 to n, and take the limit of these finite sums as n tends to ∞. According to (9.9) these finite sums are $t_n(x)$, and we saw before by using (9.10)' that if (9.11) holds, the sequence $\{t_n(x)\}$ tends to $f(x)$ as n tends to infinity. This completes the proof of Taylor's theorem. $\qquad\qquad\qquad\qquad\qquad\qquad\qquad\qquad\qquad\qquad\qquad\square$

Here are some illustrations of this theorem:

Example 2. Let $f(x) = 1/x$ in the interval $x > 0$. Then

(9.13) $$f^{(n)}(x) = \frac{(-1)^n n!}{x^{n+1}}.$$

Each $f^{(n)}(x)$ is monotonic, so its oscillation C_n on $[a, b]$ is just

$$|f^{(n)}(b) - f^{(n)}(a)| = \frac{n!}{a^{n+1}} - \frac{n!}{b^{n+1}}.$$

Clearly

$$C_n < \frac{n!}{a^{n+1}}.$$

To see when condition (9.11) is satisfied, we note that

$$\frac{C_n}{n!}(b - a)^n \le \frac{(b-a)^n}{a^{n+1}} = \frac{1}{a}\left(\frac{b}{a} - 1\right)^n;$$

as n tends to ∞, the quantity on the right tends to zero if $(b/a) - 1$ is less than 1, i.e., if $b < 2a$. So Taylor's theorem is applicable for $a = 1$, $x = 1 + y$, $0 \le y < 1$. Setting these values into (9.12) and using Formula (9.13) to evaluate $f^{(n)}(a)$, we get

$$\frac{1}{1 + y} = \sum_0^\infty (-1)^k y^k.$$

This is not a very surprising result since the right side is an old friend, a convergent geometric series. However the method of proof, unlike the classical one for finding the sum of an infinite geometric series, is more generally applicable. For instance, let us look at

Example 3. $f(x) = x^l$, $x > 0$, l any rational number, positive or negative. The n-th derivative of f is

(9.14) $$f^{(n)}(x) = l(l - 1) \cdots (l - n + 1)x^{l-n}.$$

Again, $f^{(n)}$ is monotonic on $[a, b]$, so

$$C_n = |f^{(n)}(b) - f^{(n)}(a)|.$$

We investigate now when the limit relation (9.11) holds. We note that for $n > l$, $|f^{(n)}(x)|$ is a decreasing function of x; also $f^{(n)}(a)$ and $f^{(n)}(b)$ have the same sign. Therefore,

(9.15) $$C_n < |f^{(n)}(a)| = \left|\frac{l(l-1)\cdots(l-n+1)}{a^{n-l}}\right|.$$

Denote by L a positive integer $\ge |l|$; by the triangle inequality, $|l - k| \le |l| + |k| = L + k$ for every positive integer k, so

$$|l(l - 1) \cdots (l - n + 1)| \le L \cdot (L + 1) \cdots n(n + 1) \cdots (L + n - 1).$$

Therefore, using inequality (9.15), we get that

$$\frac{C_n}{n!}(b-a)^n < \frac{L\cdots(L+n-1)}{n!}\frac{(b-a)^n}{a^{n-l}}.$$

For $n > L$, we may cancel the factor $L(L+1)\cdots n$ from numerator and denominator, getting

$$\frac{C_n}{n!}(b-a)^n < \frac{(n+1)\cdots(n+L-1)}{L!}a^l\left(\frac{b}{a}-1\right)^n.$$

For $n > L$, each of the L factors in the numerator is less than $2n$; therefore we get

$$\frac{C_n}{n!}(b-a)^n < \frac{(2n)^L}{L!}a^l\left(\frac{b}{a}-1\right)^n.$$

We claim that, for $|(b/a)-1| < 1$, the expression on the right tends to zero as n tends to ∞. Accepting this result (see Exercise 9.2), we conclude that Taylor's theorem is applicable for $a = 1$, $x = 1 + y$, $0 \le y \le 1$. Setting these values into (9.12) and using (9.14) to evaluate $f^{(n)}(a)$, we get

$$(9.16) \qquad (1+y)^l = \sum_{k=0}^{\infty} \binom{l}{k}y^k$$

which holds for $0 \le y < 1$ and where $\binom{l}{k}$ abbreviates the *binomial coefficients*

$$(9.17) \qquad \binom{l}{k} = \frac{l(l-1)\cdots(l-k+1)}{k!}.$$

In Chapter 5 we shall define x^l for arbitrary real exponent l and show that the usual rule for differentiating x^l is valid for all l. It follows then that (9.16), called the *generalized binomial theorem*, is true for all real l. Using the result of Exercise 9.1 one can show that (9.16) holds for all y, $|y| < 1$.

Example 4. Suppose we are told of a function $e(x)$ that has derivatives of all orders, that the value of all its derivatives at $x = 0$ equals 1, and that on the interval $[0, 1]$ all its derivatives lie between 1 and 3. Let us find the value of $e(1)$ with an error $< 10^{-3}$.

According to inequality (9.10) applied to $f(x) = e(x)$, with $a = 0$, $x = 1$, and $C_n = 3 - 1 = 2$, the value of $e(1)$ and of the n-th degree Taylor polynomial $t_n(1)$ differ by less than $2/n!$. Therefore $t_n(1)$ approximates the value of $e(1)$ within an error of 10^{-3} provided that n is so large that $n! > 2000$. The list

$$
\begin{aligned}
1! &= 1 \\
2! &= 2 \\
3! &= 6 \\
4! &= 24 \\
5! &= 120 \\
6! &= 720 \\
7! &= 5040
\end{aligned}
$$

shows that $n = 7$ is large enough. So

$$|e(1) - t_7(1)| < \frac{2}{5040} < \frac{1}{1000}.$$

Since $e^{(j)}(0) = 1$ for $j = 1, 2, \ldots$, we have from (9.9) that

$$t_7(1) = 1 + 1 + \frac{1}{2!} + \frac{1}{3!} + \cdots + \frac{1}{7!}.$$

A brief calculation gives the first 3 digits of $t_7(1)$ as

$$t_7(1) = 2.718.$$

Does this number look familiar? Its various appearances will be tied together in Chapter 5.

EXERCISES

9.1 Prove the validity of (9.8), Taylor's formula with a remainder, when $b < a$. [*Hint*: Apply (9.8) to the function $g(x) = f(a + b - x)$ over the interval $[b, a]$.]

9.2 Let y denote a positive number < 1, L any positive integer. Prove that the sequence $\{n^L y^n\}$ tends to zero as n tends to ∞. [*Hint*: Show that for n large enough, the sequence decreases monotonically.]

9.3 Consider the function $f(x) = \sqrt{x}$ over the interval $1 < x < 1 + d$. Find values of d small enough so that $t_2(x)$, the second-degree Taylor polynomial at $x = 1$, approximates $f(x)$ on $[1, 1 + d]$ with an error of at most
(a) $1/10$, (b) $1/100$; (c) $1/1000$.

9.4 Answer the question posed in Exercise 9.3 for the third-degree Taylor polynomial t_3 in place of t_2.

9.5 Let $s(x)$ be a function with the following properties:
  (i) s has derivatives of all order.
  (ii) All derivatives of s lie between -1 and 1.

  (iii) $s^{(j)}(0) = \begin{cases} 0 & \text{if } j \text{ is even} \\ (-1)^{(j-1)/2} & \text{for } j \text{ odd.} \end{cases}$

Determine a value of n so large that the n-th degree Taylor polynomial $t_n(x)$ approximates $s(x)$ with an error $< 10^{-3}$ on the interval $0 \le x \le 1$. Determine the value of $s(0.7854)$ with an error $< 10^{-3}$.

9.6 Let $c(x)$ be a function which has properties (i) and (ii) of Exercise 9.5, and satisfies

$$c^{(j)}(0) = \begin{cases} (-1)^{j/2} & \text{for } j \text{ even} \\ 0 & \text{for } j \text{ odd.} \end{cases}$$

Using a Taylor polynomial of appropriate degree, determine the value of $c(0.7854)$ with an error $< 10^{-3}$.

3.10* Newton's method for finding the zeros of a function

Many mathematical problems are of the following form: we are seeking a number, called "unknown" and denoted by, say, the letter z, which has some desirable property expressed in an equation. Such an equation can be written in the form

$$f(z) = 0,$$

where f is some function. Very often additional restrictions are placed on the number z; in many cases these restrictions require z to lie in a certain interval. So the task of "solving an equation" is really nothing but finding a number z where a given function f vanishes. Such a number z is called a *zero* of the function f. In some problems we are content to find *one* zero of f in a specified interval, in other problems we are interested in finding *all* zeros of f in an interval.

What does it mean "to find" a zero of a function? It means to devise a procedure which gives as good an approximation as desired of a number z where the given function f vanishes. There are two ways of measuring the goodness of an approximation z_{approx}: one is to demand that z_{approx} differ from an exact zero z by, say, less than $1/100$, or $1/1000$, or 10^{-m}. Another way of measuring the goodness of an approximation is to insist that the value of f at z_{approx} be very small, say less than $1/100$, $1/1000$ or, in general, less than 10^{-m}. Of course these notions go hand in hand: if z_{approx} is close to the true zero z, then $f(z_{approx})$ will be close to $f(z) = 0$, provided that the function f is continuous.

In this section we shall describe a method for finding approximations to zeros of functions f which are not only continuous but differentiable, preferably twice differentiable. The basic step of the method is this: starting with some fairly good approximation to a zero of f we produce a much better one. If the outcome is not yet good enough an approximation, we repeat the basic step as often as necessary to produce an approximation which is good enough according to either of the two criteria mentioned earlier. There are two ways of describing the basic step, geometrically and analytically. We start with the geometric description.

Denote by z_{old} the starting approximation; *we assume—and this is crucial for the applicability of this method—that* $f'(z_{old}) \neq 0$. This guarantees that the line tangent to the graph of f at $(z_{old}, f(z_{old}))$ (see Figure 3.22) is not parallel to the x-axis, and so intersects the x-axis at some point. This point of intersection is our new approximation z_{new}. We now calculate z_{new}; since the slope of the tangent is $f'(z_{old})$, we have

$$f'(z_{old}) = \frac{f(z_{old})}{z_{old} - z_{new}}.$$

---

* This section may be omitted on the first reading.

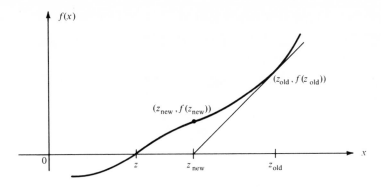

Figure 3.22

From this relation we can determine z_{new}:

$$(10.1) \qquad z_{new} = z_{old} - \frac{f(z_{old})}{f'(z_{old})}.$$

The rationale behind this procedure is this: if the graph of f were a straight line, z_{new} would be an exact zero of f. In reality the graph of f is not a straight line; but if f is differentiable, its graph over a short enough interval is *nearly* straight, and so z_{new} can reasonably be expected to be a good approximation to the exact zero z, provided that the interval (z, z_{old}) is short enough.

We now present this derivation in analytic terms, without any reference to the graph of f. We approximate f by the linear function

$$(10.2) \qquad f(x) \simeq f(z_{old}) + f'(z_{old})(x - z_{old}).$$

We define z_{new} as the exact zero of the approximating linear function on the right, in the hope that this is a good approximation to the exact zero of f. Clearly z_{new} is given by Formula (10.1).

The advantage of a purely analytic derivation over a geometric approach is that it is easier to generalize to more complicated situations involving many variables. The method described above has been devised by Newton and is therefore called *Newton's method*.

Example 1. $f(x) = x^2 - 2$; we seek a positive solution of

$$f(z) = z^2 - 2 = 0.$$

Clearly, $z = \sqrt{2}$; let us see how closely we can approximate this exact solution: For f as above, $f'(x) = 2x$, so if z_{old} is an approximation to $\sqrt{2}$, Newton's recipe (10.1) yields

$$(10.3) \qquad z_{new} = z_{old} - \frac{z_{old}^2 - 2}{2z_{old}} = \frac{z_{old}}{2} + \frac{1}{z_{old}}.$$

151

Notice that this relation is just (4.2) of Chapter 1, revisited. Let us take $z_{old} = 2$ as first approximation to $\sqrt{2}$; using Formula (10.3) we get

$$z_{new} = 1.5.$$

We then repeat the process, with $z_{new} = 1.5$ now becoming z_{old}. Thus, we construct a sequence $z_1, z_2, \ldots$ of (hopefully) better and better approximations to $\sqrt{2}$ by choosing $z_1 = 2$, and setting

$$z_{n+1} = \frac{z_n}{2} + \frac{1}{z_n}.$$

The first six approximations are

$$z_1 = 2.0$$
$$z_2 = 1.5$$
$$z_3 = 1.4166\ldots$$
$$z_4 = 1.4142157\ldots$$
$$z_5 = 1.41421356\ldots$$
$$z_6 = 1.41421356\ldots\ .$$

Since z_5 and z_6 agree up to the first 8 digits after the decimal point, we surmise that z_5 gives the first 8 digits of $\sqrt{2}$ correctly. Indeed,

$$(1.41421356)^2 = 2.00000005\ldots$$

is very near and slightly greater than 2, while

$$(1.41421355)^2 = 1.999999965\ldots$$

is very near, but slightly less than 2. It follows from the intermediate value theorem that $z^2 = 2$ at some point between these numbers, i.e., that

$$1.41421355 < \sqrt{2} < 1.41421356.$$

We remind the reader that he has encountered the sequence $z_1, z_2, \ldots$ before in Section 1.4, where it was constructed in a somewhat ad hoc fashion; the sequence was shown to converge.

The beauty of Newton's method is its universality; it can be used to find the zeros of not only quadratic functions, but functions of all sorts.

Example 2. $f(x) = x^3 - 2$; we seek a sequence of approximations to a solution of

$$z^3 - 2 = 0,$$

i.e., to the number $\sqrt[3]{2}$. Since $f'(x) = 3x^2$, Newton's recipe gives the following sequence of approximations:

(10.4)
$$z_{n+1} = z_n - \frac{z_n{}^3 - 2}{3z_n{}^2} = \frac{2z_n}{3} + \frac{2}{3z_n{}^2}.$$

Starting with $z_1 = 2$ as first approximation, we have

$$z_1 = 2.0$$
$$z_2 = 1.5$$
$$z_3 = 1.296\ldots$$
$$z_4 = 1.2609\ldots$$
$$z_5 = 1.25994\ldots$$
$$z_6 = 1.259921\ldots\ .$$

Since z_5 and z_6 agree up to the 4th digit after the decimal, we surmise that

$$\sqrt[3]{2} = 1.259921\ldots\ .$$

Indeed

$$(1.259921)^3 = 1.9999998\ldots, \qquad \text{while} \qquad (1.259922)^3 = 2.000004\ldots,$$

so that

$$1.259921 < \sqrt[3]{2} < 1.259922.$$

Example 3. Find all zeros of

$$f(x) = x^3 - 6x^2 - 2x + 12.$$

Since f is a polynomial of degree 3, an odd number, $f(x)$ is very large positive when x is very large positive, and very large negative when x is very large negative. So by the intermediate value theorem $f(x)$ is zero somewhere. To get a better idea where the zero, or zeros, might be located, we calculate the value of f at integers ranging from $x = -2$ to $x = 6$:

x	-2	-1	0	1	2	3	4	5	6
$f(x)$	-16	7	12	5	-8	-21	-28	-23	0

This table shows that f has a zero at $z = 6$, and since the value of f at $x = -2$ is negative, at $x = -1$ positive, f has a zero in the interval $(-2, -1)$; similarly f has a zero in the interval $(1, 2)$.

According to a simple theorem of elementary algebra, if z is a zero of a polynomial, $x - z$ is a factor. Indeed, we can write our f in the factored form

$$f(x) = (x - 6)(x^2 - 2).$$

This form for f shows that its other zeros are $z = \pm\sqrt{2}$, and that there are no others.

Let us ignore this exact knowledge of the zeros of f (which after all was due to a lucky accident). Let's see how well Newton's general method works in this case. The formula, for this particular function, reads

$$z_{n+1} = z_n - \frac{z_n^3 - 6z_n^2 - 2z_n + 12}{3z_n^2 - 12z_n - 2}.$$

153

Starting with $z_1 = 5$ as first approximation to the exact root 6, we get the following sequence of approximations:

$$z_1 = 5$$
$$z_2 = 6.77\ldots$$
$$z_3 = 6.147\ldots$$
$$z_4 = 6.0426\ldots$$
$$z_5 = 6.00715\ldots\ .$$

Similar calculations show that if we start with a guess z, *close enough* to one of the other two zeros $\sqrt{2}$ and $-\sqrt{2}$, we get a sequence of approximations which converge *rapidly* to the exact zeros.

How rapid is rapid, and how close is close? In the last example, starting with an initial guess which was off by 1, we obtained, after 4 steps of the method, an approximation which differs from the exact zero $z = 6$ by 0.007. Compare this to the method of bisection, described in Section 2.7, where each step cuts in half the interval in which the zero is located. Starting with an interval of length 1, 4 steps of bisection puts the zero in an interval of length $\frac{1}{16} = 0.0625$, not nearly as small an interval as that given by Newton's method. Furthermore, perusal of the examples presented so far indicates that Newton's method works faster the closer z_n gets to the zero! We shall analyze Newton's method to explain its extraordinary efficiency and also to determine its limitations.

The analytic derivation of Newton's method was based on an approximation (10.2) of f by a linear function. If there had been no error in this approximation—i.e., if f had been a linear function—then Newton's method would have furnished in one step the exact zero of f. Therefore in analyzing the error inherent in Newton's method we must start with the deviation of the function f from its linear approximation (10.2). The deviation is described by the quadratic approximation theorem, Formula (8.16):

$$(10.5) \qquad f(x) = f(z_{\text{old}}) + f'(z_{\text{old}})(x - z_{\text{old}}) + \tfrac{1}{2}f''(c)(x - z_{\text{old}})^2,$$

where c is some number between z_{old} and x. Let us introduce for simplicity the abbreviation,

$$f''(c) = s,$$

and let x be the exact zero z of f. Then (10.5) yields

$$f(z) = 0 = f(z_{\text{old}}) + f'(z_{\text{old}})(z - z_{\text{old}}) + \tfrac{1}{2}s(z - z_{\text{old}})^2,$$

from which we deduce that

$$(10.6) \qquad f(z_{\text{old}}) = -f'(z_{\text{old}})(z - z_{\text{old}}) - \tfrac{1}{2}s(z - z_{\text{old}})^2.$$

We take now Newton's recipe for the next approximation z_{new}:

$$z_{\text{new}} = z_{\text{old}} - \frac{f(z_{\text{old}})}{f'(z_{\text{old}})}$$

154

and replace $f(z_{old})$ in the numerator by the right side of (10.6); we get

$$z_{new} = z_{old} + (z - z_{old}) + \frac{1}{2}\frac{s}{m}(z - z_{old})^2,$$

where

$$m = f'(z_{old}).$$

We can rewrite the above relation as

(10.7) $$z_{new} - z = \frac{1}{2}\frac{s}{m}(z_{old} - z)^2.$$

We are interested in finding out under what conditions z_{new} is a better approximation to z than z_{old}. Formula (10.7) is ideal for deciding this, since it asserts that $(z_{new} - z)$ is the product of $(z - z_{old})$ by $(s/2m)(z - z_{old})$. Clearly there is an improvement if and only if that second factor is less than 1 in absolute value, i.e., if

(10.8) $$\frac{1}{2}\left|\frac{s}{m}\right||z - z_{old}| < 1.$$

Suppose now that $f'(z) \neq 0$; then f' is bounded away from zero at all points close to z, and clearly (10.8) holds if z_{old} is close enough to z; in fact for z_{old} close enough,

(10.9) $$\frac{1}{2}\left|\frac{s}{m}\right||z - z_{old}| < \frac{1}{2}.$$

If (10.9) holds we deduce from (10.7) that

(10.10) $$|z_{new} - z| \leq \tfrac{1}{2}|z_{old} - z|.$$

Now let $z_1, z_2, \ldots$ be a sequence of approximations generated by repeated applications of Newton's recipe. Suppose z_1 is so close to z that (10.9) holds for $z_{old} = z_1$ and for all z_{old} which are as close or closer to z than z_1. Then it follows from (10.10) that z_2 is closer to z than z_1 and, in general, that each z_{n+1} is closer to z than the previous z_n, and so (10.9) holds for all subsequent z_n. Repeated application of (10.10) shows that

(10.11) $$|z_{n+1} - z| \leq (\tfrac{1}{2})|z_n - z| \leq (\tfrac{1}{2})^2|z_{n-1} - z| \leq \cdots \leq (\tfrac{1}{2})^n|z_1 - z|.$$

This proves the

Convergence theorem for Newton's method. *Let f be a twice differentiable function, z a zero of f such that*

(10.12) $$f'(z) \neq 0.$$

Then repeated applications of Newton's recipe

(10.13) $$z_{n+1} = z_n - \frac{f(z_n)}{f'(z_n)}$$

yields a sequence of approximations $z_1, z_2, \ldots$ which converge to z, provided that the first approximation z_1 is close enough to z.

A few comments are in order:

I. The proof that z_n tends to z is based on inequality (10.11), according to which $|z_{n+1} - z|$ is less than const $(\frac{1}{2})^n$. This is a gross overestimate; to understand the true rate at which z_n converges to z, we have to go back to relation (10.7); for z_{old} close to z, the numbers m and s differ little from $f'(z)$ and $f''(z)$ respectively, so that (10.7) asserts that $|z_{new} - z|$ is practically a constant multiple of $(z_{old} - z)^2$. Now if $|z_{old} - z|$ is small, its square is enormously small! To give an example, suppose that $|f''(z)/2f'(z)| \leq 1$ and that $|z_{old} - z| \leq 10^{-3}$; then by (10.7) we conclude that

$$|z_{new} - z| \simeq (z_{old} - z)^2 = 10^{-6}.$$

In words: If the first approximation lies within one thousandth of an exact zero, and if $|f''(z)/2f'(z)| < 1$, Newton's method takes us *in one step* to a new approximation which lies within one millionth of that exact zero. The method of bisection would have required 10 steps.

II. It is necessary to start close enough to z, not only to achieve rapid convergence, but to achieve convergence at all. Figure 3.23 shows an example where Newton's method doesn't get us any closer to a zero. The points z_{old} and z_{new} are so chosen that the tangent to the graph of f at the point $(z_{old}, f(z_{old}))$ intersects the x-axis at z_{new}, and the tangent to the graph of f at $(z_{new}, f(z_{new}))$ intersects the x-axis at z_{old}. Newton's recipe brings us from z_{old} to z_{new}, then back to z_{old}, etc., without getting any closer to the zero between z_{old} and z_{new}.

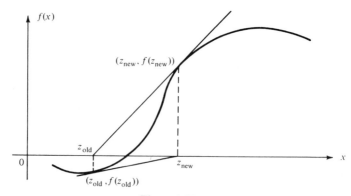

Figure 3.23

III. Our analysis indicates difficulty with Newton's method when $f'(z) = 0$ at the zero z. Here is an example: the function $f(x) = (x - 1)^2$ has a double zero at $z = 1$; therefore $f'(z) = 0$. Newton's method yields the following sequence of iterates:

$$z_{n+1} = z_n - \frac{f(z_n)}{f'(z_n)} = z_n - \frac{(z_n - 1)^2}{2(z_n - 1)} = \frac{z_n + 1}{2}.$$

Subtracting 1 from both sides, we get

$$z_{n+1} - 1 = \frac{z_n - 1}{2}.$$

Using this relation repeatedly we get

$$z_{n+1} - 1 = \frac{1}{2}(z_n - 1) = \frac{1}{4}(z_{n-1} - 1) \cdots = \left(\frac{1}{2}\right)^n (z_1 - 1).$$

Thus z_n approaches the zero $z = 1$ at the rate constant $\times \left(\frac{1}{2}\right)^n$, precisely the same rate at which the bisection method converges, and not the superfast rate at which Newton's method converges in the case where $f'(z) \neq 0$.

In the collection of FORTRAN programs at the end of this text, Program 3 is Newton's method.

EXERCISES

10.1 Determine, using Newton's method, the first 4 digits after the decimal point of the following numbers:

(a) $\sqrt{3}$ (b) $\sqrt{5}$ (c) $\sqrt{7}$

(d) $\sqrt[3]{3}$ (e) $\sqrt[3]{5}$ (f) $\sqrt[3]{7}$

(g) $3^{1/4}$ (h) $5^{1/6}$ (i) $6^{1/7}$.

[*Hint*: Evaluating $c^{1/q}$ is equivalent to finding the zero of $z^q - c$.]

10.2 Find all zeros of the following functions in the indicated domain:

(a) $f(x) = 1 + x^{1/3} - x^{1/2}$, $x \geq 0$

(b) $f(x) = x^3 - 3x^2 + 1$, $-\infty < x < \infty$

(c) $f(x) = \dfrac{x}{x^2 + 1} + 1 - \sqrt{x}$, $x \geq 0$.

10.3 (a) Let z be a zero of the function f, and suppose that neither f' nor f'' vanishes in an interval I containing z. Show that, if the first approximation z_1 lies in I, then all subsequent approximations $z_2, z_3, \ldots$ generated by Newton's method are
   (i) greater than z if $f'(z)$ and $f''(z)$ have like signs,
   (ii) less than z if $f'(z)$ and $f''(z)$ have opposite signs.
   [*Hint*: use Formula (10.7).]
   (b) Verify the truth of these assertions for the 3 examples presented in this section.

10.4 In this exercise we ask the reader to investigate the following method designed for getting a sequence of better and better approximations to a zero of a function f:

(10.14) $z_{new} = z_{old} - af(z_{old}).$

Here a is a number to be chosen in some suitable way. Clearly, if z_{old} happens to be the exact root z, then $z_{new} = z_{old}$. The question is: if z_{old} is a good approximation to z, will z_{new} be a better approximation, and how much better?

(a) Use this method to construct a sequence $z_1, z_2, \ldots$ of approximations to the positive root of

$$f(z) = z^2 - 2 = 0,$$

starting with $z_1 = 2$. Observe that

(i) For $a = 1/2$, $z_n \to \sqrt{2}$, but the z_n are alternately less than and greater than $\sqrt{2}$.

(ii) For $a = 1/3$, $z_n \to \sqrt{2}$ monotonically.

(iii) For $a = 1$, the sequence $\{z_n\}$ diverges.

(b) Prove, by using the mean value theorem, that

(10.15) $$z_{new} - z = (1 - am)(z_{old} - z),$$

where m is the value of f' somewhere between z and z_{old}. Prove that if a is so chosen that

(10.16) $$|1 - af'(z)| < 1,$$

then $z_n \to z$ provided that z_1 is taken close enough to z.

Can you explain your findings under (a) in light of Formula (10.15)?

(c) What would be the best choice for a, i.e., one that would yield the fastest converging sequence of approximants?

(d) Try this method on the problems discussed in Examples 2 and 3.

3.11 Economics and the derivative

Econometrics deals with measurable and measured quantities in economics. The basis of *econometric theory*, as of any theory, is the relation between these quantities. These relations can be expressed in functional form, that is, by expressing certain quantities as functions of others. There are three kinds of functional relations: empirical, analytical, and those imposed by law. This section contains some brief remarks on the concepts of calculus applied to the functions that occur in economic theory.

Example 1. Denote by $C(q)$ the total cost of producing q units of a certain commodity. Many ingredients make up the total cost; some, like raw materials needed, are proportional to the amount q produced. Others, like investment in plant, are independent of q. Still others, like labor costs, might increase more than proportionally to q due to inefficiencies of a large scale operation. In a word, $C(q)$ can be a pretty complicated function of q. A manager who is faced with the decision whether or not to increase production has to know how much the additional production of h units will cost. The cost per additional unit is

$$\frac{C(q + h) - C(q)}{h};$$

for reasonably small h this is well approximated by dC/dq, the derivative of C with respect to q. This is called the *marginal cost of production*.

Example 2. Let $G(L)$ be the amount of goods produced by a labor force of size L. A manager, in order to decide whether to hire more workers, wants to know how much additional goods will be produced by h additional laborers. The gain in production per laborer added is

$$\frac{G(L + h) - G(L)}{h};$$

for reasonably small h this is well approximated by dG/dL, the derivative of G with respect to L. This is called the *marginal productivity of labor*.

Example 3. Let $P(e)$ be the profit realized after the expense of e dollars. The added profit per dollar where h additional dollars are spent is

$$\frac{P(e + h) - P(e)}{h},$$

well approximated for small h by dP/de, called the *marginal profit of expenditure*.

Example 4. Let $T(I)$ be the tax imposed on a taxable income I. The increase in tax per dollar on h additional dollars of taxable income is

$$\frac{T(I + h) - T(I)}{h}.$$

For moderate h this is well approximated by dT/dI, called the *marginal rate of taxation*.

These examples illustrate two facts:

1. The rate at which functions change is as interesting in economics, business, and finance as in every other kind of quantitative description.
2. In economics, the rate of change of a function $y(x)$ is not called the derivative of y with respect to x but the marginal y of x.

Here are some primitive examples of the uses to which the notion of derivative can be put in economic thinking. The managers of a firm would not hire additional workers when the going rate of pay exceeds the marginal productivity of labor; for the firm would be losing money thereby. Thus, declining productivity places a limitation on the size of a firm.

Actually, one can argue persuasively that efficiently run firms will stop hiring even before the situation indicated above is reached. The most efficient mode for a firm is one in which the cost of producing a unit of commodity is minimal. The cost of a unit commodity is

$$\frac{C(q)}{q}.$$

The derivative must vanish at the point where this is minimum; using the rule for differentiating a quotient, we get

$$\frac{q(dC/dq) - C(q)}{q^2} = 0,$$

which implies that, at the point q_{max} of maximum efficiency,

(11.1) $$\frac{d}{dq} C(q_{max}) = \frac{C(q_{max})}{q_{max}}.$$

In words: *At the peak of efficiency, the marginal cost of production equals the average cost of production.* The firm would still make more money by expanding production, but would not be as efficient as before and so its relative position would be weakened.

Equation (11.1) has the following geometric interpretation: The line connecting $(q_{max}, C(q_{max}))$ to the origin is tangent to the graph of $C(q)$; see Figure 3.24. Such a point does not exist for all functions, but does exist for functions $C(q)$ for which $C(q)/q$ tends to ∞ as q tends to ∞. It has been remarked that the nonmonopolistic capitalistic system is possible precisely because the cost functions in capitalistic production have this property.

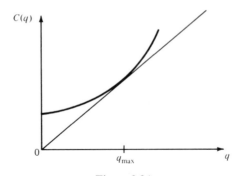

Figure 3.24

We conclude this brief section by pointing out that, to be realistic, economic theory has to take into account the enormous diversity and interdependence of economic activities. Therefore any half-way useful models deal typically with functions of very many variables. These functions are not derived *a priori* from detailed theoretical considerations, but are empirically determined. For this reason these functions are usually taken to be of very simple form, linear or quadratic; the coefficients are then determined by making a best fit to observed data. The fit that can be obtained this way is as good as could be obtained by taking functions of more complicated forms. Therefore there is no incentive or justification to consider more complicated functions. The mathematical theory of economics is thus preoccupied with statistical techniques for fitting linear and quadratic functions of many variables to recorded data and with maximizing or minimizing such functions when the variables are subject to realistic restrictions.

Integration

4

The *total amount* of some quantity is an important and useful concept. Some examples are: the total amount of energy contained in a given amount of gas; the total amount of water contained in a reservoir; the mass of earth in a hill. In this chapter, we introduce the concept of the integral, the precise mathematical expression for total amount.

4.1 Examples of integrals

Determining mileage from a speedometer

In the beginning of Chapter 3 we investigated the relation between the mileage meter and the speedometer. We have shown there that if the speedometer were broken, it would still be possible to determine the speed of the moving car from readings of the mileage meter and a clock. In this chapter, we shall investigate the inverse problem: how to determine the total mileage given the speedometer readings at various times. We assume that we have at our disposal the *total record* of speedometer readings throughout the trip, i.e., that we know the velocity of the car as a function of time. In the graph shown in Figure 4.1, t is measured in hours, velocity in miles per hour (mph). Our problem can be formulated so:

Given velocity as function f of time, determine the distance covered during the time interval S.

Let's denote the distance covered by

$$D(f, S).$$

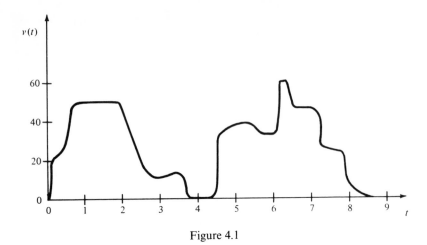

Figure 4.1

This notation emphasizes that D depends on f and S. How does D depend on S? Suppose we divide S into two disjoint intervals,

$$S = S_1 + S_2,$$

as indicated in Figure 4.2. Clearly, the distance covered during the total time interval S is the sum of the distances covered during the intervals S_1 and S_2. This property is called

Additivity

(1.1) $$D(f, S_1 + S_2) = D(f, S_1) + D(f, S_2)$$

for S_1 and S_2 disjoint.

How does D depend on f? Suppose that the velocity f exceeds some minimum speed m throughout the time interval S; clearly a car travelling with speed f covers more ground than one travelling with the minimum speed m. Likewise, if the velocity f stays less than some maximum speed M throughout the time interval S, the car travelling with speed f covers less ground than one travelling with the maximum speed M. The distance covered by a car travelling with constant velocity is

$$\text{distance} = \text{velocity} \times \text{time}.$$

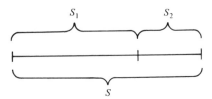

Figure 4.2

Let us denote the length of the interval S by $|S|$; cars travelling with speeds m and M would cover the distances $m|S|$ and $M|S|$, respectively, during the time interval S. So the principles described above can be expressed as the

Lower and upper bound property. *If f satisfies the inequalities*

(1.2) $$m \le f(t) \le M, \qquad t \text{ in } S$$

then

(1.3) $$m|S| \le D(f, S) \le M|S|.$$

Filling a reservoir

A reservoir is being filled with water through a pipe at a rate which varies with time. The rate of influx of water at any given time, measured in thousands of gallons per hour, is measured by a meter located in the pipe, and this influx is plotted continually as function of time in Figure 4.3. The problem is:

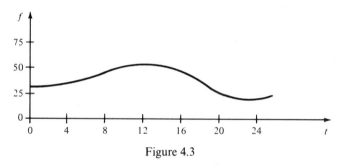

Figure 4.3

Given the influx as a function f of time, determine the total amount of water which has entered the reservoir during some time interval S. We shall denote this amount as

$$R(f, S),$$

again emphasizing that this amount depends on the function f and the interval S.

How does R depend on the time interval S? Clearly, if we break up S as a sum of two disjoint time intervals S_1 and S_2, we have

(1.4) $$R(f, S_1 + S_2) = R(f, S_1) + R(f, S_2),$$

since any water that entered during the time interval S entered either during S_1 or during S_2. This property which R shares with D was called additivity.

Next we show that R also has the lower and upper bound property, i.e., if

(1.5) $$m \le f(t) \le M \qquad \text{for all } t \text{ in } S$$

then

(1.6) $$m|S| \le R(f, S) \le M|S|.$$

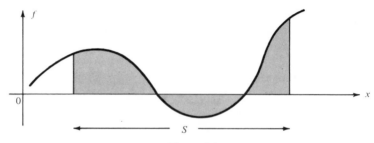

Figure 4.4

For clearly, if the rate at which water enters the reservoir does not exceed M gallons per hour at any time, then the total amount that has entered the reservoir during $|S|$ hours does not exceed $M|S|$. Likewise, if the rate exceeds m gallons per hour, at least $m|S|$ must have entered the reservoir.

Area under a curve

Let f be a function, its graph shown in Figure 4.4. We wish to calculate the area, grey-shaded in the figure, contained between the graph of f and the x-axis, and lying above the interval S, i.e., the set of points (x, y) satisfying

$$x \text{ in } S, \quad 0 \le y \le f(x).$$

Denote this area by

$$A(f, S).$$

Divide S into two disjoint subintervals: $S = S_1 + S_2$. Since the area of a pointset composed of two disjoint pointsets is the sum of the areas of the components, A has the *additive property*

(1.7) $$A(f, S_1 + S_2) = A(f, S_1) + A(f, S_2)$$

for S_1, S_2 disjoint.

We investigate the dependence of A on f; suppose that the values f takes on in S lie between m and M:

(1.8) $$m \le f(x) \le M \qquad \text{for } x \text{ in } S.$$

Then, as Figure 4.5 indicates, the pointset in question *contains* the rectangle with base S and height m, and *is contained in* the rectangle with base S and

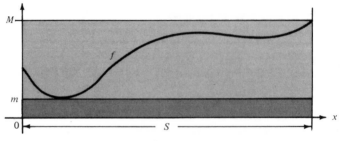

Figure 4.5

height M. Now the area of a pointset contained in another is less than the area of the latter; furthermore the area of a rectangle is the product of the lengths of two adjacent sides. Therefore we conclude that if (1.8) holds, then

$$(1.9) \qquad m|S| \leq A(f, S) \leq M|S|,$$

i.e., that A has the lower and upper bound property.

Work

In Section 8.2 we shall show that *work* W has the lower and upper bound property in its dependence on force, and that W depends additively on the spatial interval over which the work is done.

So far we have tacitly assumed that the function f is positive; we shall show now that D, R and A also make sense for functions f which take on negative values.

The notion of positive and negative distance travelled by a car along a road can be defined in the same way as positive and negative numbers are defined on the number line: The starting point divides the road into two parts, one of which is arbitrarily labelled positive. Locations on the positive side have positive distance from the starting point, while locations on the negative side are defined to have negative distance from the starting point. Velocity f is defined to be the derivative with respect to t of signed distance as defined here. It is easy to verify, and we leave it to the reader to do it, that even if f and D are allowed to take on negative values, D depends additively on S, and the lower and upper bound property (1.3), where m or M may well be negative numbers, remains true.

The interpretation of negative influx is even simpler than that of negative velocity; it means that water is flowing out of, instead of into, the reservoir. $R(S)$ negative means that during S the amount in the reservoir has decreased. It is easy to show, and we leave it to the reader to verify, that even if f and R are permitted to take on negative values, R depends additively on S, and the lower and upper bound property (1.6) is true. (Assume that the initial amount is so large that there is always some water left.)

What should the interpretation of $A(f, S)$ be when f takes on negative values, as pictured in Figure 4.4? We propose to interpret the area above the x-axis as positive, and the area below the x-axis as negative, A being the algebraic sum of these positive and negative quantities. There are two reasons for this interpretation:

1. In many applications of area the underground positions, i.e., those below the x-axis, have to be interpreted in a sense opposite to the portions above ground.
2. Only with this interpretation are the lower and upper bounds (1.9) valid. We leave it to the reader to verify this last point for himself.

We have shown that all three functions D, R and A have the additive property with respect to S, and the lower and upper bound property with

165

respect to f. We shall show in the next section that these two properties completely characterize these functions. To put it sensationally, if the reader knew no more about D, R, and A than what he has learned so far, and if he were transported to a desert island, equipped only with pencil, paper, and a hand calculator (a computer and electric outlet would be more practical), he could calculate the values of D, R, and A for any function f and any interval S. The next section is devoted to explaining how.

EXERCISES

1.1 During a hot 24 hour period S, a reservoir loses water to evaporation and community use. Suppose that, on such a day, the bounds on the rate of evaporation $f_e(t)$ in gallons per hour are

$$250 \le f_e(t) \le 350, \quad 0 \le t \le 24;$$

and the bounds on the rate of loss due to community use, $f_c(t)$, also measured in gallons per hour, are

$$2000 \le f_c(t) \le 4500, \quad 0 \le t \le 24.$$

Find bounds on the depletion $R(f, S)$ on that day. Here $f = f_e + f_c$, t is in S.

1.2 A cueball is propelled perpendicularly toward a cushion from which it rebounds. Suppose the velocity imparted by the cue stick is M_1, the velocity with which it strikes the cushion is m_1; its rebound velocity is $m_2 = -m_1$. Find upper and lower bounds on the distances traversed by the cue ball before and after rebound.

4.2 The integral

In Section 4.1 we have shown that all three quantities, $D(f, S)$, $R(f, S)$ and $A(f, S)$, are additive with respect to S and have the upper-lower bound property with respect to f. In this section we shall show that using only these two properties we can calculate D, R, and A with as great an accuracy as desired. This shows in particular that although D, R, and A have entirely different physical and geometric interpretations, they have the same value when f and S are the same. Anticipating for given f and S, that D, R, and A are identical, we shall call this quantity *the integral of f over S* and denote it by

$$I(f, S).$$

The only two properties of I that we shall use are *additivity with respect to S*:

(2.1) $$I(f, S_1 + S_2) = I(f, S_1) + I(f, S_2)$$

for S_1 and S_2 disjoint, and *the lower–upper bound property with respect to f*: if

(2.2) $$m \le f(t) \le M \quad \text{for all } t \text{ in } S,$$

then

(2.3) $$m|S| \leq I(f, S) \leq M|S|.$$

The usual notation for the integral $I(f, S)$ is

$$\int_a^b f(t)dt,$$

where S is the interval $[a, b]$. The pronunciation of this formula is "integral of $f(t)$, from a to b, with respect to t." The notation $I(f, S)$ was introduced to emphasize that integration is an operation whose *inputs* are *a function and an interval* and whose *output* is a *number*. Since the $I(f, S)$ notation is typographically simpler, we are retaining it up to Section 4.4 of this chapter; we shall explain the origin of the classical notation below. There is nothing wrong with having two different notations for the integral; each has its use in the proper place. After all, we have at least two different notations for the derivative of a function F, namely F' and dF/dx. Next we derive some important consequences of the two basic properties of integrals.

We divide (2.3) by $|S|$ and get

$$m \leq \frac{1}{|S|} I(f, S) \leq M.$$

In the usual notation this would be written as

$$m \leq \frac{1}{b-a} \int_a^b f(t)dt \leq M.$$

The quantity in the middle,

$$\frac{1}{b-a} \int_a^b f(t)dt,$$

is called the *mean value of f over the interval* $[a, b]$. Since the only restriction on m and M is inequality (2.2), we may choose m to be the *minimum* and M to be the *maximum* of f over $[a, b]$. So the above inequality can be restated in words as follows: *The mean value of a function over an interval lies between the minimum and maximum value of the function over that interval.* It follows from the intermediate value theorem in Section 2.7 that, in any interval, a continuous function f takes on all values between its minimum and maximum over that interval. Therefore, in particular, f takes on its mean value over that interval. This result is called the

Mean value theorem for integrals. *Given a function f continuous in an interval* $[a, b]$, *there is a point t_0 in the interval* $[a, b]$:

$$a < t_0 < b$$

167

such that

$$f(t_0) = \frac{1}{b-a} \int_a^b f(t)dt.$$

To derive the next consequence of the two basic properties, we multiply inequality (2.2) by $|S|$ and obtain

$$m|S| \le f(t)|S| \le M|S|.$$

This and inequality (2.3) say that the quantities $f(t)|S|$ and $I(f, S)$ lie in the interval $[m|S|, M|S|]$. Clearly, the difference of two quantities located in an interval cannot exceed the length of the interval; so

(2.5) $$|I(f, S) - f(t)|S|| \le (M - m)|S|$$

for all t in S. We can choose for M and m the maximum and minimum, respectively, of f over S. The difference $M - m$ is called *the oscillation of f over S*, and will be denoted by $\mathrm{Osc}(f, S)$,

$$\mathrm{Osc}(f, S) = M - m.$$

We can now state inequality (2.5) in the following words: *For any t in S, $f(t)|S|$ approximates $I(f, S)$ with an error not exceeding the length of S times the oscillation of f over S.*

This approximation is a good one if the oscillation of f over S is small, that is, if the values of the function f do not vary too much over the interval S. If the function f is *continuous*, its values do not vary too much over a sufficiently short interval; this is the definition of continuity, as explained in Section 2.7. So we can say: *For any t in S, $f(t)|S|$ is a good approximation to $I(f, S)$ if f is continuous and S short enough.*

We are interested in getting good approximations to $I(f, S)$ for f continuous, but S *not* necessarily short. What to do? The saving idea is to subdivide the interval S into many small pieces, each of which is short enough.

By the additivity property (2.1), we know that, if we divide S into disjoint intervals S_1 and S_2 so that $S = S_1 + S_2$, then

$$I(f, S) = I(f, S_1) + I(f, S_2).$$

Similarly, divide S into disjoint intervals S_1, S_2 and S_3 so that $S = S_1 + S_2 + S_3$; then applying the additive property twice we get

$$I(f, S_1 + S_2 + S_3) = I(f, S_1) + I(f, S_2 + S_3)$$
$$= I(f, S_1) + I(f, S_2) + I(f, S_3).$$

Generally, if we divide S into n disjoint intervals, we find that repeated application of the additive property (2.1) gives the *compound* additive property

(2.1)' $$I(f, S_1 + S_2 + \cdots + S_n) = I(f, S_1) + I(f, S_2) + \cdots + I(f, S_n).$$

Formula (2.1)′ reduces the problem of calculating $I(f, S)$ to calculating each of the n quantities $I(f, S_j)$, $j = 1, \ldots, n$, and adding the results. If n is large enough, each S_j can be made very short, and so $f(t_j)|S_j|$, where t_j is any point of S_j, is a good approximation to $I(f, S_j)$. According to (2.5) the error in this approximation does not exceed

$$(2.6) \qquad |S_j|\mathrm{Osc}_j,$$

where Osc_j is the oscillation of f over S_j. It follows from this estimate that the sum

$$(2.7) \qquad \sum_{j=1}^{n} f(t_j)|S_j|$$

differs from the sum

$$(2.8) \qquad \sum_{j=1}^{n} I(f, S_j)$$

by at most the sum of the errors (2.6) of the individual terms, i.e., by at most

$$\sum_{j=1}^{n} |S_j|\mathrm{Osc}_j.$$

Denote by Osc *the largest of the oscillations* Osc_j; the above sum is not greater than

$$(2.9) \qquad \sum_{j=1}^{n} |S_j|\mathrm{Osc}.$$

The S_j constitute a *partition* of S into disjoint intervals; the sum of the lengths of S_j is just the total length of S;

$$\sum_{j=1}^{n} |S_j| = |S|,$$

so that the sum (2.9) is $|S|\mathrm{Osc}$. We can now write our *error estimate* in the form

$$(2.10) \qquad \left| I(f, S) - \sum_{j=1}^{n} f(t_j)|S_j| \right| \le \sum_{1}^{n} |S_j|\mathrm{Osc} \le \left(\sum_{1}^{n} |S_j| \right)\mathrm{Osc} = |S|\mathrm{Osc}.$$

Thus the additivity and lower–upper bound properties of $I(f, S)$ have enabled us to derive an approximation to $I(f, S)$. We now state it in the form of an

Approximation theorem for the integral. *Let*

$$(2.11) \qquad S = \sum_{1}^{n} S_j$$

be any subdivision of S into disjoint subintervals S_j; denote by t_j any point of S_j and denote by I_{approx} the quantity

$$(2.12) \qquad I_{\mathrm{approx}}(f, S) = \sum_{1}^{n} f(t_j)|S_j|.$$

Then $I_{\text{approx}}(f, S)$ differs from $I(f, S)$, the integral of f over S, by at most

(2.13) $$|S|\,\text{Osc},$$

where

(2.14) $$\text{Osc} = \max_{j} \text{Osc}(f, S_j).$$

The quantities I_{approx} are called *approximating sums*; see Figure 4.6.

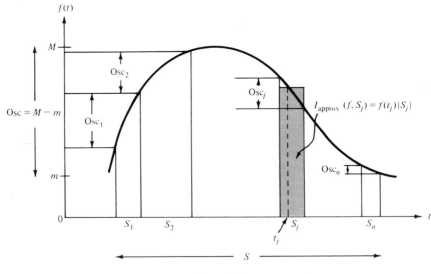

Figure 4.6

Sometimes the length S_j of the j-th interval is denoted by $d_j t$, so that the approximating sums have the form

$$\sum_{j=1}^{n} f(t_j) d_j t.$$

The classical notation

$$\int_{a}^{b} f(t)\,dt$$

for the integral has been introduced because of its resemblance to this formula; the integral sign $\int$ replaces the summation sign $\sum$, and the subscript j is dropped.

The approximation theorem can be used to determine the value of $I(f, S)$, f continuous, with any desired accuracy. All we have to do is to choose the subdivision (2.11) so fine, i.e., each subinterval S_j so short, that the oscillation of f over each is less than (permissible error)/$|S|$, something we can do for any continuous function f. This fact has important theoretical and practical uses; we discuss the theoretical ones first.

170

I Uniqueness. *The value of $I(f, S)$ is uniquely determined once the continuous function f and the interval S are specified.*

PROOF. For any given continuous function f over an interval S, the value of $I_{\text{approx}}(f, S)$ is uniquely determined by Formula (2.12) once the subdivision and the choice of the points t_j are specified. Since $I(f, S)$ can be approximated arbitrarily closely by $I_{\text{approx}}(f, S)$, it follows that the value of $I(f, S)$ is uniquely determined as well. □

II Linearity. *$I(f, S)$ depends linearly on f. This means that I satisfies*

(2.15) $$I(f + g, S) = I(f, S) + I(g, S)$$

for any pair of continuous functions f and g, and

(2.16) $$I(af, S) = aI(f, S)$$

for any real number a.

PROOF. Choose a subdivision (2.11), that is, choose a set $\{S_j\}$, and pick the points t_j; $I_{\text{approx}}(f, S)$, as defined by (2.12), depends linearly on f, i.e.,

$(2.15)_{\text{approx}}$ $\qquad I_{\text{approx}}(f + g, S) = I_{\text{approx}}(f, S) + I_{\text{approx}}(g, S)$

and

$(2.16)_{\text{approx}}$ $\qquad I_{\text{approx}}(af, S) = aI_{\text{approx}}(f, S).$

These relations follow by direct substitution into (2.12). Since $I(f, S)$, $I(f + g, S)$ and $I(af, S)$ can be approximated arbitrarily closely by $I_{\text{approx}}(f, S)$, $I_{\text{approx}}(f + g, S)$ and $I_{\text{approx}}(af, S)$ respectively, relations (2.15) and (2.16) follow from $(2.15)_{\text{approx}}$ and $(2.16)_{\text{approx}}$. □

III Monotonicity. *Suppose that the continuous functions f and g satisfy*

(2.17) $$f(t) \leq g(t) \quad \text{for all } t \text{ in } S;$$

then

(2.18) $$I(f, S) \leq I(g, S).$$

In words: If f is less than or equal to g on S, then the integral of f over S is less than or equal to the integral of g over S.

PROOF. From the definition (2.12) of I_{approx}, it follows that if $f \leq g$ on S, then $I_{\text{approx}}(f, S) \leq I_{\text{approx}}(g, S)$, provided both approximations employ the same subdivisions and the same set of points t_j. The conclusion (2.18) about the integrals themselves follows again from the approximation theorem. □

IV Approximation. *Suppose that the function a approximates f on S so that the values of $a(t)$ and $f(t)$ agree in the first m digits for all t in S. That is, we assume that for all t in S*

(2.19) $$|f(t) - a(t)| < 10^{-m}.$$

We claim that then the integral of a over S approximates the integral of f over S:

(2.20) $$|I(f, S) - I(a, S)| < 10^{-m}|S|.$$

PROOF. Inequality (2.19) means that

(2.21) $$a(t) - 10^{-m} < f(t) < a(t) + 10^{-m}.$$

According to inequality (2.18) and the linearity condition (2.15)

$$I(a, S) - I(10^{-m}, S) < I(f, S) < I(a, S) + I(10^{-m}, S).$$

Since the integral of 10^{-m} over S is $10^{-m}|S|$, we have inequality (2.20). ☐

Let f_n be a sequence of functions which converge on S to f; as explained in Section 2.8, this means that for any m, $f_n(t)$ and $f(t)$ differ by less than 10^{-m} for all t on S, provided that n is large enough. It follows from inequality (2.20) that, for all n large enough, the integrals of f_n and f over S differ by less than $10^{-m}|S|$. This means that $I(f_n, S)$ tends to $I(f, S)$ as n tends to ∞. We summarize this result as the

Convergence theorem for integrals. *If a sequence of functions f_n converges on an interval S to f, then the sequence of integrals of f_n over S converges to the integral of f over S:*

$$\lim_{n \to \infty} I(f_n, S) = I(f, S).$$

A very important extension of the convergence theorem for integrals deals with functions which depend on an additional parameter, p; we shall indicate the dependence of $f(t)$ on the parameter p by using the notation $f[p]$. For example, the linear function

$$l(x) = kx$$

depends on the constant of proportionality k, so we denote it also by $l[k]$. There are many examples of such parametric dependence. We encountered one in Chapter 3. Formula (6.11), describing the motion

$$x(t) = -\tfrac{1}{2}gt^2 + bt + c$$

of a particle under a constant gravitational force contains not one, but three parameters: g, the gravitational constant; b, the initial velocity; and c, the initial position.

If a function f depends on a parameter p, so does its integral over a fixed interval S; we indicate this dependence explicitly by

(2.22) $$I(f[p], S) = I(p).$$

In many situations it is very important to learn how $I(p)$ varies with p. If, for example, the integral of $f[p]$ represents some kind of a total payoff to be gained from the function f, we would try to choose the parameter p so that $I(p)$ is as large as possible. We shall use the tools of differential calculus to study how, for a given function $f[p]$, the integral $I(p)$ varies with p. The following theorem shows how to differentiate $I(p)$ with respect to p:

Differentiation theorem for the integral. *Suppose that, on an interval S, $f[p]$ depends differentiably on the parameter p in the sense that the difference quotients*

(2.23)
$$\frac{f[p + h] - f[p]}{h}$$

converge on S as h tends to 0. We denote the limit of these difference quotients by f_p. Then I(p), defined by (2.22), depends differentiably on p, and

(2.24)
$$\frac{d}{dp} I(p) = I(f_p, S).$$

PROOF. This result is a corollary of previously derived properties; for, using the linearity of the integral, we can write

(2.25)
$$\frac{I(p + h) - I(p)}{h} = \frac{I(f[p + h], S) - I(f[p], S)}{h}$$

$$= I\left(\frac{f[p + h] - f[p]}{h}, S\right).$$

Since we have assumed that the difference quotients (2.23) tend to f_p, we conclude from the convergence theorem for integrals that the right side of (2.25) tends to the right side of (2.24). But then the left side of (2.25) tends to the left side of (2.24), as asserted. ☐

This result is enormously useful. We shall meet several applications of it in Chapter 6; first note the following examples.

Example 1. $f[p] = (t + p^2)^3$, $\qquad f_p = 6p(t + p^2)^2$.

Example 2. $f[p] = t/(t + p)$, $\qquad f_p = - t/(t + p)^2$.

Example 3. $f[p] = \sqrt{t^2 + p^2}$, $\qquad f_p = p/\sqrt{t^2 + p^2}$.

Example 4. $f[p] = t^p$, $\qquad f_p = ?$.

We now turn to the practical application of the approximation theorem, i.e., to its use in calculating integrals approximately. Among all possible approximation formulas we shall single out three classes for numerical study: those where t_j in (2.12) is taken as the left endpoint, the right endpoint, and the midpoint, respectively, of S_j. We shall denote these collectively as I_{left}, I_{right} and I_{mid}.

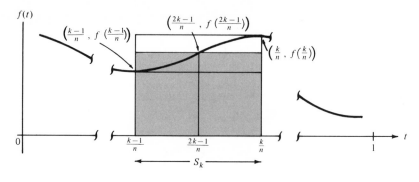

Figure 4.7

In all the examples below, the interval of integration S is taken to be $[0, 1]$, which shall be divided into n subintervals of *equal length;* see Figure 4.7. For this choice of subdivision, $|S_j| = 1/n$, so the approximation formula (2.12) reads:

$$
(2.26) \quad
\begin{cases}
I_{\text{left}}(f, S) = \dfrac{1}{n} \displaystyle\sum_{k=1}^{n} f\!\left(\dfrac{k-1}{n}\right) \\[2ex]
I_{\text{right}}(f, S) = \dfrac{1}{n} \displaystyle\sum_{k=1}^{n} f\!\left(\dfrac{k}{n}\right) \\[2ex]
I_{\text{mid}}(f, S) = \dfrac{1}{n} \displaystyle\sum_{k=1}^{n} f\!\left(\dfrac{2k-1}{2n}\right).
\end{cases}
$$

Below we tabulate the values of I_{left}, I_{right} and I_{mid} for various values of n, for a variety of functions f.

Example 5. $f(t) = t$.

n	I_{left}	I_{right}	I_{mid}
1	0	1.0	0.5
2	0.25	0.75	0.5
3	0.333...	0.666...	0.5
4	0.375	0.625	0.5
5	0.400	0.600	0.5
10	0.450	0.550	0.5
100	0.495	0.505	0.5

According to the calculation in Exercise 2.7, $I(t, [0, 1]) = 0.5$; observe that the midpoint rule gives the same exact answer, while I_{left} and I_{right} deviate from the exact value.

The tables in the next two examples have been compiled by an electronic computer:

Example 6. $f(t) = t^2$.

n	I_{left}	I_{right}	I_{mid}
1	0.0000000	1.0000000	0.2500000
5	0.2400000	0.4400000	0.3300000
10	0.2850000	0.3850000	0.3325000
100	0.3283500	0.3383500	0.3333250

According to the calculation in Exercise 2.8, the exact value of $I(t^2, [0, 1])$ is $0.3333\ldots$. Notice that for all n listed, I_{mid} gives a better approximation to the exact value than either I_{left} or I_{right}.

Example 7. $f(t) = t/(1 + t^2)$.

n	I_{left}	I_{right}	I_{mid}
1	0.0	0.5	0.4
5	0.2932233	0.3932233	0.3482551
10	0.3207392	0.3707392	0.3469912
100	0.3440653	0.3490653	0.3465778

In Section 4.4 we shall show that the exact value of $I(t/(1 + t^2), [0, 1])$ is $\frac{1}{2}\log 2 = 0.3465735 \ldots$. Observe that I_{mid} again provides a consistently better approximation to the value of the integral than either I_{left} or I_{right}. In Section 4.6 we shall explain why this is so, and derive even better approximations than the midpoint rule.

EXERCISES

2.1 Compute, using a pocket calculator, the values of I_{left}, I_{right} and I_{mid} in the following cases:
(a) $f(x) = x^3$, $S = [1, 2]$, $n = 1, 2, 4$.
(b) $f(x) = \sqrt{1 - x^2}$, $S = [0, 1/\sqrt{2}]$, $n = 1, 2, 4$.
(c) $f(x) = 1/(1 + x^2)$, $S = [0, 1]$, $n = 1, 2, 4$.

2.2 Using an electronic computer this time, calculate the same approximate sums in Exercise 2.1, for $n = 100, 150, 200$.

2.3 Prove, using Formulas (2.26), that

$$I_{left}(t, [0, 1]) = \frac{1}{2}\left(1 - \frac{1}{n}\right), \qquad I_{right}(t, [0, 1]) = \frac{1}{2}\left(1 + \frac{1}{n}\right)$$

$$I_{mid}(t, [0, 1]) = \frac{1}{2}.$$

2.4 Let S be any interval $[c, d]$, a any real number. Let S_a denote the interval obtained when S is shifted to the right by the amount a; that is

$$S_a = [c + a, d + a].$$

Let f be any continuous function on S. Denote by f_a the function obtained when f is shifted to the right by a. That is, f_a is defined on S_a according to the rule

$$f_a(t) = f(t - a).$$

The relation between f and f_a is most easily stated in terms of their graphs: to obtain the graph of f_a on S_a, shift the graph of f on S to the right by the amount a; see Figure 4.8. Prove that

(2.27) $$I(f_a, S_a) = I(f, S).$$

This property of the integral is called *translation invariance*. [*Hint*: Show that approximating sums are translation invariant.]

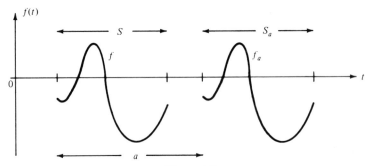

Figure 4.8

2.5 Let S be an interval $[c, d]$; the *reflected* interval is defined as $[-d, -c]$ and denoted by S_-. Let f be some continuous function on S; its *reflection*, denoted by f_-, is defined on S_- as follows:

$$f_-(t) = f(-t).$$

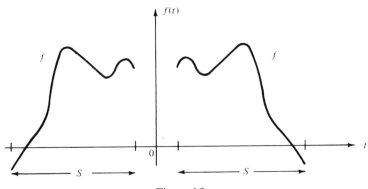

Figure 4.9

The graph of f_- is obtained from the graph of f by reflection across the f-axis; see Figure 4.9. Prove that

(2.28) $$I(f_-, S_-) = I(f, S).$$

This property of the integral is called *invariance under reflection.* [*Hint*: Show that approximating sums are invariant under reflection.]

2.6 Let S be an interval $[c, d]$, k some positive number. We denote by kS the interval obtained from S by stretching in the ratio $1:k$, i.e., $kS = [kc, kd]$. Let f be any continuous function on S; we denote by $f_{(k)}$ the function defined on kS obtained from f by stretching; i.e.,

$$f_{(k)}(t) = f(t/k);$$

see Figure 4.10. Prove that

(2.29) $$I(f_{(k)}, kS) = kI(f, S).$$

[*Hint*: Show that a similar relation holds between approximating sums.]

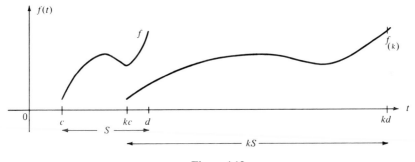

Figure 4.10

2.7 Let S be the unit interval $[0, 1]$, f any function. Reflection carries S into $[-1, 0]$, and translation to the right by 1 carries this interval back to S; see Figure 4.11. Reflection and translation carry the function $f(t)$ into $f(1 - t)$.

(a) Show, using translation and reflection invariance of the integral, that

$$I(f(t), [0, 1]) = I(f(1 - t), [0, 1]).$$

(b) Show that $I(t, [0, 1]) = I(1 - t, [0, 1])$; using this relation and linearity (2.16), show that

$$I(t, [0, 1]) = \tfrac{1}{2}.$$

(c) Show that $I(t^2, [0, 1]) = I((1 - t)^2, [0, 1])$; use this relation and linearity (2.16) to show that

$$I(t, [0, 1]) = \tfrac{1}{2}.$$

Interpret this formula geometrically in terms of the area of a triangle.

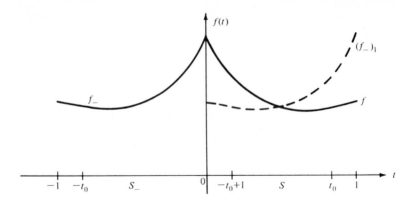

Figure 4.11

2.8 (a) Show, using the result of Exercise 2.6, that

$$I(t^2, [0, 2]) = 8I(t^2, [0, 1]).$$

(b) Show, using additivity, that

$$I(t^2, [0, 2]) = I(t^2, [0, 1]) + I(t^2, [1, 2]).$$

(c) Show, using translation invariance and linearity, that

$$I(t^2, [1, 2]) = I((1 + t)^2, [0, 1]) = I(1, [0, 1]) + 2I(t, [0, 1]) + I(t^2, [0, 1]).$$

(d) Combine (a), (b), and (c) to show that

$$I(t^2, [0, 1]) = \tfrac{1}{3}.$$

2.9 Show, using the method of Exercise 2.8, that
(a) $I(t^3, [0, 1]) = \tfrac{1}{4}$, (b) $I(t^4, [0, 1]) = \tfrac{1}{5}$.
Can you guess what $I(t^k, [0, 1])$ is for any positive integer k?

2.10 Show that

(a) $I(t, [0, d]) = \dfrac{d^2}{2}$ (b) $I(t, [c, d]) = \dfrac{d^2}{2} - \dfrac{c^2}{2}$

(c) $I(t^2, [0, d]) = \dfrac{d^3}{3}$ (d) $I(t^2, [c, d]) = \dfrac{d^3}{3} - \dfrac{c^3}{3}$.

[*Hint*: Use the results of Exercises 2.6, 2.7, 2.8, and additivity.]

2.11 (a) Prove that the integral of a linear function $l(t) = mt + b$, m, b constants, over an interval $S = [c, d]$ is given by

(2.30) $$I(l, S) = \frac{l(c) + l(d)}{2} |S|.$$

(b) Using (2.30), verify that, for linear functions, I depends additively on S and has the lower–upper bound property with respect to l.
(c) Interpret Formula (2.30) geometrically as the formula for the area of a trapezoid.

4.3* Existence of the integral

In the last section we have shown, on the basis of the approximation theorem, that as the partition of S gets more and more refined, the approximating sums $I_{approx}(f, S)$ tend to the integral I of f over S for any continuous f. In our proof we assumed that f has an integral over each subinterval. In this section we shall prove the convergence of suitable approximating sums *without* such an assumption. By defining the integral to be the limit of approximating sums *we will prove the existence of the integral.*

We know from Section 4.2 how to integrate some simple functions. In particular, Formula (2.30) in Exercise 2.11 gives the integral of all *linear* functions over any interval; furthermore we have verified there that for this class of functions I depends additively on the interval and has the lower–upper bound property with respect to the function. We introduce next a larger class of functions, the *continuous, piecewise linear functions*, defined as follows.

Partition an interval S into disjoint subintervals:

$$S = \sum_1^n S_j.$$

A function f is called a piecewise linear, continuous function if it is linear on each subinterval, i.e.,

$$(3.1)_j \qquad\qquad f(t) = l_j(t) = m_j t + b_j, \quad t \text{ in } S_j,$$

and if it is continuous in the whole interval S, so that $l_j(t)$ and $l_{j+1}(t)$ have the same value at the common endpoint of the adjoining intervals S_j and S_{j+1}.

The graph, see Figure 4.12, of a continuous piecewise linear function consists of straight line segments joined at the endpoints. Since we know the

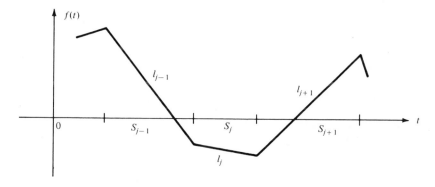

Figure 4.12

* This section may be omitted on the first reading.

integral of linear functions, we can define the integral of piecewise linear functions $(3.1)_j$ as

(3.2)
$$I(f, S) = \sum_{j=1}^{n} I(l_j, S_j).$$

As we have seen earlier, the integral of linear functions is additive in S and has the lower–upper bound property with respect to f. It follows therefore that $I(f, S)$ as defined by (3.2) for piecewise linear functions has these same properties.

Let f and g be two piecewise linear functions, f being linear on each subinterval of the partition $S = \sum_{j=1}^{n} S_j$, g linear on each subinterval of the partition $S = \sum_{k=1}^{n} T_k$ of the same interval S. Denote by R_{jk} the intersection $S_j \cap T_k$ (common part) of the intervals S_j and T_k. Clearly, R_{jk} is again an interval (possibly empty) and

$$S = \sum_{\substack{j=1 \\ k=1}}^{\substack{m \\ n}} R_{jk}$$

is again a partition of S, called the intersection of the two partitions. In Figure 4.13, the intersection of the 2 partitions consists of 8 subintervals. Clearly, the intersection of the two partitions is again a partition of S. Since each R_{jk} lies in a subinterval of both partitions, both f and g are linear over each R_{jk}; therefore, so are their sum and difference. This shows that *the sum and difference of 2 piecewise linear functions are piecewise linear.*

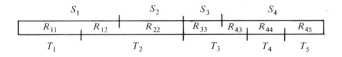

Figure 4.13

We shall introduce a single symbol, P, to denote a particular partition

(3.3)
$$S = \sum_{j=1}^{n} S_j$$

of an interval S into disjoint subintervals.

Let f be a continuous function defined over an interval S, P the partition (3.3) of S. We define the continuous piecewise linear function f_P over S as follows:

$$\left. \begin{array}{l} \text{(i) } f_P(t) = f(t) \text{ at the endpoints of } S_j, \\ \text{(ii) } f_P(t) \text{ is linear in } S_j. \end{array} \right\} \quad j = 1, \dots, n.$$

Since a linear function is uniquely determined if its value is prescribed at two distinct points, the above two conditions determine f_P over each S_j, and

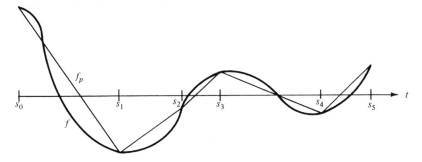

Figure 4.14

therefore over all of S. f_P is called the *piecewise linear approximation to f*, *corresponding to the partition P.*

Given the graph of f, the graph of f_P consists of the straight line segments connecting the pairs of points $(s_{j-1}, f(s_{j-1}))$, $(s_j, f(s_j))$, $j = 1, \ldots, n$, where s_j ($j = 0, 1, \ldots, n$) denotes the endpoints of the intervals S_j arranged in increasing order; see Figure 4.14. From this geometrical description of the construction of f_P it is evident that, if f is continuous and if the partition P is fine enough, f_P will lie pretty close to f. We formulate this as the

Piecewise linear approximation theorem. *Let f be a continuous function on S, P a partition of S, f_P the piecewise linear approximation to f corresponding to the partition P. Define the quantity*

$$\text{Osc}(f, P)$$

[as in Section 4.2, Formula (2.14)] as the largest of the oscillations of f on any of the subintervals S_j of the partition. Then for all t in S

(3.4) $|f(t) - f_P(t)| \le 2 \operatorname{Osc}(f, P)$.

PROOF. To prove that, at every point of S, the values of f and f_P are as close as (3.4) claims, we shall

(i) Compare the values of f at a point t in S_j and at the right endpoint s_j of S_j;
(ii) compare the values of f_P at the same two points;
(iii) use the monotonicity of f_P in each S_j;
(iv) use the agreement of f_P and f at the endpoint s_j of each S_j.

We spell out the details below.

Suppose t lies in the j-th subinterval $S_j = [s_{j-1}, s_j]$. By definition of oscillation, for every t in S_j

$$|f(t) - f(s_j)| \le \text{Osc}(f, S_j) \le \text{Osc}(f, P).$$

A linear function is monotonic; since f_P is linear on S_j, for every t in S_j

$$|f_P(t) - f_P(s_j)| \leq |f_P(s_{j-1}) - f_P(s_j)|.$$

By definition of piecewise linear approximation, the values of f_P at the endpoints of S_j agree with those of f, so we can rewrite this last inequality as

$$|f_P(t) - f(s_j)| \leq |f(s_{j-1}) - f(s_j)| \leq \text{Osc}(f, P).$$

Since both $f(t)$ and $f_P(t)$ differ from $f(s_j)$ by at most $\text{Osc}(f, P)$, it follows that they can differ from each other at most by $2\,\text{Osc}(f, P)$, as asserted in (3.4). $\square$

Consider now an infinite sequence of partitions P_n such that, as n tends to ∞, the length of the largest subinterval of P_n tends to zero; e.g., we may choose P_n to be the partition of S into n subintervals of equal length. It follows from the continuity of f over S that

$$\text{Osc}(f, P_n)$$

tends to zero as n tends to infinity.

Assertion. *For any function f continuous over S, the infinite sequence*

$$I(f_{P_n}, S)$$

converges.

PROOF. Let us compare two terms of the sequence. Since the difference of two piecewise linear functions is piecewise linear, and since the integral is additive in its dependence on f, see (2.15), we have for any two partitions P and Q

(3.5) $$I(f_P, S) - I(f_Q, S) = I(f_P - f_Q, S).$$

According to the piecewise linear approximation theorem, f_P and f_Q each differ from f by at most $2\,\text{Osc}(f, P)$ and $2\,\text{Osc}(f, Q)$ respectively; it follows that f_P and f_Q differ from each other at most by the sum of these:

(3.6) $$|f_P(t) - f_Q(t)| \leq 2\,\text{Osc}(f, P) + 2\,\text{Osc}(f, Q).$$

Since the integral of piecewise linear functions has the upper–lower bound property, it follows from (3.6) that

$$|I(f_P - f_Q, S)| \leq 2[\text{Osc}(f, P) + \text{Osc}(f, Q)]|S|.$$

Thus from relation (3.5), we may conclude that

(3.7) $$|I(f_P, S) - I(f_Q, S)| \leq 2[\text{Osc}(f, P) + \text{Osc}(f, Q)]|S|.$$

We saw before that $\text{Osc}(f, P_n)$ tends to zero as $n \to \infty$; therefore it follows from (3.7) that

$$|I(f_{P_n}, S) - I(f_{P_m}, S)|$$

tends to zero as n and m tend to ∞.

According to the convergence criterion of Section 1.4, a sequence a_n for which $|a_n - a_m| < 10^{-k}$ for $n, m > N$, N suitably chosen, converges to a limit. The foregoing shows that the sequence

$$a_n = I(f_{P_n}, S)$$

satisfies the convergence criterion; therefore it tends to a limit. We define this limit to be the integral of f over S:

$$I(f, S) = \int_a^b f(t)dt = \lim_{\substack{n \to \infty \\ |S_j| \to 0}} I(f_{P_n}, S), \qquad S = [a, b]. \qquad \square$$

We saw earlier that for piecewise linear f, $I(f, S)$ depends additively on S and has the lower–upper bound property with respect to f. $I(f, S)$ is defined for arbitrary continuous f as a limit of integrals of piecewise linear approximations f_P. It follows from (3.4) in the piecewise linear approximation theorem that Max f_P tends to Max f, Min f_P to Min f. It follows that $I(f, S)$, thus extended, retains both aforementioned properties. This completes the proof of the following

Existence theorem for the integral. *Every function f continuous on an interval S has a well determined integral $I(f, S)$. I depends additively on S and has the lower–upper bound property with respect to f.*

The integral was defined as the limit of a sequence of exact integrals of piecewise linear approximations to f. It is instructive to look at the form of these integrals $I(f_P, S)$. According to Formula (2.30), see Exercise 2.11, for l linear,

(3.8)
$$I(l, [c, d]) = \frac{l(c) + l(d)}{2}(d - c).$$

Apply this to the interval $[c, d] = [s_{j-1}, s_j] = S_j$ and to the linear function $l = f_P$; since by definition $f_P = f$ (see Figure 4.15) at the endpoints of S_j,

$$I(f_P, S_j) = \frac{f(s_{j-1}) + f(s_j)}{2}|S_j|.$$

Summing with respect to j we get

(3.9)
$$I(f_P, S) = \sum_1^n \frac{f(s_{j-1}) + f(s_j)}{2}|S_j|.$$

Since (3.8) expresses the area of a trapezoid (average of bases $\times$ altitude), (3.9) is called the *trapezoid rule* for approximating an integral.

183

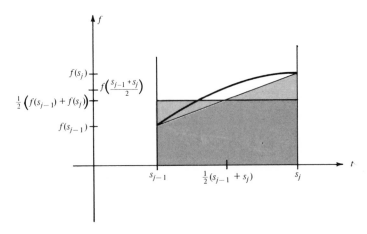

Figure 4.15

4.4 The fundamental theorem of calculus

At the beginning of this chapter, we posed this problem: Determine the *total distance D* that a moving vehicle has covered during a time interval S from the knowledge of the *velocity v* of the vehicle at each instant of S. The answer we found was that the total distance D is the integral

$$(4.1) \qquad\qquad D = I(v, S)$$

of the velocity as function of time over the interval S. Formula (4.1) expresses the total distance covered during the whole trip. A similar formula holds, of course, for the distance $D(t)$ covered during the part of the journey up to time t. This formula is

$$(4.1)_t \qquad\qquad D(t) = I(v, S(t)),$$

where $S(t)$ is the interval between the starting time and the time t.

In Section 3.1, we discussed the *converse problem*: If we know the distance $D(t)$ of a moving vehicle from its starting point for all values of t, how can we determine its velocity as function of time t? The answer we found was that *velocity is the derivative of D as function of time*:

$$(4.2) \qquad\qquad v(t) = \frac{dD(t)}{dt}.$$

Now compare the two answers $(4.1)_t$ and (4.2). We can summarize them in the following words: *If the function D(t) is the integral of v over the interval S(t), then v is the derivative of D.* Omitting all qualifying phrases, we can express this statement as an epigram.

Differentiation is the inverse of integration.

The argument presented in favor of this proposition was based on physical intuition concerning the way the velocity and distance covered by a moving vehicle are related to each other. We proceed to give a purely mathematical proof. This will require a modest amount of notation.

Let f be any continuous function defined on an interval $S = [a, b]$. Denote the subinterval $[a, t]$ by $S(t)$, where t is any value between a and b. We define the function $F(t)$ to be the integral of f over $S(t)$:

$$(4.3) \qquad F(t) = I(f, S(t)).$$

We assert:

$$(4.4) \qquad \frac{dF(t)}{dt} = f(t).$$

Like all our proofs concerning properties of the integral, the proof of (4.4) too will be based on the two basic properties of the integral. Assertion (4.4) means that

$$(4.5) \qquad \lim_{h \to 0} \frac{F(t + h) - F(t)}{h} = f(t).$$

By the definition (4.3),

$$F(t + h) = I(f, S(t + h)).$$

Let h be positive; then the interval $S(t + h)$ can be broken up into the two subintervals $[a, t]$ and $[t, t + h]$, so that

$$S(t + h) = S(t) + [t, t + h].$$

Using the additive property of the integral, we find that

$$\begin{aligned} F(t + h) = I(f, S(t + h)) &= I(f, S(t)) + I(f, [t, t + h]) \\ &= F(t) + I(f, [t, t + h]); \end{aligned}$$

so the difference quotient in (4.5) is

$$(4.6) \qquad \frac{I(f, [t, t + h])}{h}.$$

Denote by m_h and M_h the minimum and maximum of f on $[t, t + h]$, respectively. The length of this interval is h. The lower–upper bound property of integrals states that

$$m_h h \leq I(f, [t, t + h]) \leq M_h h;$$

the equivalent inequality

$$m_h \leq \frac{I(f, [t, t + h])}{h} \leq M_h$$

places the difference quotient (4.6) between m_h and M_h. Now the function f is continuous; therefore over a short interval, the maximum and minimum of f differ very little from the value of f at any point of that interval. In particular, for h small enough, m_h and M_h differ little from $f(t)$; this proves that as h tends to zero, the quantity (4.6) tends to $f(t)$, as asserted in (4.5). The case where h is negative can be treated in a similar fashion. All this leads up to the most fundamental of theorems of calculus, aptly called

The fundamental theorem of calculus

(a) *Every continuous function f defined on some interval S is the derivative of a differentiable function F defined on the same interval S.*

(b) *The function F is determined up to an additive constant.* This means that two functions F_1 and F_2 which have f as their derivative differ on S by a constant.

(c) *Let F be a differentiable function on an interval S, f its derivative. Then for any two points x and y of S, $x < y$,*

(4.7)$_I$

$$F(y) - F(x) = I(f, [x, y]).$$

PROOF. We have just shown that, given any continuous f, the function F_1 defined by (4.3),

$$F_1(t) = I(f, S(t)),$$

has f as its derivative. This completes the proof of part (a).

Let F_2 be any function whose derivative is f; then

$$(F_1 - F_2)' = F_1' - F_2' = f - f = 0,$$

so that the function $(F_1 - F_2)$ has derivative 0 in S; see Figure 4.16. But we have shown in Section 3.3 that a function whose derivative is zero on an interval is constant there; this completes the proof of part (b).

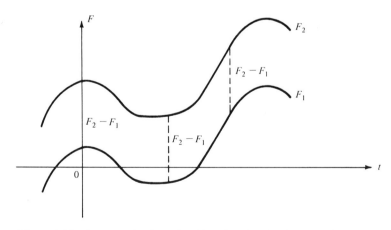

Figure 4.16 An example of a pair of functions having the same derivative.

We use the above definition of F_1 to write

(4.8) $$F_1(y) - F_1(x) = I(f, S(y)) - I(f, S(x)).$$

Now the interval $S(y)$ is the union of the nonoverlapping intervals $S(x)$ and $[x, y]$ (Figure 4.17), that is,

$$S(y) = S(x) + [x, y].$$

So by additivity of the integral,

$$I(f, S(y)) = I(f, S(x)) + I(f, [x, y]).$$

Substituting this for $I(f, S(y))$ on the right side of (4.8) we obtain $(4.7)_I$ for the function F_1 defined by (4.3). Let F be any function which has f as its derivative; according to part (b), F differs from the function F_1 by a constant. It follows that $(4.7)_I$ is valid for any F which has f as its derivative. This completes the proof of the fundamental theorem.   □

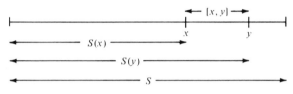

Figure 4.17

This is a good time to switch to the classical notation:

$$I(f, [x, y]) = \int_x^y f(t)dt.$$

In terms of this notation, the fundamental relation $(4.7)_I$ reads

$(4.7)_f$
$$F(y) - F(x) = \int_x^y \frac{dF(t)}{dt} dt.$$

The symbol dt occurs twice on the right side. If we succumbed to a deeply rooted impulse and cancelled dt, we would obtain

$$\int_x^y dF(t).$$

If we think of $dF(t)$ as an increment in F, and of $\int_x^y$ as the sum of all such increments as t goes from x to y, we obtain the total increment of the function F in $[x, y]$; this is just what the left side of $(4.7)_f$ says. Thus the classical notation for the integral forces the fundamental theorem to our attention.

The fundamental theorem deserves its honorific name; it has (at least) three important uses. First and foremost, it is the *fundamental existence theorem of analysis*; in Chapters 5 and 7 we shall exploit its power in our treatment of the exponential, logarithmic, and trigonometric functions.

187

Its second use lies in uncovering new, important properties of the integral; these will be discussed in the next section. Its third use lies in furnishing a constructive method for evaluating the integral of any function we recognize to be the derivative of a known function $F: f = F'$. We shall illustrate this on some examples.

Example 1. $f(t) = t^n$, n any integer except -1; find $\int_a^b f(t)dt$.
We notice that $f(t) = t^n$ is the derivative of

$$F(t) = \frac{t^{n+1}}{n+1};$$

therefore by the fundamental theorem,

$$\int_a^b t^n \, dt = F(b) - F(a) = \frac{b^{n+1}}{n+1} - \frac{a^{n+1}}{n+1}.$$

In particular, if we take $a = 0$, $b = 1$, we obtain, for $n = 1$, that

$$\int_0^1 t \, dt = \frac{1}{2},$$

in agreement with Exercise 2.7; for $n = 2$ we get

$$\int_0^1 t^2 \, dt = \frac{1}{3},$$

in agreement with Exercise 2.8.

Example 2. $f(t) = t/\sqrt{1 + t^2}$; find $\int_a^b f(t)dt$.
We notice that $f(t)$ is the derivative of

$$F(t) = \sqrt{1 + t^2};$$

therefore, by the fundamental theorem,

$$\int_a^b \frac{t}{\sqrt{1 + t^2}} \, dt = \sqrt{1 + b^2} - \sqrt{1 + a^2}.$$

In particular, let $[a, b] = [0, 1]$. Then

$$\int_0^1 \frac{t}{\sqrt{1 + t^2}} \, dt = \sqrt{2} - 1.$$

Example 3. $f(t) = t^2\sqrt{1 + t^3}$, and $S = [0, 1]$; find $\int_0^1 f(t)dt$.
Here we observe that f is the derivative of

$$F(t) = \tfrac{2}{9}(1 + t^3)^{3/2}.$$

Therefore by the fundamental theorem,

$$\int_0^1 t^2\sqrt{1 + t^3} \, dt = (2)^{3/2} - \tfrac{2}{9}.$$

By now the reader must have noticed that the key to evaluating integrals by the fundamental theorem lies in an ability to notice that the function f presented for integration is the derivative of another handy function F. How does one acquire such an uncanny ability? It comes with the experience of differentiating lots of functions; in addition, the search for F can be systematized with the aid of a few basic rules. These will be presented in the next section.

We end this section with a theoretical application of the fundamental theorem. According to $(4.7)_f$, for any function F, differentiable on an interval $[x, y]$,

$$F(y) - F(x) = \int_x^y \frac{dF(t)}{dt} \, dt.$$

According to the mean value theorem for integrals (see Section 4.2) applied to $f = dF/dt$, there is a t_0 between x and y such that

$$(4.9) \qquad \frac{dF}{dt}(t_0) = \frac{1}{y - x} \int_x^y \frac{dF(t)}{dt} \, dt.$$

Combining (4.9) with $(4.7)_f$, we obtain that

$$(4.10) \qquad \frac{F(y) - F(x)}{y - x} = \frac{d}{dt} F(t_0)$$

for some t_0. This is the mean value theorem for derivatives, proved in Chapter 3, and seen here as the consequence of the mean value theorem for integrals.

EXERCISE

4.1 For the following functions $f(t)$, find a function F such that $dF/dt = f$.

(a) $t/\sqrt{1 - t^2}$ (b) $(1 + t)^{1/3}$
(c) $t^2(1 + t^3)^{1/3}$ (d) $2 + 4t^{-2} - 8t^{-3}$.

4.5 Rules of integration and how to use them

Section 3.2 contains the rules for differentiation. These rules specify how to express the derivatives of the sum, product, and composite of two functions in terms of the derivatives of those functions themselves. Using the fundamental theorem we shall convert each of these rules into a rule for integration. Let f and g be the derivatives of F and G, respectively. The sum rule says that

$$(F + G)' = F' + G' = f + g.$$

Applying relation $(4.7)_f$ of the fundamental theorem to the function $f + g$, we obtain

$$\int_a^b (f + g) dt = (F(b) + G(b)) - (F(a) + G(a)).$$

On the other hand, again by (4.7)$_f$,

$$\int_a^b f\,dt = F(b) - F(a) \quad \text{and} \quad \int_a^b g\,dt = G(b) - G(a);$$

comparing the first with the sum of these two, we deduce the *sum rule*

$$\int_a^b (f + g)dt = \int_a^b f\,dt + \int_a^b g\,dt$$

for integrals. This rule is not new to our readers; they have already en-countered it under the name of *linearity* in Section 4.2, where it was deduced from the linearity of approximating sums. We have shown there that for any constant k

$$\int_a^b kf\,dt = k\int_a^b f\,dt.$$

We now recall the product rule

$$(fg)' = f'g + fg'$$

of differentiation. Integrate each side over an interval $[a, b]$ and apply relation (4.7)$_f$ of the fundamental theorem to the left to obtain

(5.1) $$\int_a^b (f'g + fg')dt = f(b)g(b) - f(a)g(a).$$

The linearity of the integral on the left yields, after rearrangement, the following relation:

(5.2) $$\int_a^b f'g\,dt = f(b)g(b) - f(a)g(a) - \int_a^b fg'\,dt.$$

The process of expressing the integral on the left with the aid of another integral on the right as in Formula (5.2) is called *integration by parts*. The reader will appreciate that integration by parts is helpful if we know more about the integral on the right than about the integral on the left. "Knowing more" could mean knowing the exact value of the integral on the right, or it could mean that the integral on the right is easier to evaluate approximately than the one on the left. In the examples below, we illustrate both possibilities.

Example 1. Find

$$\int_a^b x^3\sqrt{1 + x^2}\,dx.$$

We factor the function to be integrated as follows:

$$x^3\sqrt{1 + x^2} = (x\sqrt{1 + x^2})x^2 = f'g,$$

where

$$f(x) = \tfrac{1}{3}(1 + x^2)^{3/2}, \qquad g(x) = x^2.$$

We integrate by parts, and get

(5.3)
$$\int_a^b x^3 \sqrt{1 + x^2}\, dx$$

$$= \tfrac{1}{3}(1 + b^2)^{3/2}b^2 - \tfrac{1}{3}(1 + a^2)^{3/2}a^2 - \int_a^b \tfrac{2}{3}(1 + x^2)^{3/2}x\, dx.$$

We scrutinize the integral on the right and notice that the function to be integrated, $\tfrac{2}{3}(1 + x^2)^{3/2}x$, is the derivative of the function

$$\tfrac{2}{15}(1 + x^2)^{5/2}.$$

Therefore it follows from (5.3) that

$$\int_a^b x^3 \sqrt{1 + x^2}\, dx = H(b) - H(a),$$

where the function H is

$$H(x) = \tfrac{1}{3}(1 + x^2)^{3/2}x^2 - \tfrac{2}{15}(1 + x^2)^{5/2}.$$

Example 2. Find

$$\int_a^b \frac{x^5}{\sqrt{1 + x^2}}\, dx.$$

We factor the integrand as follows:

$$\frac{x^5}{\sqrt{1+x^2}} = \left(\frac{x}{\sqrt{1+x^2}}\right)x^4 = f'g,$$

with

$$f(x) = (1 + x^2)^{1/2}, \qquad g(x) = x^4.$$

We integrate by parts:

$$\int_a^b \frac{x^5}{\sqrt{1 + x^2}}\, dx = f(b)g(b) - f(a)g(a) - \int_a^b fg'\, dx$$

$$= (1 + b^2)^{1/2}b^4 - (1 + a^2)^{1/2}a^4 - \int_a^b 4(1 + x^2)^{1/2}x^3\, dx.$$

Notice that the integrand in the integral on the right is 4 times the function occurring in Example 1; therefore the integral is 4 times the integral of Example 1 which we have already evaluated.

There are many ways of factoring a given integrand into a product $f'g$. Though our choice may seem arbitrary to the inexperienced reader, he will soon see why some factorizations are more helpful than others. More will be said about this after Examples 3 and 4.

Example 3. Find $\int_a^b x^5 \sqrt{1 + x^3}\, dx$. We choose factors f' and g of the integrand in such a way that the corresponding fg', which will appear as new integrand after integration by parts, is the derivative of a function. We write

$$x^5 \sqrt{1 + x^3} = (x^2 \sqrt{1 + x^3})x^3 = f'g,$$

where

$$f(x) = \tfrac{2}{9}(1 + x^3)^{3/2}, \qquad g(x) = x^3.$$

Now integrate by parts obtaining

$$\int_a^b x^5 \sqrt{1 + x^3}\, dx = f(b)g(b) - f(a)g(a) - \int_a^b \tfrac{2}{9}(1 + x^3)^{3/2} 3x^2\, dx,$$

where the integrand on the right,

$$\tfrac{2}{3}(1 + x^3)^{3/2}x^2,$$

is indeed the derivative of the function

$$\tfrac{4}{45}(1 + x^3)^{5/2}.$$

So we can express that integral by using the fundamental theorem. Putting everything together we find that

$$\int_a^b x^5 \sqrt{1 + x^3}\, dx = H(b) - H(a),$$

where

$$H(x) = \tfrac{2}{9}(1 + x^3)^{3/2}x^3 - \tfrac{4}{45}(1 + x^3)^{5/2}.$$

Example 4. Find $\int_0^1 \sqrt{x^2 - x^3}\, dx$. Factor the integrand:

$$\sqrt{x^2 - x^3} = (\sqrt{1 - x})x = f'g,$$

where

$$f(x) = -\tfrac{2}{3}(1 - x)^{3/2}, \qquad g(x) = x.$$

The function f vanishes at the endpoint $x = 1$, g at the other endpoint $x = 0$. So integration by parts gives

$$\int_0^1 \sqrt{1 - x}\, x\, dx = - \int_0^1 -\tfrac{2}{3}(1 - x)^{3/2}\, dx.$$

We recognize that $-\tfrac{2}{3}(1 - x)^{3/2}$ is the derivative of

$$\tfrac{4}{15}(1 - x)^{5/2}.$$

So, using the fundamental theorem, we find that

$$\int_0^1 -\tfrac{2}{3}(1-x)^{3/2} = -\tfrac{4}{15};$$

therefore, using the result of integration by parts, we get that

$$\int_0^1 \sqrt{x^2 - x^3}\, dx = \tfrac{4}{15}.$$

These examples illustrate the mechanics of applying integration by parts to an integral $I(H, S)$; the first step is to factor the integrand H into a product of the derivative f' of some function f and another function g:

$$H = f'g.$$

This first step presents no problem; in fact we may choose f arbitrarily, and set g equal to the quotient H/f'. Integrating by parts expresses then the original integral in terms of the integral

$$\int_a^b fg'\, dx.$$

If we are clever or lucky in our choice of f, we shall recognize fg' as the derivative of some function G; if not, we have 2 choices:

1. Start all over with a different factorization leading to another function f and apply integration by parts using the new factorization.
2. Give up.

In Chapters 5 and 7 we shall introduce the basic transcendental functions of analysis, the logarithm, the exponential function, sine, cosine, and their inverses. Having that many new functions at our disposal we can play the game of integration by parts much more efficiently, and we shall achieve astonishing feats of expressing integrals as products, quotients, and composites of polynomials, powers, and the transcendental functions just mentioned. Yet the importance of integration by parts is not limited to these nearly miraculous cases of explicit integration; the examples below illustrate a different point of view.

Example 5. Find

$$\int_0^1 x^2\sqrt{1-x^2}\, dx.$$

We factor the integrand as follows:

$$x^2\sqrt{1-x^2} = (x\sqrt{1-x^2})x = f'g,$$

where

$$f(x) = -\tfrac{1}{3}(1-x^2)^{3/2}, \qquad g(x) = x.$$

193

Note that the function f is 0 at $x = 1$, and that g is 0 at $x = 0$; so integrating by parts we obtain

(5.4) $$\int_0^1 x^2 \sqrt{1 - x^2}\, dx = \int_0^1 \tfrac{1}{3}(1 - x^2)^{3/2}\, dx.$$

In contrast to the previous examples, no matter how we wrack our brains, we fail to recognize any simple (or even complicated) function which would have $\tfrac{1}{3}(1 - x^2)^{3/2}$ as its derivative. So seemingly our strategy of integrating by parts has merely reduced the original task of finding the value of an integral to the problem of finding the value of another integral. We shall now show that something useful has been achieved. The second integral is easier to approximate than the first one. To convince the reader of this, we plot the graphs of both integrands in Figure 4.18.

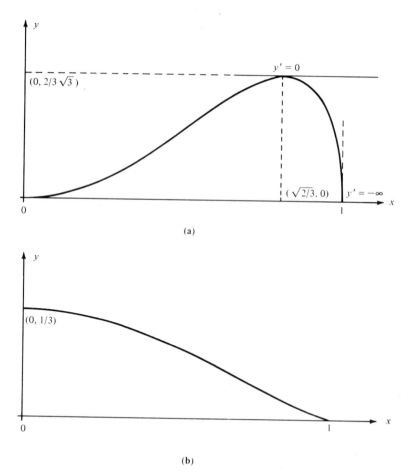

(a)

(b)

Figure 4.18   (a): $y = x^2 \sqrt{1 - x^2}$; (b): $y = \tfrac{1}{3}(1 - x^2)^{3/2}$.

The derivative of the first function, $x^2\sqrt{1-x^2}$, is

$$2x\sqrt{1-x^2} - \frac{x^3}{\sqrt{1-x^2}},$$

which tends to $-\infty$ as x tends to the right endpoint $x = 1$; this fact causes the graph of the function to have a vertical tangent at $x = 1$. A function which has a large derivative is changing very fast. In contrast the derivative of the second function, $\frac{1}{3}(1-x^2)^{3/2}$, is $-x\sqrt{1-x^2}$, which stays decently bounded throughout the whole interval $[0, 1]$. A function with such a derivative changes slowly. We claim that *it is easier to evaluate approximately the integral of a slowly changing function than of one which changes fast.* To see this, examine the approximation theorem stated in Section 4.2; the estimate (2.13) of the amount by which the approximating sum differs from the exact integral is proportional to the maximum oscillation of the integrand over the intervals of the subdivision. For a given subdivision, this quantity is much larger for a rapidly varying function than for a slowly varying one.

A word of caution: just because *our estimate* for the difference between approximating sum and exact integral is large, it doesn't follow that the difference itself is large; after all, our estimate could have been very crude! It turns out, in this case, that our estimate is realistic.

The midpoint rule, introduced at the end of Section 4.3, gives a far better approximation to the integral for a slowly varying integrand than for one that varies fast. These questions will be discussed in Section 4.6; in the meantime, we content ourselves with some numerical evidence. The table below lists, for various values of n, approximating sums for both integrals in (5.4), evaluated according to the midpoint rule for subdivisions of the interval $[0, 1]$ into n equal subintervals. According to (5.4), these integrals are equal.

n	$I_{mid}(x^2\sqrt{1-x^2})$	$\frac{1}{3}I_{mid}((1-x^2)^{3/2})$
1	0.21650635	0.21650635
2	0.21628707	0.19951025
5	0.20306231	0.19664488
10	0.19890221	0.19640021
100	0.19643512	0.19634969
1000	0.19635226	0.19634954
10000	0.19634963	0.19634954

The last two entries in the second column have the same first 8 digits; so we are reasonably certain that the first seven digits of the integral are

$$0.1963495.$$

In the second column, $n = 2$ gives the first 2, $n = 5$ the first 3, and $n = 10$ the first 4 digits accurately. In the first column $n = 10$ gives only 2 digits accurately.

195

To clinch the claim that it is easier to approximate the second integral than the first one, we show that it takes the same time to evaluate the n-th approximating sum for each. The arithmetic operations involved in calculating I_{mid} are: evaluating the integrand at n points, adding the values of these functions and dividing the sum by n.

We shall now show that, if we use a little trick, then it takes the same number of arithmetic operations to evaluate both functions. This becomes evident if we write the second function in the form $(1 - x^2)\sqrt{1 - x^2}$. Clearly, in order to evaluate either function we have to perform these steps:

1. Calculate x^2,
2. Calculate $1 - x^2$.
3. Calculate $\sqrt{1 - x^2}$.
4. Calculate the product $\sqrt{1 - x^2}$ by x^2 for the first, and by $(1 - x^2)$ for the second function.

For the second integral we have to multiply the total result by $\frac{1}{3}$, which is a negligible contribution to the operation count.

We now turn to an entirely different kind of use of integration by parts:

Example 6. Let $x(t)$ be any twice differentiable function, and S an interval at whose endpoints x' is zero. Can one say something about the integral of the product xx'' over $S = [t_1, t_2]$, that is, about $\int_{t_1}^{t_2} xx'' \, dt$? To answer this question we factor the integrand as follows:

$$x''x = f'g,$$

where

$$f = x', \qquad g = x.$$

S was so chosen that f is 0 at both endpoints. Integration by parts gives

$$\int_{t_1}^{t_2} x''x \, dt = -\int_{t_1}^{t_2} x' \cdot x' \, dt = \int_{t_1}^{t_2} -x'^2 \, dt.$$

Since the integrand on the right is the negative quantity $-x'^2$, then so is the value of the integral on the right; this shows that the left side is negative. This example demonstrates that integration by parts can sometimes reveal a quality of an integral such as negativity. There are some situations when this is all we want to know.

We now deduce an important integration formula from the chain rule. Let F and g be differentiable functions. Again $F \circ g$ denotes the *composite function* whose value at t is $F(g(t))$. Denote by f the derivative of F:

(5.5)
$$F' = f.$$

196

According to the chain rule,

(5.5)' $$(F \circ g)' = (F' \circ g)g' = (f \circ g)g'.$$

Let S be an interval on which g' is positive; then g is strictly increasing. Denote by $g(S)$ the image of S under g; the endpoints of $g(S)$ are the images of the endpoints of S under g, i.e., if S is $[a, b]$, $g(S)$ is $[g(a), g(b)]$.

According to the fundamental theorem, the integral of the derivative of a function over an interval equals the difference of the values of that function at the endpoints of the interval. We apply this to the function F and the interval $g(S)$, as well as to the function $F \circ g$ and the interval S. Using (5.5) and (5.5)' to express the derivatives of these functions, we get

(5.6) $$\int_{g(a)}^{g(b)} f \, dg = F(g(b)) - F(g(a))$$

and

(5.7) $$\int_{a}^{b} (f \circ g)g' \, dt = (F \circ g)(b) - (F \circ g)(a).$$

By definition of composition the right sides of (5.6) and (5.7) are equal; therefore so are the left sides:

(5.8) $$\boxed{\int_{g(a)}^{g(b)} f \, dg = \int_{a}^{b} (f \circ g)g' \, dt}.$$

This is the celebrated *formula for changing variables in an integral*. Let us illustrate its use in the examples below.

Example 7. Consider the integral $\int_{1}^{2} \sqrt{1 + x^2} \, dx$. Let $f(x) = \sqrt{1 + x^2}$ and make the "change of variable" $x = g(t) = \sqrt{t}$; then $f \circ g = \sqrt{1 + t}$. Moreover, $g'(t) = 1/(2\sqrt{t})$. Also, when $x = 1 = \sqrt{t}$, $t = 1$, and when $x = 2 = \sqrt{t}$, $t = 4$. So g maps $[1, 4]$ into $[1, 2]$ and Formula (5.8) gives

$$\int_{1}^{2} \sqrt{1 + x^2} \, dx = \int_{1}^{4} \frac{\sqrt{1 + t}}{2\sqrt{t}} \, dt = \frac{1}{2} \int_{1}^{4} \sqrt{\frac{1 + t}{t}} \, dt.$$

Example 8. Let f be any continuous function, and let g be the linear function

$$g(t) = kt + c, \qquad k \text{ and } c \text{ constants.}$$

We get from (5.8) that

(5.9) $$\int_{ka+c}^{kb+c} f(x)dx = \int_{a}^{b} f(kt + c)k \, dt = k \int_{a}^{b} f(kt + c)dt.$$

Notice that relation (5.9) embodies two rules derived in Section 4.2; setting $k = 1$ we obtain (2.27), the translation invariance of an integral, and setting $c = 0$ we obtain (2.29), which describes the effect of stretching on integrals. Thus we see that (5.8) is a powerful generalization of the simple rules we found in Exercises 2.4 and 2.6, showing the effect of shifting and stretching functions on their integrals.

The proofs suggested in Section 4.2 about shifting and stretching were based on properties of approximating sums; we now give a proof of our general result (5.8) for changing variables in an integral based on these properties. The general result is important enough to merit two distinct proofs. The proof based on approximating sums can be extended to yield a change of variables theorem for integrals of functions of several variables; however, a proof based on the fundamental theorem has no counterpart for functions of several variables.

We start with any partition

$$S = \sum S_j$$

of $S = [a, b]$ into disjoint subintervals S_j. We pick points t_j in each S_j; denote as before the length of S_j by $|S_j|$. The sum

(5.10) $$\sum f(g(t_j))g'(t_j)|S_j|$$

is an approximating sum for the integral

(5.11) $$\int_a^b (f \circ g)g' \, dt.$$

Let us denote by X the image of S under g, and by X_j the image of S_j under g; i.e.,

$$X = g(S), \quad X_j = g(S_j).$$

We recall that the function g was assumed to have a positive derivative. It follows that it is a monotonic continuous function. Therefore the X_j are nonoverlapping distinct intervals which cover the interval $X = g(S)$, so that

$$X = \sum X_j$$

is a disjoint partition of X.

We pick points x_j in each X_j; the sum

(5.12) $$\sum f(x_j)|X_j|$$

is an approximating sum for the integral

(5.13) $$\int_{g(a)}^{g(b)} f(x)dx.$$

In particular, since t_j is located in S_j, picking

(5.14) $$x_j = g(t_j)$$

198

guarantees that x_j lies in X_j. This choice of x_j makes

(5.15) $$f(x_j) = f(g(t_j)).$$

We now show that we can choose t_j so that

(5.16) $$g'(t_j)|S_j| = |X_j|.$$

If this is so, each term in the approximating sum (5.10) is equal to the corresponding term in (5.12), so that the two sums are equal.

To see what needs to be done to accomplish (5.16), we recall that the intervals

$$S_j = [a_j, b_j] \quad \text{and} \quad X_j = g(S_j) = [g(a_j), g(b_j)]$$

have lengths

$$|S_j| = b_j - a_j \quad \text{and} \quad |X_j| = g(b_j) - g(a_j),$$

respectively. Thus relation (5.16) can be written as

(5.17) $$g'(t_j) = \frac{|X_j|}{|S_j|} = \frac{g(b_j) - g(a_j)}{b_j - a_j}.$$

Is there a value t_j in S_j for which (5.17) is true? The mean value theorem, Section 3.8, says yes, and (5.16) is satisfied for this choice of t_j.

We have explained in Section 4.2 that if the subdivision $S = \sum S_j$ is fine enough, i.e., if the length of each S_j is small, then any approximating sum (5.10) differs very little from the integral (5.11). Since g is continuous, making each S_j very small also makes each image $X_j = g(S_j)$ small, so that the approximating sum (5.12) differs little from the integral (5.13). Since for the proper choice of t_j and x_j these approximating sums are equal, it follows that the integrals themselves are equal; the equality of these integrals is just formula (5.8) for changing variables in an integral. In the case $g'(t) < 0$, that is, for a monotonically decreasing function g, the change of variable formula is given in (5.21), see Exercise 5.4.

An easy way to remember the change of variable Formula (5.8) is to recall how useful the dF/dx notation is for formulating the chain rule; in this notation the chain rule (5.5) for $F(x(t))$ appears as

$$\frac{dF}{dt} = \frac{dF}{dx}\frac{dx}{dt},$$

merely a cancellation law. This formula is quite unforgettable. We shall now show that the change of variable formula for integrals is just as unforgettable. Writing $g = x(t)$ and $g' = dx/dt$ in the integrands of Formula (5.9) gives

(5.18) $$\boxed{\int_{x(a)}^{x(b)} f(x)dx = \int_a^b f(x(t))\frac{dx}{dt}dt}$$

Equality of the two sides is suggested by cancelling dt and observing that if t varies between a and b, $x = g(t)$ varies between $g(a)$ and $g(b)$. Formula (5.19) has the virtue of being more easily remembered than (5.8).

Example 9. Transform the integral

$$\int_{\sqrt{a}}^{\sqrt{b}} \frac{1}{x^2 + 1}\,dx$$

with respect to x by the change of variable $x = \sqrt{t}$ to an integral with respect to t. Let

$$f(x) = \frac{1}{x^2 + 1}, \quad x = \sqrt{t}; \quad \text{so} \quad \frac{dx}{dt} = \frac{1}{2\sqrt{t}}.$$

Then by Formula (5.18),

$$\int_{\sqrt{a}}^{\sqrt{b}} \frac{1}{x^2 + 1}\,dx = \int_a^b \frac{1}{t + 1}\frac{dx}{dt}\,dt = \frac{1}{2}\int_a^b \frac{1}{t + 1}\frac{1}{\sqrt{t}}\,dt.$$

What are the uses of the formula for changing variables in an integral? There are many; certainly it holds out the possibility of transforming an integral into another one which can be expressed in terms of known functions; we shall give a number of examples of this in Chapter 5. Often the transformed integral cannot be *evaluated* explicitly but is easier to *estimate* than the original integral. Here is an example:

Example 10. Estimate

$$I = \int_1^4 \frac{1}{1 + x}\,dx.$$

The length of the interval of integration is $4 - 1 = 3$. The integrand is a decreasing function; its largest value, taken on at $x = 1$, is 0.5; its smallest value, taken on at $x = 4$, is 0.2. According to the upper–lower bound property for the integral, I lies between 0.2×3 and 0.5×3:

(5.19) $$0.6 \leq I \leq 1.5.$$

Now let us change variables; set $x = t^2$, then

$$\int_1^4 \frac{1}{1 + x}\,dx = \int_1^2 \frac{1}{1 + t^2}\frac{dx}{dt}\,dt = \int_1^2 \frac{2t}{t + 1^2}\,dt.$$

Now the length of the interval of integration is $2 - 1 = 1$. We claim that the new integrand, $2t/(1 + t^2)$, is a decreasing function of t; we see this by differentiating the above function:

$$\frac{d}{dt}\left(\frac{2t}{1 + t^2}\right) = \frac{2}{1 + t^2} - \frac{4t^2}{(1 + t^2)^2} = \frac{2(1 - t^2)}{(1 + t^2)^2}.$$

200

The expression on the extreme right shows that, for $t > 1$, this derivative is negative; by the monotonicity principle we conclude that in the interval $[1, 2]$, the integrand is a decreasing function of t, as asserted. Its maximum, taken on at $t = 1$, is 1 and its minimum, taken on at $t = 2$, is 0.8. Therefore, according to the lower–upper bound property, I lies between these limits:

(5.20)
$$0.8 \leq I \leq 1.$$

Notice how much narrower the bounds in (5.20) are than in (5.19). [*Note*. The value of the integral, to 4 figures, is log $2.5 = 0.9163$.]

The most important use by far of the formula for changing variables in an integral comes from frequent encounters with integrals where the integrand is not handed down from on high, but is a function which arises in another part of a larger problem. It often happens that it is necessary, or convenient, to change variables in the other part of the problem; it is then necessary to perform that change of variable in the integral, too.

EXERCISES

5.1 Evaluate the integral

$$\int_0^1 \sqrt{1 + \sqrt{x}}\, dx$$

by integration by parts. [*Hint*: Write the integrand as the product $\sqrt{x} \cdot (\text{something}).$]

5.2 Evaluate the integral in Exercise 5.1 by introducing $\sqrt{x} = t$ as new variable of integration.

5.3 Evaluate approximately the integral

$$\int_1^2 \sqrt{x^2 - 1}\, dx.$$

(a) Apply the midpoint rule directly to the above integral.
(b) Express, using integration by parts, the integral as another one, whose integrand varies more slowly near $x = 1$ than our present integrand; then apply the midpoint rule to the new integral.
(c) Introduce $\sqrt{x - 1} = t$ as the new variable of integration; then apply the midpoint rule to the resulting integral.

5.4 Let g be a function whose derivative on the interval $[a, b]$ is negative. Prove that in this case we have the following formula for changing variables in an integral:

(5.21)
$$\int_{g(b)}^{g(a)} f(x)dx = \int_a^b f(x(t))|g'(t)|dt.$$

[*Hint*: Use the invariance of the integral under reflection as expressed in (2.28).]

5.5 Define the number $B(n, m)$ by the definite integral

$$B(n, m) = \int_0^1 x^n(1 - x)^m \, dx, \qquad n > 0, \quad m > 0.$$

(a) Integrating by parts, show that

(5.22)
$$B(n, m) = \frac{n}{m + 1} B(n - 1, m + 1).$$

(b) For positive integers n and m, show that repeated application of the recursion relation derived in part (a) yields

$$B(n, m) = \frac{n!m!}{(n + m + 1)!}.$$

Remark. The result in part (a) is valid for any positive real numbers n and m. We shall see in Section 5.1 that $n!$ can be defined also for noninteger n. With this definition, the result in (b) remains valid for arbitrary real, positive n and m.

4.6 The approximation of integrals

A basic property of the integral is that $I(f, S) = \int_a^b f(t)dt$ is bracketed between $m|S| = m(b - a)$ and $M|S| = M(b - a)$, where m and M are the minimum and maximum values, respectively, of f over S. According to the intermediate value theorem, a continuous function takes on all values between its minimum and maximum. Since according to the foregoing,

$$\frac{I(f, S)}{|S|} = \frac{1}{b - a} \int_a^b f(t)dt$$

is such a value, it follows that it equals the value of f at some point t_0 of S:

$$f(t_0)(b - a) = \int_a^b f(t)dt;$$

see Figure 4.19 and the mean value theorem for integrals, Section 4.2. Of course this formula cannot be used directly to evaluate the integral of f, since we do not know where to choose the point t_0. On the other hand, if f

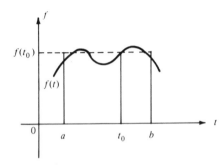

Figure 4.19

is continuous and S short enough, the value of f does not change much from point to point, so that it doesn't matter crucially where we choose t_0; this is the basis of the approximation formulas

$$I_{\text{approx}}(f, S) = \sum_j f(t_j)|S_j|$$

which we have employed so successfully in Section 4.2. We saw there, however, that if we chose each t_j to lie at the midpoint of the interval S_j rather than at one of the endpoints, we obtained a significantly better approximation to the integral. In this section we shall analyze why this is so, and, using the fruits of that analysis, derive an approximation formula which for smooth functions is even better than the midpoint rule.

Suppose f is monotonic on S. Then its maximum and minimum are reached at the endpoints, so that the left and right point rules furnish the upper and lower bounds $M|S|$ and $m|S|$, whereas the midpoint rule furnishes something in between. To see how good this "in between value" is, we recall that the graph of a differentiable function f over short intervals deviates very little from the tangent line; see Figure 4.20, where the tangent to the graph of f

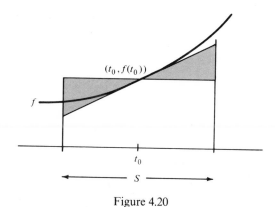

Figure 4.20

has been drawn through the point $(t_0, f(t_0))$, and t_0 is the midpoint of the interval S. The two triangles shown shaded in the diagram are congruent and so have the same area. This shows that the area of the rectangle whose height is $f(t_0)$ and whose base is $|S|$ equals the area of the trapezoid contained between the tangent line to the graph of f and the interval S. The area of the rectangle is $f(t_0)|S|$, just the midpoint approximation to the integral. Since the tangent line is close to the graph of f over S, for a differentiable function f, the midpoint approximation is very close to the integral, over a short enough interval.

We now give another, nongeometric analysis of the accuracy of the midpoint rule. We start by investigating the midpoint rule for linear functions l:

$$l(s) = k + ms.$$

Let $S = [a, c]$ be any interval; to evaluate the integral $\int_a^c l(s)ds$, we note that l is the derivative of the quadratic function

$$q(s) = ks + \frac{ms^2}{2}.$$

So, by the fundamental theorem,

$$\int_a^c l(s)ds = q(c) - q(a) = kc + \frac{mc^2}{2} - ka - \frac{ma^2}{2}.$$

We rewrite and factor the right side as follows:

$$kc - ka + \frac{mc^2}{2} - \frac{ma^2}{2} = k(c - a) + m\frac{c + a}{2}(c - a)$$

$$= \left(k + m\frac{c + a}{2}\right)(c - a).$$

The first factor, $k + m(a + c)/2$, is just the value of l at the midpoint $b = (a + c)/2$ of S; the second factor, $c - a$, is the length of S. Therefore we have shown that

$$\int_a^c l\,ds = l(b)(c - a),$$

where b is the midpoint of S. This proves that *for linear functions the midpoint formula gives the exact value of the integral*.

Let f be a differentiable function on an interval S. Differentiability means that, if S is short enough, the function f can be well approximated by a linear function l over the interval S so that

(6.1) $$f(s) - l(s) \quad \text{is small on } S.$$

Since l is close to f, its integral will be close to that of f; so

(6.2) $$\int_a^c l\,ds = \int_a^c f\,ds + \text{something small}.$$

According to what we have shown before, the integral of l is given exactly by the midpoint formula $I_{mid}(l, S)$. If l is chosen so that $l(b) = f(b)$ at the midpoint b of S, then $I_{mid}(l, S)$ is equal to $I_{mid}(f, S)$. Combining this with (6.2) we conclude that for S short enough, $I_{mid}(f, S)$ differs little from $\int_a^c f\,ds$.

The analysis so far is qualitative. To make it quantitative we would have to estimate how small the "small" is in expressions (6.1) and (6.2), i.e., how well f can be approximated by a linear function. For twice differentiable functions f, we can get such an estimate from the quadratic approximation theorem in Section 3.8; we shall not carry out the details of estimating the error committed by using the midpoint rule but merely state the outcome:

Let f be twice differentiable on an interval $S = [a, c]$, and let M be an upper bound for $|f''(s)|$ over S. Consider a partition of S into subintervals S_j, the largest of which has length h. Denote by $I_{mid}(f, S)$ the midpoint approximation using this partition; then

(6.3)
$$\left| \int_a^c f \, ds - I_{mid}(f, S) \right| \leq \frac{M}{24} h^2 |S|.$$

Example 1. Find the error committed when the midpoint rule is used with a subdivision into 100 equal intervals in approximating the value of $I = \int_0^1 s^2 \, ds$. In this case $f''(s) = 2$, so we may choose $M = 2$. Dividing $[0, 1]$ into a hundred equal subintervals means that $h = 1/100$. Applying (6.3) we get

$$|I - I_{mid}| \leq \tfrac{2}{24} 10^{-4} \simeq 8.3 \times 10^{-6}.$$

Now

$$I_{mid} = h \sum_{j=1}^{100} \left[\left(j - \frac{1}{2} \right) h \right]^2 = 0.33332500 \ldots .$$

Therefore $I - I_{mid} = \frac{1}{3} - 0.3333250 \ldots \simeq 8.3 \times 10^{-6}$; so in this case (6.3) gives a precise estimate.

The analysis just presented traces the success of the midpoint formula to two sources:

1. Every differentiable function can be well approximated over short intervals by linear functions.
2. The midpoint formula is exact for linear functions.

In our study of the approximation of functions in Section 3.8 we found that, over short intervals, twice-differentiable functions can be exceedingly well approximated by quadratic functions. Therefore, if there were an approximation formula which is exact for quadratic functions, such a formula, when applied over a sufficiently fine partition, would yield an excellent approximation to the integral of any twice differentiable function.

We shall next construct an approximation to the integral which is exact for any quadratic function over any interval. For the sake of simplicity we take the underlying interval to be $[-1, 1]$. A quadratic function q is of the form

(6.4)
$$q(s) = ms^2 + ns + k.$$

Its integral is easily computed; q is the derivative of

$$\frac{m}{3} s^3 + \frac{n}{2} s^2 + ks.$$

205

Therefore, by the fundamental theorem,

$$(6.5) \quad \int_{-1}^{1} q \, ds = \frac{m}{3}(1)^3 + \frac{n}{2}(1)^2 + k(1) - \frac{m}{3}(-1)^3 - \frac{n}{2}(-1)^2 - k(-1)$$

$$= \frac{2m}{3} + 2k.$$

It follows from the definition (6.4) of q that k is the value of q at $s = 0$:

$$(6.6) \quad\quad\quad\quad\quad\quad k = q(0).$$

Since q is quadratic, its second derivative q'' has the same value,

$$(6.7) \quad\quad\quad\quad\quad\quad q'' = 2m,$$

at every point in $[-1, 1]$. We saw earlier (Section 3.8 and Exercise 8.5) that the second derivative can be approximated by the second difference quotient

$$\frac{q(1) - 2q(0) + q(-1)}{1^2} = \frac{1}{2}q''(c) + \frac{1}{2}q''(d),$$

where c and d are points of the interval $[-1, 1]$. Since in this case q'' has the same value at every point, given by (6.7), we have

$$(6.8) \quad\quad\quad q(1) - 2q(0) + q(-1) = 2m.$$

Now for k and $2m$ in (6.5) substitute the expressions (6.6) and (6.8), respectively, and obtain

$$(6.9) \quad\quad\quad \int_{-1}^{1} q \, ds = \frac{q(1) + q(-1)}{3} + \frac{4}{3}q(0).$$

Denote by a, b and c the left endpoint, midpoint and right endpoint of S, and by $|S|$ its length. For the interval $S = [-1, 1]$, $a = -1$, $b = 0$, $c = 1$ and $|S| = 2$, so (6.9) can be rewritten as follows:

$$\int_{a}^{c} q \, ds = [(\tfrac{1}{6}q(a) + \tfrac{2}{3}q(b) + \tfrac{1}{6}q(c)](c - a).$$

Let f be any function defined on an interval S; we denote by I_{Simp} the following approximation to the integral of f over $S = [a, c]$:

$$(6.10) \quad\quad I_{\text{Simp}} = [\tfrac{1}{6}f(a) + \tfrac{2}{3}f(b) + \tfrac{1}{6}f(c)](c - a).$$

The subscript "Simp" stands for Simpson, who invented this particular approximation. Let's see how well Simpson's rule works in a few examples.

Example 2. Take $f(s) = 1$ on the interval $S = (0, c)$. The integral of a constant function is just the product of the function value and the length of the interval of integration:

$$\int_{0}^{c} 1 \, ds = c.$$

Simpson's rule gives $(\tfrac{1}{6} + \tfrac{2}{3} + \tfrac{1}{6})c = c$, the exact value of the integral.

206

Example 3. Consider the function $f(s) = s$ on the interval $S = [0, c]$. The exact value of the integral (obtained by the fundamental theorem) is

$$\int_0^c s\, ds = \frac{c^2}{2}.$$

Simpson's rule gives

$$\left[\frac{1}{6}(0) + \frac{2}{3}\left(\frac{c}{2}\right) + \frac{1}{6}(c)\right]c = \left[\frac{1}{3} + \frac{1}{6}\right]c^2 = \frac{c^2}{2},$$

which is the exact value of the integral.

Example 4. Suppose $f(s) = s^2$ and $S = [0, c]$; the exact value of the integral is

$$\int_0^c s^2\, ds = \frac{c^3}{3}.$$

Simpson's rule gives

$$\left[\frac{1}{6}(0) + \frac{2}{3}\left(\frac{c}{2}\right)^2 + \frac{c^2}{6}\right]c = \left[\frac{1}{6} + \frac{1}{6}\right]c^3 = \frac{c^3}{3},$$

again the exact value of the integral.

That Simpson's rule gives exact answers for linear and quadratic functions is not surprising; after all, the approximation was designed to do just that.

Example 5. Now take $f(s) = s^3$ on the same interval as in the preceding examples. The exact integral is

$$\int_0^c s^3\, ds = \frac{c^4}{4};$$

Simpson's rule gives

$$\left[\frac{1}{6}(0) + \frac{2}{3}\left(\frac{c}{2}\right)^3 + \frac{c^3}{6}\right]c = \left[\frac{1}{12} + \frac{1}{6}\right]c^4 = \frac{c^4}{4}.$$

Simpson's rule strikes again and furnishes the exact answer.

Example 6. For $f(s) = s^4$ and $S = [0, c]$, the exact integral is

$$\int_0^c s^4\, ds = \frac{c^5}{5}.$$

However, Simpson's rule gives

$$\left[\frac{1}{6}(0) + \frac{2}{3}\left(\frac{c}{2}\right)^4 + \frac{c^4}{6}\right]c = \left[\frac{1}{24} + \frac{1}{6}\right]c^5 = \frac{5}{24}c^5.$$

At last we have found a case where the rule fails to give the exact value. The failure is not excessive; the percentage error, $[(\frac{5}{24} - \frac{1}{5})/\frac{1}{5}] \times 100$, is about 4.1%.

Consider $f(s)$, an arbitrary polynomial of degree 3:

(6.11) $$f(s) = a_0 + a_1 s + a_2 s^2 + a_3 s^3.$$

We have seen earlier that the integral depends linearly on the function f. Therefore for f of form (6.11),

(6.12) $$\int_a^c f \, ds = a_0 \int_a^c 1 \, ds + a_1 \int_a^c s \, ds + a_2 \int_a^c s^2 \, ds + a_3 \int_a^c s^3 \, ds.$$

Similarly, $I_{\text{Simp}}(f, S)$ defined by (6.10) depends linearly on f. Therefore for f of form (6.11),

(6.13) $$I_{\text{Simp}}(f, S) = a_0 I_{\text{Simp}}(1, S) + a_1 I_{\text{Simp}}(s, S) \\ + a_2 I_{\text{Simp}}(s^2, S) + a_3 I_{\text{Simp}}(s^3, S).$$

We have shown in Examples 2 to 5 that for the functions 1, s, s^2, and s^3, Simpson's rule furnishes the exact value for the integral. So it follows by comparing (6.12) and (6.13) that *Simpson's rule applied to any cubic polynomial furnishes the exact value of the integral*.

How well does Simpson's rule work for other functions? That depends on how closely these functions can be approximated by cubic polynomials. In Section 3.9 we have estimated how well n-times differentiable functions are approximated by Taylor polynomials of degree $n - 1$; that estimate can be used to judge the accuracy of Simpson's rule for 4-times differentiable functions. Rather than carry out the details we shall content ourselves with presenting a few numerical examples; many more will occur in Chapters 5 and 7.

Example 7. Evaluate the integral of $f(s) = s/(1 + s^2)$ over the interval $S = [0, 1]$, using Simpson's rule applied over the unit interval S without subdivision. Formula (6.10) gives the approximate value

$$\left[\frac{1}{2}(0) + \frac{2}{3}\left(\frac{\frac{1}{2}}{1 + \frac{1}{4}}\right) + \frac{1}{6}\left(\frac{1}{2}\right)\right] = \frac{7}{20} = 0.35$$

for the integral. Below we display the results of approximating this integral $I = \int_0^1 s/(1 + s^2) ds$ by Simpson's rule but with n equal subdivisions.

n	I_{Simp}
1	0.35
5	0.346577...
10	0.3465739...
100	0.3465735903....

Notice that the difference between the value of I_{Simp} for $n = 100$ and for $n = 1$ is about 1%. The exact value of I is $\frac{1}{2}\log 2 \simeq 0.346573590$.

Example 8. Using Simpson's rule, evaluate $\int_0^1 \sqrt{2 + s^2}\, ds$. The exact value of this integral can be expressed in terms of logarithms; its first 8 digits are

$$\int_0^1 \sqrt{2 + s^2}\, ds = 1.5245043\ldots.$$

The following table gives approximations using Simpson's rule with partition into n equal subintervals:

n	I_{Simp}
1	1.5243 ...
2	1.524495 ...
4	1.5245038 ...
8	1.5245043 ....

Thus we see that with only 4 subdivisions, Simpson's rule gives correctly the first 6 digits of the integral.

Example 9. Let $f(s) = \sqrt{1 - s^2}$ and compute $\int_0^{1/\sqrt{2}} f(s)ds$. The geometrical meaning of the integral

$$I = \int_0^{1/\sqrt{2}} \sqrt{1 - s^2}\, ds$$

is the area of the grey-shaded region shown in Figure 4.21. This grey-shaded area is the sum of the area marked A of a square of side length $1/\sqrt{2}$ and of a piece marked B bounded by two perpendicular segments and

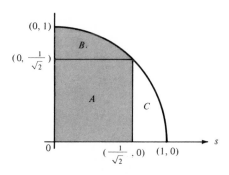

Figure 4.21

209

part of a circular arc of radius 1. The quarter circle is the union of A, B, and C. Since B and C are congruent, and the area of A is $\frac{1}{2}$, we get that

$$\text{area of quarter circle} = \frac{\pi}{4} = \text{area } (A) + \text{area } (B) + \text{area } (C)$$

$$= \tfrac{1}{2} + 2 \text{ area } (B).$$

Hence

$$\pi = 2 + 8 \text{ area } (B).$$

But

$$I = \text{area } (A) + \text{area } (B) = \tfrac{1}{2} + \text{area } (B),$$

so that

$$\text{area } (B) = I - \tfrac{1}{2}.$$

Thus

$$\pi = 2 + 8(I - \tfrac{1}{2}) = 8I - 2.$$

The table below, generated with the fourth computer program in the back of the book, lists various approximations to I and to $8I - 2$, using Simpson's rule with a partition of S into n equal subintervals:

n	I_{Simp}	$8I_{Simp} - 2 \simeq \pi$
1	0.6421 ...	3.137 ...
5	0.642697 ...	3.14158 ...
10	0.64269897 ...	3.141591 ...
100	0.642699081688 ...	3.14159265351 .... .

Observe that for $n = 10$, we get the first 6 digits of π correctly.

EXERCISES

6.1 Evaluate approximations to the integrals

(a) $\int_0^1 \frac{x}{\sqrt{1+x^2}}\, dx$
(b) $\int_0^1 \frac{1}{1+x^2}\, dx$

(c) $\int_0^2 \frac{1}{(1+x)^2}\, dx$
(d) $\int_{-1}^1 \frac{(1+x)^2}{\sqrt{1+x^2}}\, dx$

using the midpoint rule and then using Simpson's rule employing partitions into n equal subintervals, $n = 1, 5, 10, 100$. Use an electronic computer. Calculate, beg, borrow, or steal the exact values of these integrals, and compare the results of your numerical calculations with the exact answers.

6.2 Looking at Figure 4.21, we see that

$$\int_0^1 \sqrt{1-s^2}\, ds = \frac{\pi}{4}.$$

Approximate, using Simpson's rule, the integral on the left. You will observe that even if a large number of subintervals is used, the approximation to $\pi/4$ is poor. Can you explain why Simpson's rule works so poorly in this case?

6.3 Suppose that the function f is convex over $S = [a, b]$, i.e., that its second derivative is positive there. Using inequality (8.17) of Section 3.8, show that

$$\int_a^b f\, dx \geq I_{\mathrm{mid}}(f, S).$$

Can you give a geometric interpretation of this inequality?

6.4 Suppose that the function f is concave over $S = [a, b]$, i.e., its second derivative is negative there. Show that

$$\int_a^b f\, dx \leq I_{\mathrm{mid}}(f, S).$$

Interpret your result geometrically.

4.7* Improper integrals

In this section we study improper integrals. The first kind of "improper" integral arises if the interval of integration becomes infinite, e.g., if the endpoint b of $S = [0, b]$ tends to infinity.

The second kind of "improper" integral arises when a function becomes unbounded at some point P of the interval of integration. As we shall see later in this section, an "improper" integral of such a function f may nevertheless be defined, provided f is continuous in every closed subinterval of S not containing the point P.

Both kinds of "impropriety" of integrals will be illustrated in the examples that follow, and some of their uses will become apparent.

Example 1. Consider the integral

$$\int_1^b \frac{1}{x^2}\, dx.$$

The function $1/x^2$ is the derivative of $-1/x$; so the integral can be evaluated with the aid of the fundamental theorem:

$$\int_1^b \frac{1}{x^2}\, dx = -\frac{1}{b} - \left(-\frac{1}{1}\right) = -\frac{1}{b} + 1.$$

* This section may be omitted on first reading.

Let b tend to ∞; $1/b$ tends to zero, so we get the result

$$\lim_{b \to \infty} \int_1^b \frac{1}{x^2}\, dx = 1.$$

The limit on the left is denoted, not unreasonably, as

$$\int_1^\infty \frac{1}{x^2}\, dx.$$

More generally, whenever $\int_a^b f(x)dx$ tends to a limit as b tends to ∞, that is, whenever

(7.1)
$$\lim_{b \to \infty} \int_a^b f(x)dx$$

exists, that limit is denoted by

(7.2)
$$\int_a^\infty f(x)dx.$$

Such an integral is called an *improper integral*, and the function f is said to be *integrable on* $[a, \infty)$. If the limit (7.1) does not exist, we say that the function f is not integrable on $[a, \infty)$.

Example 2. We shall calculate the improper integral

$$\int_1^\infty \frac{1}{x^3}\, dx.$$

By the fundamental theorem,

$$\int_1^b \frac{1}{x^3}\, dx = -\frac{1}{2b^2} + \frac{1}{2}.$$

As b tends to ∞, $1/2b^2$ tends to zero, so we conclude that

$$\int_1^\infty \frac{1}{x^3}\, dx = \frac{1}{2}.$$

Example 3. Let n be any number greater than 1; since $1/x^n$ is the derivative of $-1/(n-1)x^{n-1}$, we get from the fundamental theorem that

$$\int_1^b \frac{1}{x^n}\, dx = -\frac{1}{n-1}\frac{1}{b^{n-1}} + \frac{1}{n-1}.$$

Letting $b \to \infty$ we deduce that, for $n > 1$,

$$\int_1^\infty \frac{1}{x^n}\, dx = \frac{1}{n-1}.$$

Example 4. In this example we show that the improper integral

$$\int_1^\infty \frac{1}{x} dx$$

does not exist. To see this we rely on the lower and upper bound property of the integral

$$\int_a^b f(x)dx > (b - a)m,$$

where m is the minimum of f in $[a, b]$. Apply this to the integral

$$\int_a^{2a} \frac{1}{x} dx;$$

clearly, the minimum of $1/x$ occurs at the right endpoint of the interval of integration, so

$$\int_a^{2a} \frac{1}{x} dx > a\frac{1}{2a} = \frac{1}{2}, \quad \int_{2a}^{4a} \frac{1}{x} dx > 2a\frac{1}{4a} = \frac{1}{2}, \quad \text{etc.}$$

This, together with the additivity of integrals, allows us to write the integral $1/x$ over the interval $[1, 2^k]$ as the following sum:

$$\int_1^{2^k} \frac{1}{x} dx = \int_1^2 + \int_2^4 + \int_4^8 + \cdots + \int_{2^{k-1}}^{2^k} \frac{1}{x} dx$$

$$> \frac{1}{2} + \frac{1}{2} + \frac{1}{2} + \cdots + \frac{1}{2} = \frac{k}{2}.$$

This estimate shows that as k tends to ∞,

$$\int_1^{2^k} \frac{1}{x} dx$$

also tends to ∞, and therefore does not have a limit. Thus the function $1/x$ is not integrable on $[1, \infty)$.

The adjective in the phrase "improper integral" is not pejorative. Let us revert for a moment to our earlier notation for the integral:

$$\int_a^b f(x)dx = I(f, S), \quad S = [a, b].$$

One should think of an improper integral as $I(f, S)$, where at least one endpoint of S is infinitely far away. It follows from the definition given above that $I(f, S)$ has all those properties of the integral which make sense when S is an unbounded interval:

(i) *Additivity.* If $S = S_1 + S_2$, $I(f, S) = I(f, S_1) + I(f, S_2)$.
(ii) *Linearity.* $I(f + g, S) = I(f, S) + I(g, S)$, $I(kf, S) = kI(f, S)$.
(iii) *Monotonicity.* If $f(x) \leq g(x)$ for all x in S, then $I(f, S) \leq I(g, S)$.

We leave it to the reader to verify that improper integrals have these properties.

Now we present a very useful criterion, based on the monotonicity of the integral, for deciding which functions are integrable and which are not. We start by examining what it means that the limit

$$\lim_{b \to \infty} \int_a^b f(x)dx$$

exists. The meaning is that for b large enough, the value of the integral on $[a, b]$ hardly depends on b; more precisely, for any choice of m

(7.3) $$\left| \int_a^{b_1} f(x)dx - \int_a^{b_2} f(x)dx \right| < 10^{-m}$$

for b_1 and b_2 large enough; how large depends of course on m (cf. Section 1.5).
    The first basic property of the integral is additivity. Suppose $b_2 > b_1$, so that $[a, b_2] = [a, b_1] + [b_1, b_2]$. Then

$$\int_a^{b_2} = \int_a^{b_1} + \int_{b_1}^{b_2}.$$

Therefore the above difference of integrals can be rewritten as a single integral over $[b_1, b_2]$, and (7.3) becomes

(7.4) $$\left| \int_{b_1}^{b_2} f(x)dx \right| < 10^{-m}.$$

Suppose g is a positive function bigger than $|f|$ for all $x > a$:

(7.5) $$|f(x)| < g(x) \quad \text{for } x > a.$$

According to the monotonicity of the integral,

(7.6) $$\left| \int_{b_1}^{b_2} f(x)dx \right| < \int_{b_1}^{b_2} g(x)dx.$$

Suppose now that g is integrable from a to ∞. Then according to the analysis previously presented, for any choice of m, the right hand side of (7.6) satisfies

$$\int_{b_1}^{b_2} g(x)dx < 10^{-m},$$

for b_1, b_2 large enough. Then it follows from (7.6) that also

$$\left| \int_{b_1}^{b_2} f(x)dx \right| < 10^{-m},$$

which means that f is integrable on $[a, \infty)$. This result is called the

Comparison theorem for improper integrals. *If* $|f(x)| \le g(x)$ *on* $[a, \infty)$, *and if* g *is integrable on* $[a, \infty)$, *then so is* f.

This theorem can be used as an integrability criterion for f, or a non-integrability criterion for g. We shall illustrate both applications in a few examples.

Example 5. Is the function $1/(1 + x^2)$ integrable on $[1, \infty)$? To answer this question, observe that the function under discussion is obviously less than $1/x^2$:

$$\frac{1}{1 + x^2} < \frac{1}{x^2}.$$

According to the calculations in Example 1, the function $1/x^2$ is integrable. Therefore it follows from the comparison theorem that $1/(1 + x^2)$ is also integrable.

Example 6. Is the function $1/x\sqrt{1 + x}$ integrable on $[1, \infty)$? It is obvious that

$$\frac{1}{x\sqrt{1 + x}} < \frac{1}{x\sqrt{x}} = \frac{1}{x^{3/2}}.$$

According to Example 3, the function $x^{-3/2}$ is integrable on $[1, \infty)$. Therefore the function $1/x\sqrt{1 + x}$ is also integrable.

Example 7. Is the function $x/(1 + x^2)$ integrable on $[1, \infty)$? Since $x^2 \ge 1$ on the interval in question, we have the inequality

$$\frac{x}{1 + x^2} \ge \frac{x}{x^2 + x^2} = \frac{1}{2x} \qquad \text{for } x \ge 1.$$

The function $1/2x$ is just half the function $1/x$ which, according to Example 4, is not integrable; therefore it follows from the comparison theorem that neither is the larger function $x/(1 + x^2)$.

There is another variant of the comparison theorem which can be used as a criterion for the convergence or divergence of infinite series discussed in Chapter 1:

Integral test for the convergence of a series

(i) *Let* $f(x)$ *be a positive, decreasing function defined and integrable on* $[1, \infty)$. *Let* $\sum_1^\infty a_n$ *be an infinite series whose terms satisfy the inequalities*

(7.7) $$|a_n| \le f(n).$$

Then the series $\sum_1^\infty a_n$ *converges.*

215

(ii) *Let* $f(x)$ *be a positive, decreasing function defined on* $[1, \infty)$. *Let* $\sum_1^\infty a_n$ *be a convergent infinite series whose terms satisfy the inequalities*

(7.8) $$a_n \geq f(n).$$

Then f *is integrable on* $[1, \infty)$.

PROOF. Since $f(x)$ is assumed to be decreasing, its minimum on the interval $[n-1, n]$ occurs at the right endpoint n. By the lower bound property of the integral,

$$f(n) \leq \int_{n-1}^n f(x)dx.$$

Using assumption (7.7) we conclude therefore that

$$a_n \leq \int_{n-1}^n f(x)dx.$$

Add these inequalities for all n between j and k; using the additive property of the integral, we get

(7.9) $$|a_{j+1}| + \cdots + |a_k| \leq \int_j^{j+1} f(x)dx + \cdots + \int_{k-1}^k f(x)dx = \int_j^k f(x)dx.$$

The function f is assumed to be integrable on $[1, \infty)$. That means that inequality (7.4) holds for b_1, b_2 large enough. Taking j to be b_1, k to be b_2 we see that the right side of (7.9) is less than 10^{-m} for k, j large enough. So it follows from (7.9) that no matter how large m,

(7.10) $$|a_{j+1}| + \cdots + |a_k| \leq 10^{-m}$$

for j, k large enough. According to the convergence criterion for series, see Section 1.5, (7.10) implies that the series $\sum a_j$ converges. This proves part (i).

To prove part (ii), we observe that, since f is decreasing, its maximum on the interval $(n, n+1)$ is reached at n. By the upper bound property of the integral

$$f(n) \geq \int_n^{n+1} f(x)dx.$$

Using assumption (7.8) we conclude that

$$a_n \geq \int_n^{n+1} f(x)dx.$$

Add these inequalities for all n between j and k. Using the additive property of the integral, we get

(7.11) $$a_j + \cdots + a_{k-1} \geq \int_j^{j+1} f(x)dx + \cdots + \int_{k-1}^k f(x)dx = \int_j^k f(x)dx.$$

We have assumed that the series $\sum a_n$ converges; that means, as we have seen in Chapter 1, that no matter how large m,

$$a_j + \cdots + a_{k-1} \leq 10^{-m}$$

for j, k large enough. It follows then from (7.11) that

$$\int_j^k f(x)dx \leq 10^{-m}.$$

According to (7.4) this implies that f is integrable on $[1, \infty)$. This completes the proof of the integral test for convergence. □

The integral test is enormously useful in applications. Here are some examples.

Example 8. $f(x) = 1/x^p, p > 1$. In this case we see from Example 4 that f is integrable on $[1, \infty)$. Set

$$a_n = f(n) = \frac{1}{n^p}.$$

It follows from part (i) of the integral test for convergence that $\sum 1/n^p$ converges for $p > 1$.

Example 9. Take $f(x) = 1/x$, and set $a_n = f(n) = 1/n$. We claim that the infinite series

$$\sum_1^\infty \frac{1}{n}$$

diverges. For if it converged, then according to part (ii) of the integral test for convergence the function $f(x) = 1/x$ would be integrable over $[1, \infty)$, whereas according to Example 4 it is not. Another convergence proof for the harmonic series $\sum 1/n$ was given in Example 6, Section 1.5.

Next we show how the introduction of $z = 1/x$ as new variable of integration can change an improper integral into a proper one. According to the change of variable formula for integrals described in Section 4.5,

$$I = \int_a^b f(x)dx = \int_{z(a)}^{z(b)} f[x(z)] \frac{dx}{dz} dz.$$

Taking $x = 1/z$, we have $dx/dz = -1/z^2$, so we obtain

$$I = \int_{1/a}^{1/b} f[x(z)] \left(-\frac{1}{z^2}\right) dz,$$

and (see Exercise 5.4) we have

(7.12) $$I = \int_{1/b}^{1/a} f\left(\frac{1}{z}\right) \frac{1}{z^2} dz.$$

Suppose that the function $f(x)$ is such that, as x tends to ∞, $x^2 f(x)$ tends to a finite limit L. Then the change of variable $x = 1/z$ leads to the function $f(1/z)/z^2$ which tends to the same limit L as z tends to zero. It follows that, as b tends to ∞, the integral (7.12) tends to the proper integral

$$\int_0^{1/a} \frac{1}{z^2} f\left(\frac{1}{z}\right) dz .$$

The next example makes use of such a change of variable.

Example 10. Evaluate

$$\int_1^\infty \frac{1}{1 + x^2} dx.$$

Putting $f(x) = 1/(1 + x^2)$ into (7.12), we get

$$\int_1^b \frac{1}{1 + x^2} dx = \int_{1/b}^1 \frac{1}{1 + (1/z)^2} \frac{1}{z^2} dz = \int_{1/b}^1 \frac{1}{z^2 + 1} dz.$$

As b tends to ∞, the integral on the right tends to the perfectly proper integral

$$\int_0^1 \frac{1}{z^2 + 1} dz.$$

Example 11. Make the improper integral

$$\int_1^\infty \frac{x}{x^3 + 1} dx$$

proper by introducing, as before, $z = 1/x$ as a new variable of integration. Using (7.12), we get

$$\int_1^b \frac{x}{x^3 + 1} dx = \int_{1/b}^1 \frac{1/z}{(1/z)^3 + 1} \frac{1}{z^2} dz = \int_{1/b}^1 \frac{1}{z^3 + 1} dz.$$

As b tends to ∞, this last integral tends to the very proper integral

$$\int_0^1 \frac{1}{z^3 + 1} dz .$$

We turn now to another class of integrals also called "improper," with the feature that the integrand is not bounded over the interval of integration.

Example 12. Consider the integral

$$\int_a^1 \frac{1}{\sqrt{x}} dx .$$

Since the function $1/\sqrt{x}$ is the derivative of $2\sqrt{x}$, the fundamental theorem yields

$$\int_a^1 \frac{1}{\sqrt{x}}\,dx = 2 - 2\sqrt{a}\ .$$

Now let a tend to zero; then $\sqrt{a}$ also tends to zero, and

$$\lim_{a\to 0} \int_a^1 \frac{1}{\sqrt{x}}\,dx = 2\ .$$

The limit on the left is denoted, not unreasonably, by

$$\int_0^1 \frac{1}{\sqrt{x}}\,dx.$$

Although the function $1/\sqrt{x}$ becomes unbounded as x tends to zero, its integral exists.

More generally, let f be a function defined on a half open interval $S = (a, b]$, such that f is continuous on every subinterval $(a + h, b]$, $h > 0$, but not on the interval S itself. The function f is called *integrable on* $(a, b]$ if the limit

$$\lim_{h\to 0} \int_{a+h}^b f(x)dx$$

exists; this limit is denoted by

$$\int_a^b f(x)dx$$

and is called an *improper integral*. If the limit does not exist, we say that f is not integrable on $(a, b]$.

Example 13. Consider $\int_0^1 dx/x$; the function $1/x$ is not defined at the left endpoint $x = 0$ of the interval of integration $S = [0, 1]$. We cut off the point $x = 0$ by introducing some positive number $h < 1$, and we integrate over the modified interval $S_h = [h, 1]$; that is, we consider the integral $\int_h^1 dx/x$.

Now the function $1/x$ is the derivative of $\log x$ (as we shall show in a later chapter), so that the fundamental theorem yields

$$\int_h^1 \frac{dx}{x} = \log 1 - \log h = \log \frac{1}{h}.$$

As $h \to 0$, $\log(1/h)$ becomes arbitrarily large, so the integral *diverges*. Figure 4.22 shows the graph of the integrand, a hyperbola. The shaded area is the integral over S_h.

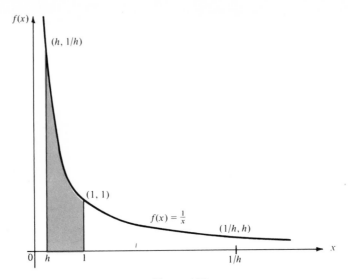

Figure 4.22

Example 14. Evaluate the improper integral

$$\int_0^1 \frac{1}{x^p}\,dx, \qquad 0 < p < 1.$$

Since the function $1/x^p$ is the derivative of $x^{1-p}/(1-p)$, it follows from the fundamental theorem that

$$\int_a^1 \frac{1}{x^p}\,dx = \frac{1}{1-p} - \frac{a^{1-p}}{1-p}.$$

As a tends to zero, so does a^{1-p} for $p < 1$; therefore

$$\int_0^1 \frac{1}{x^p}\,dx = \frac{1}{1-p}, \quad \text{for all } p < 1.$$

The comparison criterion applies also to improper integrals of this kind: If $|f(x)| \le g(x)$ on S, and if g is integrable on S, then f is also integrable on S.

Example 15. Consider the improper integral $\int_0^1 dx/\sqrt{x+x^2}$. The integrand satisfies the inequality $1/\sqrt{x+x^2} < 1/\sqrt{x}$. Since, according to Example 12, $1/\sqrt{x}$ is integrable on $(0, 1]$, so is $1/\sqrt{x+x^2}$.

Example 16. Consider the integral $I = \int_1^3 dx/(x-2)^{1/3}$, whose integrand is not defined at the interior point $x = 2$ of the interval of integration $S = [1, 3]$. Here we divide S into two subintervals, $[1, 2]$ and $[2, 3]$, and write

$$I = \int_1^2 \frac{dx}{(x-2)^{1/3}} + \int_2^3 \frac{dx}{(x-2)^{1/3}}.$$

In each of these integrals, the integrand is undefined at an endpoint of the interval, hence can be handled by the method of Example 13. Thus, for $0 < h < 1$, the fundamental theorem yields

$$\int_1^{2-h} \frac{dx}{(x-2)^{1/3}} = \tfrac{3}{2}[(2-h)-2]^{2/3} - \tfrac{3}{2}[1-2]^{2/3} = \tfrac{3}{2}[(-h)^{2/3}-1]$$

and

$$\int_{2+h}^3 \frac{dx}{(x-2)^{1/3}} = \tfrac{3}{2}[3-2]^{2/3} - \tfrac{3}{2}[(2+h)-2]^{2/3} = \tfrac{3}{2}[1-h^{2/3}].$$

When $h \to 0$, the first integral tends to $-\tfrac{3}{2}$, the second to $\tfrac{3}{2}$, so that the value of I is 0. We conclude that $(x-2)^{-1/3}$ is integrable over $[1,3]$, and we show the graph of the integrand in Figure 4.23. Observe that the integrand is

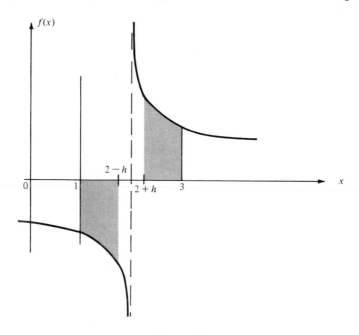

Figure 4.23

unbounded at the right endpoint of $[1, 2]$ and the left endpoint of $[2, 3]$; yet each of the integrals converges. Thus, the shaded areas in Figure 4.23 are finite, in contrast to the situation in Example 13, where the shaded area in Figure 4.22 becomes arbitrarily large as $h \to 0$.

In conclusion we show via an example how to turn improper integrals into proper ones by integrating by parts.

Example 17. Turn

$$\int_0^1 \frac{1}{\sqrt{x}} \frac{1}{x+1}\, dx$$

into a "proper" integral. Integration by parts asserts that

$$\int_a^1 f'g\, dx = f(1)g(1) - f(a)g(a) - \int_a^1 fg'\, dx.$$

Take $f' = 1/\sqrt{x}$ and $g(x) = 1/(x+1)$; then $f(x) = 2\sqrt{x}$. We obtain

$$\int_a^1 \frac{1}{\sqrt{x}} \frac{1}{x+1}\, dx = \frac{2}{2} - \frac{2\sqrt{a}}{a+1} + \int_a^1 2\sqrt{x}\frac{1}{(x+1)^2}\, dx.$$

As a tends to 0, the right side tends to the quantity

$$1 + 2\int_0^1 \frac{\sqrt{x}}{(x+1)^2}\, dx$$

whose second term is an integral of a bounded function over a finite interval. Thus integration by parts can turn an improper integral into a proper one.

EXERCISES

7.1 Which of the functions below are integrable on $(0, \infty)$?

(a) $\dfrac{1}{\sqrt{x}(1+x)}$ (b) $\dfrac{x}{1+x^2}$ (c) $\dfrac{\sqrt{x+1}-\sqrt{x}}{1+x}$

(d) $\dfrac{1}{x+x^4}$ [*Hint*: Use the comparison theorem. Treat the two endpoints separately.]

7.2 Let $q(x)$ be a polynomial of degree n, $n \geq 2$, and suppose that $q(x)$ is nonzero for $x \geq 1$. Let $p(x)$ be a polynomial of degree $\leq n-2$; show that the introduction of $z = 1/x$ as new variable of integration turns the improper integral $\int_1^\infty (p(x)/q(x))\, dx$ into a proper one.

7.3 (a) Show, by the change of variables $z = 1/x$, that

$$\int_1^\infty \frac{1}{1+x^2}\, dx = \int_0^1 \frac{1}{1+z^2}\, dz.$$

(b) Evaluate the integral on the right by Simpson's rule, $n = 1$.
(c) Evaluate approximately

$$\int_{-\infty}^\infty \frac{1}{1+x^2}\, dx.$$

How close is your approximation to π?

7.4 Integrate by parts to turn the improper integral $\int_0^1 dx/\sqrt{x+x^2}$ into a proper one.

7.5 Integrate by parts to verify that the improper integral

$$\int_1^\infty \frac{\cos x}{x}\,dx$$

exists.

7.6 Show that the following improper integrals exist, and evaluate them.

(a) $\displaystyle\int_1^\infty \frac{dx}{x^{1.0001}}$ 　　　　(b) $\displaystyle\int_0^4 \frac{dx}{\sqrt{4-x}}$

(c) $\displaystyle\int_0^1 \frac{dx}{x^{0.9999}}$ 　　　　(d) $\displaystyle\int_{-1}^1 \frac{dx}{x^{2/3}}$.

5 Growth and decay

Let us call addition, subtraction, multiplication, division, composition, and inversion of functions "algebraic" operations. All functions that can be constructed out of the identity function $f(x) = x$ by algebraic operations are called elementary, all others are called transcendental. This chapter deals with the logarithmic and exponential functions. They are transcendental from the algebraic point of view. As we shall see, they are elementary from the point of view of calculus because they satisfy extremely simple differential equations.

5.1 The exponential function

It is a well known fact that radioactive elements are not immutable, but with the passage of time change into other elements. We shall study the time history of this change. Specifically we ask this question: If we start out with a given mass of radioactive material, how much of it will be left after time t has elapsed? To give a mathematical treatment we introduce the following function:

$M(t)$ *is the mass of material left after time t of an initial supply of* 1 *unit of mass.*

Suppose we start out with an initial supply of A units; how much will be left after time t? Assuming that the various atoms have no influence on each other, we conclude that the amount left is proportional to the initial amount, so that

$$AM(t)$$

is the mass left after time t.

How much is left of 1 unit of material after time $s + t$? There are two ways of answering this question; one answer is $M(s + t)$. The other is obtained by observing that after time s we are left with the mass $A = M(s)$; after t additional units of time this amount is reduced to

$$AM(t) = M(s)M(t).$$

Since the two answers must be the same, we conclude that

(1.1)
$$\boxed{M(s + t) = M(s)M(t)}$$

for all positive values of s and t. We now turn to another example, that of bacteria multiplying.

We denote by $P(t)$ the size of a bacterial population which results when a colony of unit size is allowed to grow for time t in a surrounding of ample nutrients.

If the size of the initial colony is A, then, assuming that the bacteria do not hamper each other's rate of reproduction (it is here that the ampleness of food supply comes in), we conclude that the size of the colony after time t is proportional to the initial amount, so that it has grown to

$$AP(t).$$

What will be the size after time $s + t$ of a bacterial population whose initial size is 1 unit? There are two ways of computing the answer; the first is to call it $P(s + t)$. The second is to observe that after time s the population has grown to size $A = P(s)$, and that the additional growth during the next time interval of length t results in the total amount of

$$AP(t) = P(s)P(t).$$

Since the two answers must be the same we conclude that

(1.2)
$$\boxed{P(s + t) = P(s)P(t)}$$

for all positive values of s and t. Equations (1.1) and (1.2) relate the values of the functions M and P, respectively, at three points s, t and $s + t$. Such an equation is called a *functional equation*. Comparing (1.1) and (1.2) we see that M and P satisfy the same functional equation.

The functional equation (1.2) is pretty simple; we shall deduce from it an even simpler relation between the function P and its derivative. Subtract $P(t)$ from both sides of (1.2):

$$P(s + t) - P(t) = (P(s) - 1)P(t).$$

Divide both sides by s:

(1.3)
$$\frac{P(s + t) - P(t)}{s} = \frac{P(s) - 1}{s} P(t).$$

225

Recall that at time $t = 0$, the population had unit size, so that

(1.4) $$P(0) = 1.$$

With this initial value, (1.3) can be rewritten as

(1.3)$_0$ $$\frac{P(s + t) - P(t)}{s} = \frac{P(s) - P(0)}{s} P(t).$$

Assume that P is differentiable; then as s tends to 0, the left side of (1.3) tends to $P'(t)$, while the right side tends to $P'(0)P(t)$. Abbreviate $P'(0)$ by a; the limit of (1.3)$_0$ can then be written as

(1.5) $$\boxed{P' = aP}.$$

A relation like (1.5) between a function P and its derivative is called a *differential equation*. A specification like (1.4), fixing the value of P at $t = 0$ is called an *initial condition*.

We shall now show that *all properties of the function P are contained in the differential equation (1.5) and initial condition (1.4)*.

(i) Uniqueness

Let $Q(t)$ be another function satisfying (1.4) and (1.5); form the function

(1.6) $$F(t) = Q(t)P(u - t), \quad u \text{ an arbitrary constant.}$$

The derivative of this function is

$$F'(t) = Q'(t)P(u - t) - Q(t)P'(u - t);$$

using the differential equation satisfied by Q and P, we can rewrite this as

$$F'(t) = aQ(t)P(u - t) - Q(t)aP(u - t)$$

which is zero. Therefore the function $F(t)$ defined by (1.6) is a constant; in particular its value at $t = 0$ equals its value at $t = u$:

$$Q(0)P(u) = Q(u)P(0).$$

Since by (1.4), both $Q(0)$ and $P(0)$ are equal to 1, we see that

$$P(u) = Q(u),$$

i.e., P and Q have the same value at u. Since u is arbitrary, this shows that P and Q are the same everywhere. Thus (1.4) and (1.5) uniquely characterize the function P. Therefore all properties of the function P are logical consequences of (1.4) and (1.5).

We shall now show that the *interesting properties of the function P* can be easily extracted from (1.4) and (1.5).

(ii) The functional equation

Form the function

(1.7) $$P(t)P(u - t).$$

Its derivative is

$$P'(t)P(u - t) - P(t)P'(u - t),$$

which by the differential equation (1.5) is equal to

$$aP(t)P(u - t) - P(t)aP(u - t),$$

and this is zero. Therefore the function (1.7) is a constant; its value at any point t equals its value at $t = 0$:

$$P(t)P(u - t) = P(0)P(u).$$

Since by (1.4), $P(0) = 1$, we get that

(1.8) $$P(t)P(u - t) = P(u).$$

Setting $u = s + t$ in this relation gives the functional equation (1.2).

(iii) Positivity

We claim that $P(t)$ is positive for all values of t. First we show that $P(t)$ is never zero. Set $u = 0$ in the functional equation (1.8); from the initial condition $P(0) = 1$, we get

(1.9) $$P(t)P(-t) = 1,$$

which shows that $P(t) \neq 0$. In fact, (1.9) shows that $P(t)$ and $P(-t)$ are reciprocals of each other. To complete the proof of positivity we have to show that $P(t)$ cannot be negative. Since $P(0)$ is positive, and since $P(t)$ is continuous (even differentiable), if P were negative anywhere, it would follow from the intermediate value theorem that P is zero somewhere; but that has already been shown to be impossible.

Denote the values of P at $t = 1$ by b:

$$P(1) = b.$$

Repeated application of the functional equation (1.2) shows that for every positive integer n,

$$P(n) = P(1 + 1 + \cdots + 1) = b^n.$$

Another m-fold application of (1.2) gives

$$P(n) = P\left(\frac{n}{m} + \frac{n}{m} + \cdots + \frac{n}{m}\right) = P^m\left(\frac{n}{m}\right) = b^n.$$

Since P is always positive, the real m-th power of $P(n)$ is defined; so

(1.10) $$P(t) = b^t$$

holds for every positive rational number t. By (1.9) it holds also for negative rational numbers. So we have shown that functions P which satisfy the functional equation (1.2) are *exponential functions*.

We now consider a particular exponential function, one that satisfies the differential equation (1.5) with $a = 1$, as well as the initial condition (1.4). Denote this function by $E(t)$:

(1.11)
$$E'(t) = E(t), \qquad E(0) = 1$$

The relations (1.11) uniquely characterize the function E, called *the exponential function* and denoted in the exponential notation by

(1.12)
$$E(t) = e^t,$$

sometimes also written as

(1.12)′
$$E(t) = \exp\{t\}.$$

The *number e* occurring in (1.12) is nothing but the value of the function E at $t = 1$; in Section 5.3 we shall describe several ways of calculating its value. Figure 5.1 shows some of the basic properties of the function E.

It follows from (1.11) that the solution P of (1.4) and (1.5) can be expressed in terms of the exponential function E as follows:

(1.13)
$$P(t) = E(at).$$

We proceed to derive further properties of the exponential function.

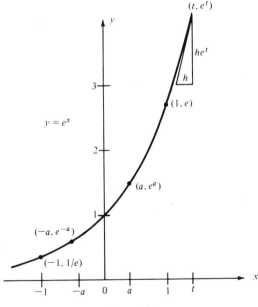

Figure 5.1

(iv) Monotonicity

According to the differential equation (1.11), $E' = E$; and according to (iii), E is positive. Therefore E has positive derivative, and so is an *increasing function*.

The function $P(t)$, measuring the growth of bacterial population, is an increasing function; it follows therefore from (1.13) that the number a is positive. On the other hand, the function $M(t)$, measuring radioactive decay, is a decreasing function. Therefore in the differential equation satisfied by M,

$$M' = bM,$$

the constant b, the value of M' at 0, is negative.

How fast does the function E increase? Pretty fast, as we shall now show.

(v) Behavior of the exponential function near ∞ and 0

As t tends to infinity, e^t increases *faster than any power of* t. The meaning of this statement is that no matter how large you take n,

(1.14)
$$\frac{e^t}{t^n}$$

tends to ∞ as $t \to \infty$.

To prove this, differentiate the function (1.14); we get

$$\frac{e^t}{t^n} - \frac{ne^t}{t^{n+1}} = \left(1 - \frac{n}{t}\right)\frac{e^t}{t^n}.$$

The first factor on the right, $1 - n/t$, is positive for $t > n$; the remaining factor is always positive for $t > 0$, so it follows that the function (1.14) has positive derivative for $t > n$. Therefore, for $t > n$, (1.14) is an increasing function; in particular, its value at t is greater than its value at n when $t > n$, so that

$$\frac{e^t}{t^n} > \frac{e^n}{n^n} \quad \text{for } t > n.$$

Multiplying this inequality by t, we see that

$$\frac{e^t}{t^{n-1}} > \frac{e^n}{n^n}t \quad \text{for } t > 0.$$

As t tends to ∞, the function $e^n t/n^n$ tends to ∞; therefore so does the function e^t/t^{n-1} on the left. Since n is an arbitrary integer, this proves the contention.

How does e^{-t} behave as t tends to ∞? It follows from (1.9), with $P(t) = e^t$, that

$$e^{-t} = \frac{1}{e^t}.$$

229

It follows therefore from property (v) that e^{-t} tends to zero faster than any negative power of t, i.e., that

(1.15)
$$t^n e^{-t}$$

tends to zero as t tends to ∞ for any n.

We now turn to integrating the function e^{-t}; since e^{-t} is the derivative of $-e^{-t}$, we deduce from the fundamental theorem of calculus that

$$\int_0^b e^{-t}\, dt = -e^{-b} + e^0 = 1 - e^{-b}.$$

Now let b tend to ∞; according to (1.15), e^{-b} tends to 0, so we deduce that

(1.16)
$$\int_0^\infty e^{-t}\, dt = 1,$$

an amusing answer. For more amusement we turn to the integral

$$\int_0^b t e^{-t}\, dt.$$

We shall integrate by parts, i.e., use the identity

$$\int_a^b f g'\, dt = f(b)g(b) - f(a)g(a) - \int_a^b f' g\, dt.$$

Setting $f(t) = t$, and $g'(t) = e^{-t}$, we obtain, with $a = 0$,

$$\int_0^b t e^{-t}\, dt = -b e^{-b} + \int_0^b e^{-t}\, dt.$$

As before, let b tend to infinity; by (1.15) the first term on the right tends to zero, so we get

$$\int_0^\infty t e^{-t}\, dt = \int_0^\infty e^{-t}\, dt.$$

Using (1.16) we deduce that

(1.17)
$$\int_0^\infty t e^{-t}\, dt = 1,$$

another amusing result.

Emboldened by this success we turn to evaluating the integrals

(1.18)
$$I_n = \int_0^b t^n e^{-t}\, dt$$

for any positive integer value of n. We integrate by parts, setting $f(t) = t^n$ and $g'(t) = e^{-t}$, so that $f'(t) = nt^{n-1}$ and $g(t) = -e^{-t}$, and obtain

$$I_n = \int_0^b t^n e^{-t}\, dt = -b^n e^{-b} + \int_0^b nt^{n-1} e^{-t}\, dt.$$

As b tends to ∞, the first term on the right tends to zero, so

$$\int_0^\infty t^n e^{-t}\, dt = n \int_0^\infty t^{n-1} e^{-t}\, dt;$$

using the abbreviation introduced in (1.18) we can write this as

(1.19) $$I_n = nI_{n-1}.$$

Since we already know from (1.17) that $I_1 = 1$, we get from (1.19), with $n = 2$, that

$$I_2 = 2I_1 = 2.$$

Setting $n = 3$ in (1.19) we get

$$I_3 = 3I_2 = 6.$$

So, using (1.19) we can, step by step, determine the value of I_n:

$$I_n = nI_{n-1} = n(n-1)I_{n-2} = \cdots = n(n-1)\cdots 2\cdot 1.$$

We recall that the product of all positive integers from 1 to n is denoted by $n!$, pronounced n factorial. So we have shown that

(1.20) $$\int_0^\infty t^n e^{-t}\, dt = n!.$$

This is a very interesting result; for instance, whereas the factorial on the right is defined only for positive integer values of n, the left side is defined for any positive rational n. Therefore Formula (1.20) serves to define the factorial function $n!$ for noninteger n, a remarkable feat.

Even for integer values of n, the integral expression on the left can be used to give a better idea of the size of $n!$ than the original definition as an n-fold product; Exercise 1.5 at the end of this section contains information on this point.

In summary, in this section we have derived a large number of properties of the exponential function e^t, such as its functional equation, how fast it grows as $t \to \infty$, how to differentiate and integrate it. What we don't yet know about e^t is: does it exist, and how do we compute its value at any particular point? The next two sections will be devoted to these problems.

EXERCISES

1.1 Differentiate the following functions:

(a) $\dfrac{1 - e^t}{1 + e^t}$

(b) $\sqrt{1 + e^t}$

(c) e^{-t^2}

(d) $e^{\sqrt{t}}$

(e) $\dfrac{e^{-t^2}}{\sqrt{t}}$

(f) $e^{f(t)}$, where f is any differentiable function.

231

1.2 We define the functions $C(t)$ and $S(t)$ as follows:

$$C(t) = \frac{e^t + e^{-t}}{2}, \qquad S(t) = \frac{e^t - e^{-t}}{2}.$$

Prove that:
(a) $C' = S, \ S' = C$ (b) $C^2 - S^2 = 1$
(c) $C(t + s) = C(t)C(s) + S(t)S(s)$
    $S(t + s) = C(t)S(s) + S(t)C(s)$.

1.3 Evaluate the following integrals:

(a) $\displaystyle\int_0^\infty t e^{-t^2/2} \, dt$ (b) $\displaystyle\int_0^\infty t^3 e^{-t^2/2} \, dt$

(c) $\displaystyle\int_0^\infty t^{2q+1} e^{-t^2/2} \, dt$,   where q is any positive integer.

1.4 Show that the function $f(t) = t^n e^{-t}$, n a positive integer, achieves its maximum at $t = n$; therefore its values at $t = n - 1$ and $t = n + 1$ are less than the maximum. From this fact deduce that

(1.21) $\left(\dfrac{n+1}{n}\right)^n < e < \left(\dfrac{n}{n-1}\right)^n.$

1.5 (a) Using (1.20) show that

$$\int_n^{n+1} t^n e^{-t} \, dt < n! \ .$$

(b) Show that

$$(n + 1)^n e^{-n-1} < n! \ .$$

(c) Find the value of both sides of the inequality in (b) for $n = 3, 7, 10$ using a table of values of e^n, or using the approximate value 2.718 for e.

1.6 Let P be a continuous function which satisfies the functional equation

(1.22) $P(t + s) = P(t)P(s)$

for all t and s.
(a) Show that, unless $P(t) = 0$ for all t, $P(0) = 1$.
(b) Let t be any positive number, n any positive integer, and N the largest integer less than nt. Show that

$$\frac{1}{n}\left[P(0) + P\left(\frac{1}{n}\right) + P\left(\frac{2}{n}\right) + \cdots + P\left(\frac{N}{n}\right)\right]$$

tends to

$$\int_0^t P(s) ds$$

as $n \to \infty$.

(c) Using the functional equation (1.22) satisfied by P show that the terms of the above sum form a geometric progression (see Section 1.5) and that the sum equals

$$\frac{1}{n}\left[P\left(\frac{N+1}{n}\right) - 1\right]\Big/\left[P\left(\frac{1}{n}\right) - 1\right].$$

(d) Show that, unless $P(t) = 1$ for all t,

$$n\left[P\left(\frac{1}{n}\right) - 1\right]$$

tends to a nonzero limit a, and that

$$\int_0^t P(s)ds = \frac{1}{a}[P(t) - 1].$$

(e) Using the result of (d), prove that $P(t)$ is differentiable, and that

$$P' = aP.$$

1.7 Let $f(x) = e^{-1/x}$ for $x > 0$.
(a) Find $f'(x)$ and show that $\lim_{x \to 0} f'(x) = 0$.
(b) Show that the n-th derivative $f^{(n)}(x)$ is of the form $f^{(n)}(x) = g(1/x)e^{-1/x}$ where g is a polynomial in $1/x$.
(c) Show that $\lim_{x \to 0} f^{(n)}(x) = 0$ for $n = 0, 1, 2, \ldots$.

1.8 (a) Show that for all y

$$1 + y \le e^y.$$

(b) Show that for every $s > 0$,

$$e^{(1-s)y} \le 1 + y$$

for $0 \le y < Y$, where Y is some number which depends on s.
(c) Show that

$$e^{(1+s)y} \le 1 + y$$

for $-Y < y \le 0$, where Y is some number which depends on s.
(d) Show that for any a

$$\lim_{n \to \infty} \left(1 + \frac{a}{n}\right)^n = e^a.$$

(e) Let $a_1, a_2 \ldots$ be a sequence of positive numbers, such that

$$\lim_{n \to \infty} \frac{a_1 + \cdots + a_n}{n} = a.$$

Show that

$$\lim_{n \to \infty} \left(1 + \frac{a_1}{n}\right)\left(1 + \frac{a_2}{n}\right)\cdots\left(1 + \frac{a_n}{n}\right) = e^a.$$

(f) Let $f_j(x)$ be a sequence of twice differentiable functions such that

$$f_j(0) = 1, \qquad f'_j(0) = a_j$$

and

$$\lim_{n \to \infty} \frac{a_1 + \cdots + a_n}{n} = a.$$

Assume further that the second derivatives $f''_j(x)$ are bounded by the same constant, i.e.,

$$|f''_j(x)| < K \quad \text{for all } j.$$

Prove that

$$\lim_{n \to \infty} f_1\left(\frac{x}{n}\right) \cdots f_n\left(\frac{x}{n}\right) = e^{ax}.$$

(g) Let g_j be a sequence of 3-times differentiable functions with the following properties:

$$g_j(0) = 1, \qquad g'_j(0) = 0, \qquad g''_j(0) = b_j,$$

and

$$\lim_{n \to \infty} \frac{b_1 + \cdots + b_n}{n} = b.$$

Assume further that the third derivatives g'''_j are bounded by some constant, the same for all j:

$$|g'''_j(x)| < K.$$

Prove that

$$\lim_{n \to \infty} g_1\left(\frac{x}{\sqrt{n}}\right) \cdots g_n\left(\frac{x}{\sqrt{n}}\right) = \exp\left(\frac{bx^2}{2}\right).$$

5.2 The logarithm

We have shown in the last section that e^t is an increasing function of t, and that, as t goes from $-\infty$ to ∞, e^t goes from 0 to ∞. This shows that $E(t) = e^t$ is *invertible*, and that its *inverse function*, let us denote it by $L(x)$, is defined for all positive x. The inverse relationship between E and L is expressed as follows:

(2.1) $$L[E(t)] = L(e^t) = t, \qquad E[L(x)] = e^{L(x)} = x.$$

According to the rule for differentiating the inverse of a function [see Chapter 3, Equation (2.7)],

$$L' = \frac{1}{E'} = \frac{1}{e^t}.$$

Setting $e^t = x$, we obtain

(2.2)
$$L'(x) = \frac{1}{x}, \qquad L(1) = 0 \quad,$$

since the value of the exponential function at $t = 0$ is 1. According to the fundamental theorem of calculus, there exists exactly one function $L(x)$ defined for all $x > 0$ whose derivative is $1/x$ and which is zero at $x = 1$. This function is called the *logarithm*, and is written as log x. In symbols:

(2.3)
$$\int_1^x \frac{1}{s}\, ds = \log x \quad.$$

Thus relations (2.2) uniquely determine the log function; we shall now extract a whole array of interesting properties of the logarithm from (2.2) and (2.3).

(*i*) *Monotonicity*

The derivative of $L(x)$, $1/x$, is positive for $x > 0$; it follows from the monotonicity criterion that the function log x is *an increasing* function of x. Since log $1 = 0$, it follows that log x is positive for $x > 1$, negative for $x < 1$.

(*ii*) *Functional equation*

Let a be any positive number, and consider the function $L(ax)$. According to the chain rule and (2.2), its derivative is

$$\frac{d}{dx} L(ax) = a\frac{1}{ax} = \frac{1}{x} \, .$$

Thus the function $L(ax)$ has the same derivative as $L(x)$. According to the fundamental theorem this can be true only if $L(ax)$ and $L(x)$ differ by a constant:

$$L(ax) - L(x) = \text{constant}$$

The value of that constant can be determined by setting $x = 1$:

$$L(ax) - L(x) = L(a) - L(1).$$

According to (2.2), $L(1) = 0$, so we get the following result, in which we write log for L:

(2.4)
$$\log(ax) = \log x + \log a \quad.$$

In words: *The logarithm of a product is the sum of the logarithms of the factors.*

235

This celebrated functional equation is the basis of using log tables and slide rule* for multiplication. If you wish to multiply a and x, you look up their logs, add them, and look up the number whose log is that sum. According to (2.4), that number is the desired product ax.

We now show that, conversely, the multiplicative rule (2.4) characterizes the logarithmic function, except for a multiplicative constant, i.e., *if $l(x)$ is any function differentiable at $x = 1$ satisfying the functional equation*

$$l(xy) = l(x) + l(y),$$

then $l(x)$ is a constant multiple of $\log x$: $l(x) = \text{constant} \times \log x$.

PROOF. We shall calculate the derivative of l. Since

$$x + h = x\left(1 + \frac{h}{x}\right),$$

we can, using the functional equation (2.4), write

$$(2.5) \qquad l(x + h) = l\left[x\left(1 + \frac{h}{x}\right)\right] = l(x) + l\left(1 + \frac{h}{x}\right).$$

Setting $h = 0$ in (2.5), we conclude that

$$(2.6) \qquad\qquad\qquad l(1) = 0.$$

Subtracting $l(x)$ and $l(1)$ from both sides of Equation (2.5) and dividing by h gives, in view of (2.6),

$$(2.7) \qquad \frac{l(x + h) - l(x)}{h} = \frac{l[1 + (h/x)] - l(1)}{h}.$$

By assumption $l(x)$ is differentiable at $x = 1$; denote the value of $l'(1)$ by A. Set $p = h/x$. Then the right member of (2.7) becomes

$$\frac{1}{x}\frac{l(1 + p) - l(1)}{p},$$

and, as $h \to 0$, also $p \to 0$, so that the limit of the right member of (2.7) is $(1/x)l'(1)$. Thus the limit, as $h \to 0$, of the left member of (2.7) exists, and we have

$$l'(x) = \frac{1}{x}l'(1) = \frac{A}{x}.$$

We conclude that l is differentiable everywhere. It has the same derivative as $A \log x$. Since $l(x)$ and $A \log x$ both vanish at $x = 1$, we deduce from the

---

* A slide rule consists of movable number lines on which the logarithm of numbers has been calibrated. Multiplication is performed by aligning one scale with respect to the other, so that the lengths corresponding to the logs of the factors are added to form the length corresponding to the log of the product.

236

fundamental theorem that

$$l(x) = A \log x$$

as asserted.
□

(*iii*) *Behavior of the logarithmic function near* ∞ *and* 0

Repeated application of the functional equation (2.4) gives

(2.8)
$$\log(a^n) = n \log a$$

for any integer n and any $a > 0$. Suppose $a > 1$; then, as we saw before, $\log a$ is positive. Now let n tend to ∞; it follows that the right side of (2.8) tends to ∞; so we have shown that $\log x$ tends to ∞ if x tends to ∞ through a special sequence of the form $x = a^n$. But since $\log x$ is an increasing function, we see that $\log x$ *tends to* ∞ *as* x *tends to* ∞.

It follows from the functional equation for the log and from $\log 1 = 0$ that

$$\log x + \log \frac{1}{x} = \log\left(x \cdot \frac{1}{x}\right) = \log 1 = 0,$$

so that

(2.9)
$$\log x = -\log \frac{1}{x}.$$

As x tends to 0, $1/x$ tends to ∞; since $\log(1/x)$ also tends to ∞, we deduce from (2.9) that as x tends to 0, $\log x$ tends to $-\infty$.

How fast does $\log x$ tend to ∞? To answer this question we take any positive number n and examine the function

$$x^{-n} \log x.$$

Its derivative is

$$-nx^{-n-1} \log x + x^{-n}\frac{1}{x} = x^{-n-1}[1 - n \log x].$$

Since $\log x$ tends to ∞ as x tends to ∞, it follows that, for any positive n, the square bracket $[1 - n \log x]$ is negative for x large enough. From this and the monotonicity criterion, we conclude that for x large enough $x^{-n} \log x$ decreases. Since n is an arbitrary positive number, it follows that $x^{-n} \log x$ decreases to 0 as x tends to ∞. In words: *As* x *tends to* ∞, $\log x$ *tends to* ∞ *more slowly than any positive power of* x. Using the relation (2.9) we conclude from the above that *as* x *tends to* 0, $|\log x|$ *tends to* ∞ more slowly *than any negative power* of x.

(*iv*) *Concavity*

The first derivative of the log function is $1/x$; therefore the second derivative of $\log x$ is $-1/x^2$. We have seen in Section 3.8 that a function whose second derivative is negative is concave. Since $-1/x^2$ is negative, we conclude that *the function* $\log x$ *is concave*.

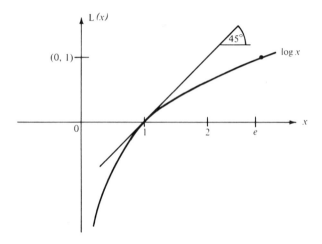

Figure 5.2

What we know so far about the log function enables us to make a sketch of its graph, see Figure 5.2. The sketch shows that as x goes from 0 to ∞, log x increases from $-\infty$ to ∞; that log $1 = 0$, and the derivative of log x at $x = 1$ equals 1. The graph is concave, i.e., it lies below the lines tangent to it.

Having shown that, as x goes from 0 to ∞, log x goes, monotonically, from $-\infty$ to ∞, we see that log x has an inverse; let us call it $E(x)$, defined on $(-\infty, \infty)$. Since log $1 = 0$,

(2.10) $$E(0) = 1.$$

The derivative of E can be determined from the rule linking the derivatives of inverse functions; differentiate

$$\log(E(x)) = x,$$

and get

$$\frac{1}{E(x)} E'(x) = 1,$$

so

(2.11) $$E'(x) = E(x).$$

This is the basic differential equation for the exponential function; as we saw in Section 5.1, this and relation (2.10) together uniquely characterize the exponential function. So we have shown: *The inverse of the logarithmic function is the exponential function.*

At this point we are exposing ourselves to the charge of redundancy. Haven't we, after all, defined the logarithmic function as the inverse of the exponential function, and doesn't this automatically make e^t the inverse of log x? Our defense of going from the logarithm to the exponential function rather than the other way around is this: we haven't yet demonstrated that there is such a function as $E(x)$, whereas the existence of log x is guaranteed by the fundamental theorem of calculus. Defining $E(x)$ as the inverse of log x, and verifying that it satisfies the differential equation (2.11) constitutes therefore an unimpeachable proof of the existence of the exponential function.

We now turn to some simple and important uses in calculus for the logarithmic function. Let f be any function whose values are positive, and consider the composite function

$$\log f(x).$$

Its derivative, by the chain rule, is

(2.12) $$[\log f(x)]' = \frac{f'(x)}{f(x)}.$$

This particular combination of f' and f is sometimes called the logarithmic derivative of f. Since f' measures the rate of change of f, the logarithmic derivative f'/f measures the fractional rate of change of f.

Example 1. Let $f(x) = 1 + x^2$; then $f'(x) = 2x$, so the logarithmic derivative of f is

$$\frac{d}{dx} \log f(x) = \frac{f'}{f} = \frac{2x}{1 + x^2};$$

when $x = 3$, $f' = 6$, $f = 10$, so the fractional rate of change of f is $f'/f = 0.6$. This means that, when $x = 3$, the function f changes at a rate which is 60% of its value.

We now turn to the power function

$$x^r.$$

For a positive integer r, x^r is defined directly in terms of multiplication. For a negative integer r, x^r is defined as the reciprocal of x^{-r}. When r is the reciprocal of a positive integer p, $x^{1/p}$ is defined for positive x as that positive number whose p-th power is x. When r is rational, i.e., $r = p/q$, p, q are integers, $q \neq 0$, x^r is defined as $x^{1/q}$ raised to the p-th power. Finally, to define x^r for an irrational r, one has to approximate r by a sequence of rational numbers r_n and define x^r as the limit of the sequence x^{r_n}.

We shall now show how this piecemeal business can be replaced at one stroke by a single definition. To arrive at such a definition, take r to be any rational number p/q, p and q integers, $q \neq 0$. Then, as we saw,

$$x^{p/q} = y$$

239

is defined through the relation

$$x^p = y^q.$$

Taking the logarithm of both sides and using the functional equation of the logarithm, we get that

$$p \log x = q \log y,$$

so that

$$\log y = \frac{p}{q} \log x.$$

Using again the inverse relation between log and exp we get

$$y = e^{\log y} = e^{(p/q)\log x}.$$

This suggests that for any r and any positive x *we define the power x^r by the formula*

(2.13)
$$\boxed{x^r = e^{r \log x} = E(r \log x)} ,$$

where the exponential function E is defined as the inverse of the log. As we have shown above, for rational r, this definition agrees with the classical one. Since the exponential function is continuous, it follows that x^r as defined by (2.13) depends continuously on r.

The beauty of Definition (2.13) is that all properties of the power function follow easily from it. By definition

$$x^{r+s} = e^{(r+s)\log x} = e^{r \log x + s \log x}.$$

The last expression could have been written in the alternate notation as $x^{r+s} = \exp\{r \log x + s \log x\}$. From the functional equation of the exponential function, we get

(2.14)
$$x^{r+s} = x^r x^s,$$

the first law of exponents. Similarly from the functional equation for the logarithm,

$$(xy)^r = e^{r \log xy} = e^{r(\log x + \log y)} = e^{r \log x + r \log y},$$

and from the functional equation of the exponential function, we obtain

(2.15)
$$(xy)^r = x^r y^r,$$

the second law of exponents.

To differentiate x^r as defined by (2.13), we appeal to the chain rule:

$$\frac{d}{dx}(x^r) = \frac{d}{dx}(e^{r \log x}) = \frac{r}{x} e^{r \log x} = \frac{r}{x} x^r = rx^{-1}x^r.$$

Using rule (2.14), we obtain the familiar formula

(2.16)
$$\frac{d}{dx}(x^r) = rx^{r-1}$$

for differentiating a power.

Let b denote any positive number different from 1; the logarithm of the positive number x to base b, denoted by

$$\log_b x,$$

is defined as that number y to which b must be raised to obtain x, i.e.,

$$x = b^y.$$

Using the power formula (2.13) we find that

$$x = e^{y \log b}.$$

Taking the logarithm of both sides we obtain

$$\log x = y \log b.$$

Solve this for $y = \log_b x$:

(2.17)
$$\log_b x = \frac{\log x}{\log b}.$$

We now turn to integrals that can be evaluated in terms of logarithms. We saw earlier that, for any positive function f, the derivative of $\log f(x)$ is $f'(x)/f(x)$. Therefore it follows from the fundamental theorem that

(2.18)
$$\int_a^b \frac{f'(x)}{f(x)} dx = \log f(b) - \log f(a) = \log \frac{f(b)}{f(a)}.$$

Now suppose f is negative in some interval $a \leq x \leq b$, say $f(x) = -g(x)$, $g(x) > 0$ in $a \leq x \leq b$. Then $f' = -g'$, so that $f'/f = g'/g$ and

$$\int_a^b \frac{f'}{f} dx = \int_a^b \frac{g'}{g} dx = \log g(b) - \log g(a)$$

$$= \log |f(b)| - \log |f(a)|$$

$$= \log \left| \frac{f(b)}{f(a)} \right|.$$

Thus, if we modify (2.18) to read

$$\int_a^b \frac{f'(x)}{f(x)} dx = \log \left| \frac{f(b)}{f(a)} \right|,$$

it is valid for positive and negative f.

Example 2. Find the value of

$$I = \int_0^1 \frac{x}{1 + x^2} \, dx.$$

Solution: The integrand is of the form $\frac{1}{2}f'(x)/f(x)$ with $f(x) = 1 + x^2$. Therefore by (2.18)

$$\int_a^b \frac{x}{1 + x^2} \, dx = \frac{1}{2} \log \frac{1 + b^2}{1 + a^2};$$

with $a = 0$ and $b = 1$, we see that the value of the integral I is $\frac{1}{2} \log 2$.

Example 3. Find the value of

$$J = \int_0^1 \frac{x^2}{1 + x^3} \, dx.$$

Solution: The integrand is of the form $\frac{1}{3}f'(x)/f(x)$, with $f(x) = 1 + x^3$. Therefore, by (2.18),

$$J = \int_0^1 \frac{x^2}{1 + x^3} \, dx = \frac{1}{3} \log 2.$$

Example 4. Find the value of

$$K = \int_0^\infty \frac{1}{x^2 + 3x + 2} \, dx.$$

Solution: We factor the denominator into linear factors:

$$x^2 + 3x + 2 = (x + 1)(x + 2).$$

We write the integrand in terms of partial fractions (see Appendix 2.1):

$$\frac{1}{x^2 + 3x + 2} = \frac{1}{x + 1} - \frac{1}{x + 2}.$$

The right side is the derivative of $\log(x + 1) - \log(x + 2)$, so according to the fundamental theorem

$$\int_0^b \frac{1}{x^2 + 3x + 2} \, dx = \log(b + 1) - \log 1 - \log(b + 2) + \log 2$$

$$= \log 2 + \log \frac{b + 1}{b + 2}.$$

As b tends to ∞, $(b + 1)/(b + 2)$ tends to 1; since $\log 1 = 0$, the limit of the right side is $\log 2$. Thus

$$K = \int_0^\infty \frac{1}{x^2 + 3x + 2} \, dx = \log 2.$$

Example 5. Let q be a polynomial of degree n which has n real and distinct zeros $s_1, \ldots, s_n$. We can write q in the factored form

$$q(x) = a_n(x - s_1) \cdots (x - s_n),$$

where a_n is the coefficient of x^n in q. Let p be a polynomial of degree $<n$; as we have shown in Appendix 2.1, the quotient

$$\frac{p(x)}{q(x)}$$

can be expanded in partial fractions; i.e., it can be written in the form

(2.19) $$\frac{p(x)}{q(x)} = \sum_{j=1}^{n} \frac{c_j}{x - s_j}.$$

Suppose that all the roots s_j are negative, and let a and b denote positive numbers. Using (2.19) we can evaluate the integral of p/q on $[a, b]$ as follows:

(2.20) $$\int_a^b \frac{p(x)}{q(x)} dx = \int_a^b \sum_{j=1}^{n} \frac{c_j}{x - s_j} dx = \sum_{j=1}^{n} c_j \log \frac{b - s_j}{a - s_j}.$$

We are emphatically *not* advocating (2.20) as a way of numerically evaluating integrals of rational functions. In all but the simplest cases the labor of factoring q is much greater than that involved in a straightforward application of Simpson's rule.

Next we consider an integral where the integrand involves the log function.

Example 6. Evaluate

$$\int_a^b \log x \, dx.$$

We shall integrate by parts, i.e., use the identity

$$\int_a^b f'g \, dx = f(b)g(b) - f(a)g(a) - \int_a^b fg' \, dx.$$

We choose $f(x) = x$, $g(x) = \log x$ and get

$$\int_a^b \log x \, dx = b \log b - a \log a - \int_a^b x \frac{1}{x} dx$$
$$= b \log b - a \log a - (b - a).$$

EXERCISES

2.1 Find the derivative of these functions:

(a) $\log(\log x)$

(b) $\log\left(x^2 + \frac{1}{x^2}\right)$

(c) $x^2 \log x$

(d) $x^n \log x$.

2.2 Find functions whose derivatives are the following functions:

(a) $x \log x$.

(b) $\dfrac{e^x - e^{-x}}{e^x + e^{-x}}$.

2.3 What is the limit of

$$x^x,$$

as x tends to zero?

2.4 Evaluate the following integrals:

(a) $\displaystyle\int_a^1 \log x \, dx$. What is the limit of this integral as a tends to 0?

(b) $\displaystyle\int_a^1 x \log x \, dx$. What limit does this integral approach as a tends to 0?

(c) $\displaystyle\int_0^2 \dfrac{x}{x^2 + 4x + 2} \, dx$

(d) $\displaystyle\int_1^n x^{n-1}(n \log x + 1) dx$.

2.5 Evaluate the integral

$$\int_1^2 \frac{1}{x^3 + 3x^2 - 2} \, dx$$

(a) using the method of Example 4;
(b) using Simpson's rule.
To calculate the answer to an accuracy of 6 decimal digits, which method is quicker?

2.6 Evaluate

$$\int_{-3}^{-2} \frac{1}{x} \, dx.$$

2.7 Let b denote any positive number, $b \neq 1$; show that for any pair of positive numbers x and a

$$\log_b ax = \log_b a + \log_b x.$$

2.8 In Section 3.8 we have shown that the graph of a concave function lies above its secant, see Figure 3.21; i.e., for p concave,

$$p(ta + (1 - t)b) \geq tp(a) + (1 - t)p(b)$$

for all t in the interval $0 \leq t \leq 1$. Apply this inequality to $p(x) = \log x$ and show that

$$ta + (1 - t)b \geq a^t b^{1-t} \quad \text{for } 0 \leq t \leq 1.$$

Note that for $t = \frac{1}{2}$, this is the inequality between the arithmetic and geometric means studied in Section 3.5.

2.9 Find the fractional rate of change of
(a) the function f of Example 1 when $x = -2$ and when $x = 10$;
(b) the function e^{kt} for any value of t;
(c) the function $\log[\log s]$.

244

5.3 The computation of logarithms and exponentials

The numerical calculation of logarithms is, at least in principle, a straight-forward affair; $\log a$ is defined as $\int_1^a 1/x \, dx$. Therefore, to find the value of $\log a$, one has to evaluate approximately an integral, which we can do with the aid of the numerical methods described in Section 4.6. For example, let us find the value of

$$\log 2 = \int_1^2 \frac{1}{x} dx$$

by Simpson's rule, applied to the interval $[1, 2]$ with a single subdivision; the approximate value is

$$\frac{1}{6} 1 + \frac{2}{3} \frac{1}{1.5} + \frac{1}{6} \frac{1}{2} = \frac{25}{36} = 0.6944 \ldots .$$

Compare this with the tabulated value of $\log 2$, whose first four digits are

$$\log 2 = 0.6931 \ldots .$$

Our crude approximation differs from this value by less than 1 part in the third decimal place. Indeed, tables of logarithms are prepared by evaluating approximately the integral defining $\log a$.

It is important to observe that extensive log tables are not required; it suffices to tabulate values of $\log a$ in a relatively small range, since values outside this range can be obtained by means of the functional equation of logarithms. Specifically, suppose we only know the value of $\log a$ in the range $1 \leq a \leq 2$. Then the value of $\log b$ for any b can be obtained as follows.

Every positive number b can be written in the form

$$b = 2^n a,$$

where n is an integer, and a lies in the range $1 \leq a < 2$. To see this, note that every positive number b can be sandwiched between two successive powers of 2:

$$2^n \leq b < 2^{n+1}.$$

It follows that

$$1 \leq \frac{b}{2^n} < 2,$$

and the number

$$a = \frac{b}{2^n}$$

lies between 1 and 2, so its logarithm is known. According to the functional equation for logarithms, we deduce from $b = 2^n a$ that

$$\log b = n \log 2 + \log a.$$

In Exercises 3.1 and 3.2 following this section, we describe a different, more practical method for calculating $\log b$.

We now turn to the problem of computing exponentials e^s for any given number s. The exponential function e^s may be defined as the inverse of the log function:

$$y = e^s \qquad \text{means} \qquad \log y = s;$$

that is, y satisfies the equation

(3.1) $$\log y - s = 0,$$

and so y is a zero of the function

$$f(y) = \log y - s.$$

We shall use Newton's method, described in Section 3.10, to solve the equation

$$f(y) = \log y - s = 0$$

by constructing better and better approximations to e^s. For the sake of definiteness, take $s = 1$, so that

$$y = e^1 = e,$$

and the above equation becomes

(3.2) $$f(y) = \log y - 1 = 0.$$

To compute e^s for arbitrary s, we apply Newton's method to the function $\log y - s$.

Given an approximation y_{old} to the solution of (3.2), Newton's method gives a new approximation in Formula (10.1), Section 3.10:

$$y_{new} = y_{old} - \frac{f(y_{old})}{f'(y_{old})} .$$

For f given in (3.2), we have $f'(y) = 1/y$, so

(3.3) $$y_{new} = y_{old} - \frac{\log y_{old} - 1}{1/y_{old}} = 2y_{old} - y_{old} \log y_{old}.$$

Suppose we choose, for y_{old}, the rather crude approximation

$$y_{old} = 2.6$$

to $e = 2.71828\ldots$. We compute $\log 2.6$ by applying Simpson's rule to the integral

$$\int_1^{2.6} \frac{1}{x}\, dx$$

defining log 2.6. Employing subintervals of length 1/10, we obtain

$$\log 2.6 = 0.955511\ldots,$$

and substituting these values into (3.3), we get

$$y_{\text{new}} = 5.2 - 2.6(0.955511) = 2.7157.$$

This value differs from e by less than 1 part in a 1000.

Next, we develop other methods for evaluating e^x approximately; these are based on the characterization of $E(x) = e^x$ not as the inverse of the log function, but as the solution of the differential equation

$$(3.4) \qquad\qquad E' = E$$

whose value at $x = 0$ is

$$(3.5) \qquad\qquad E(0) = 1.$$

$E'(x)$ is defined as the limit of difference quotients

$$\frac{E(x + h) - E(x)}{h}$$

as $h \to 0$; so for h small the derivative and the difference quotient differ by a small quantity:

$$(3.6) \qquad\qquad E'(x) = \frac{E(x + h) - E(x)}{h} + \text{s.q.},$$

where s.q. stands for a "small quantity." Substitute (3.6) into (3.4) to obtain:

$$E(x) = \frac{E(x + h) - E(x)}{h} + \text{s.q.}$$

Multiplying by h we can solve for $E(x + h)$:

$$(3.7) \qquad\qquad E(x + h) = (1 + h)E(x) - h \times (\text{s.q.}).$$

We shall now construct an approximation $F(x)$ to $E(x)$ as the solution of the equation obtained from (3.7) by omitting the small amount $h \times (\text{s.q.})$;

$$(3.8) \qquad\qquad F(x + h) = (1 + h)F(x).$$

At $x = 0$ we take F to have the same value as E:

$$(3.9) \qquad\qquad F(0) = E(0) = 1.$$

The Equation (3.8) for F is considerably simpler than the Equation (3.7) for E and, as we shall see in a moment, can be solved explicitly. The guiding idea is that, since for small h the equation satisfied by F differs little from the equation satisfied by E, we expect $F(x)$ to differ little from $E(x)$ for small h. Hence the solution of (3.7) is obtained by evaluating $F(1)$ starting with $F(0) = 1$, then stepping toward $x = 1$ by incrementing by h at each step. To determine the function F, set as a first step $x = 0$ in (3.8):

$$F(h) = (1 + h)F(0),$$

247

and since, by (3.9), $F(0) = 1$, we obtain

$(3.9)_1$ $$F(h) = 1 + h.$$

In the next step, set $x = h$ in (3.8):

$$F(2h) = (1 + h)F(h).$$

Using the value of $F(h)$ determined previously, we get

$(3.9)_2$ $$F(2h) = (1 + h)^2.$$

The next step is clear: set $x = 2h$ in (3.8) and use the previously determined value $(3.9)_2$ to get

$$F(3h) = (1 + h)^3.$$

Proceeding like this step by step, we get for any positive integer n that

$(3.9)_n$ $$F(nh) = (1 + h)^n.$$

Now choose h to be $1/n$ and introduce the abbreviation F_n by

(3.10) $$F_n = F(1) = \left(1 + \frac{1}{n}\right)^n.$$

As n tends to ∞, $h = 1/n$ tends to 0, so, according to our guiding principle, we surmise that the sequence (3.10) tends to e as n tends to ∞. We recall that, in Section 1.4, we have investigated this sequence and have calculated F_n for several values of n; we reproduce the table from Section 1.4:

(3.11)
$$F_4 = 2.44141\ldots$$
$$F_8 = 2.56578\ldots$$
$$F_{16} = 2.63793\ldots$$
$$F_{64} = 2.69734\ldots$$
$$F_{256} = 2.71299\ldots$$
$$F_{1024} = 2.71696\ldots.$$

As we said before, the correct value of the first 3 digits of e is

$$e = 2.718.$$

The numerical evidence suggests that the sequence (3.11) converges to e. Later on in this section we shall prove that it does. All the numbers in (3.11) are less than e. This suggests that the quantities (3.10) are all less than e, see (1.21) of Exercise 1.4. That this is indeed so can be deduced from the *convexity* of the function $E(x)$ noted in Section 5.1. For, according to the discussion at the end of Section 3.8, the graph of a convex function lies above

248

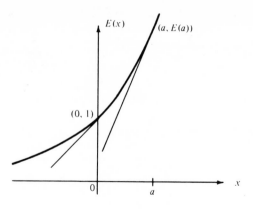

Figure 5.3

the lines tangent to it. The tangent to the graph of E at the point $(a, E(a))$, see Figure 5.3, has the equation

$$y = (x - a)E'(a) + E(a).$$

By convexity, at the point $x = a + h$, this line lies below the graph of E; analytically this is expressed by the inequality

$$E(a + h) > hE'(a) + E(a).$$

The function E satisfies the differential equation $E' = E$; so the above inequality can be rewritten as

(3.12) $$E(a + h) > (1 + h)E(a).$$

Setting $a = 0$ in (3.12) and using the fact that $E(0) = 1$, we get

$$E(h) > (1 + h).$$

Setting $a = h$ in (3.12) and using the above inequality, we get

$$E(2h) > (1 + h)^2.$$

We continue following this pattern. Setting $a = 2h, 3h$, etc., successively, we obtain inequalities

$$E(nh) > (1 + h)^n.$$

Setting $h = 1/n$, we get that

(3.13) $$e = E(1) > \left(1 + \frac{1}{n}\right)^n,$$

as foreshadowed by the numerical evidence of Table (3.11).

We return now to Equation (3.6), in which the difference quotient $[E(x + h) - E(x)]/h$ was regarded as an approximation of the value of E'

at x. One can, however, with equal justice regard it as an approximation to the value of E' at $x + h$:

$$E'(x + h) = \frac{E(x + h) - E(x)}{h} + \text{s.q.}$$

Since $E'(x + h) = E(x + h)$, this expression is equivalent to

$$E(x + h) = \frac{E(x + h) - E(x)}{h} + \text{s.q.},$$

where s.q. stands for a small quantity. We solve for $E(x + h)$ and obtain

(3.14) $$E(x + h) = \frac{1}{(1 - h)} E(x) - \frac{h}{1 - h} \times (\text{s.q.}).$$

We now construct an approximation $G(x)$ to $E(x)$ as the solution of the equation obtained from (3.14) by omitting the term containing s.q.:

(3.15) $$G(x + h) = \frac{1}{1 - h} G(x).$$

At $x = 0$, we take G to have the same value as E;

$$G(0) = 1.$$

We are guided by the same idea as before: since for small h the equation satisfied by G differs little from the equation satisfied by E, we expect G to differ little from E when h is small enough.

Next we determine the values of the function G at all integer multiples of h; setting $x = nh$ in (3.15) gives

$$G((n + 1)h) = \frac{1}{1 - h} G(nh).$$

When $n = 0$, this gives

$$G(h) = \frac{1}{1 - h} G(0) = \frac{1}{1 - h}.$$

Setting $n = 0, 1, \ldots$ successively, we obtain

$$G((n + 1)h) = \frac{1}{1 - h} G(nh) = \frac{1}{(1 - h)^2} G((n - 1)h) = \cdots = \frac{1}{(1 - h)^{n+1}}.$$

Setting $h = 1/(n + 1)$ in this relation, we evaluate $G(1)$:

$$G(1) = \left(\frac{1}{1 - 1/(n + 1)} \right)^{n+1} = \left(\frac{n + 1}{n} \right)^{n+1} = \left(1 + \frac{1}{n} \right)^{n+1}.$$

We introduce the abbreviation

(3.16) $$G_n = \left(1 + \frac{1}{n} \right)^{n+1}$$

and tabulate the values of G_n for various n:

$$G_4 = 3.05176\ldots$$
$$G_8 = 2.88651\ldots$$
$$G_{16} = 2.80280\ldots$$
$$G_{64} = 2.73949\ldots$$
$$G_{256} = 2.72359\ldots$$
$$G_{1024} = 2.71961\ldots\ .$$

(3.17)

According to our guiding principle, the smaller h is, the closer (3.16) ought to be to $e = 2.718\ldots$. That is, as n tends to ∞ we expect the sequence G_n to tend to e; our expectation is reinforced by Table (3.17).

The Table (3.17) suggests not only that the sequence G_n tends to e, but also that all the quantities G_n are greater than e. This can be deduced from the convexity of the function E in a manner analogous to the discussion which showed that the quantities F_n are less than e. In Exercise 3.7 the reader is led, in a series of steps, to prove the inequality $e < G_n$.

Tables (3.11) and (3.17) which suggested the inequalities

(3.18) $$F_n < e < G_n,$$

also suggest that *the sequence F_n is increasing and the sequence G_n is decreasing.* In what follows we shall present proofs of these statements and deduce from them that *the sequences F_n and G_n converge to a common limit.* According to Definition (2.3)

$$\log\left(1 + \frac{1}{n}\right) = \int_1^{1+1/n} \frac{1}{s}\, ds.$$

The integrand $1/s$ is a decreasing function, so in the interval of integration,

$$\frac{1}{1 + 1/n} \le \frac{1}{s} \le 1.$$

The length of the interval of integration is $1/n$; therefore, according to the lower upper bound property of the integral,

(3.19) $$\frac{1}{n+1} < \log\left(1 + \frac{1}{n}\right) < \frac{1}{n},$$

or, equivalently,

$$1 < (n+1)\log\left(1 + \frac{1}{n}\right) \quad \text{and} \quad n\log\left(1 + \frac{1}{n}\right) < 1.$$

The last inequalities may be restated in the form

$$1 = \log e < \log\left(1 + \frac{1}{n}\right)^{n+1} \quad \text{and} \quad \log\left(1 + \frac{1}{n}\right)^n < \log e = 1,$$

and since log is an increasing function, it follows that

$$e < \left(1 + \frac{1}{n}\right)^{n+1} \qquad \text{and} \qquad \left(1 + \frac{1}{n}\right)^{n} < e.$$

This furnishes another proof of inequalities (3.18).

To see that $\{F_n\}$ is an increasing sequence, differentiate the function $\log F_n = \log(1 + 1/n)^n = n \log(1 + 1/n)$ with respect to n (thinking of n now as a continuous variable in $1 \leq n < \infty$):

$$\frac{d}{dn} \log F_n = \frac{d}{dn} n \log\left(1 + \frac{1}{n}\right) = \log\left(1 + \frac{1}{n}\right) - n \frac{1}{1 + 1/n} \frac{1}{n^2}$$

$$= \log\left(1 + \frac{1}{n}\right) - \frac{1}{n + 1}.$$

By (3.19), the last expression is *positive*. It follows then from the monotonicity criterion of Section 3.3, that $\log F_n$ is an increasing function of n, and hence, that the sequence $\{F_n\}$ increases.

Analogously, the derivative of the logarithm of G_n is

$$\frac{d}{dn} \log G_n = \frac{d}{dn} (n + 1)\log\left(1 + \frac{1}{n}\right) = \log\left(1 + \frac{1}{n}\right) - (n + 1)\frac{1}{1 + 1/n} \frac{1}{n^2}$$

$$= \log\left(1 + \frac{1}{n}\right) - \frac{1}{n}$$

which, by (3.19), is *negative*. So by the monotonicity criterion, $\log G_n$ decreases, and so does the sequence $\{G_n\}$.

So far we have shown that $\{F_n\}$ increases and is bounded from above by e, and that $\{G_n\}$ decreases and is bounded from below by e. To show that both actually converge to e we must show that they have the *same* limit. To this end, we form the ratio

$$\frac{G_n}{F_n} = 1 + \frac{1}{n}$$

and observe that $\lim_{n \to \infty} (G_n/F_n) = 1$, whence

$$\lim_{n \to \infty} G_n = \lim_{n \to \infty} F_n = e.$$

The approximations (3.10) and (3.16) were obtained by regarding the difference quotient $(E(x + h) - E(x))/h$ as an approximation to the value of E' at x and $x + h$, respectively. We have shown, at the end of Section 3.8, that this difference quotient is a better approximation of the value of E' at the midpoint $x + h/2$ of the interval $[x, x + h]$ than at either endpoint; therefore,

$$\frac{E(x + h) - E(x)}{h} = E'\left(x + \frac{h}{2}\right) + \text{very small quantity}.$$

We again note that

$$E'\left(x + \frac{h}{2}\right) = E\left(x + \frac{h}{2}\right),$$

and recall from Exercise 8.3, Section 3.8 that a good approximation to the value of E' at the midpoint of an interval is the average of the values of E' at the endpoints. So we write

$$\frac{E(x + h) - E(x)}{h} = \frac{E(x) + E(x + h)}{2} + \text{v.s.q.,}$$

where v.s.q. is a very small quantity.

Multiply this relation by h and solve for $E(x + h)$:

$$(3.20) \qquad E(x + h) = \frac{1 + h/2}{1 - h/2} E(x) + \frac{h}{1 - h/2} \text{(v.s.q.).}$$

We now construct an approximation $H(x)$ to $E(x)$ as the solution of the equation obtained from (3.20) by omitting the term containing v.s.q.:

$$(3.21) \qquad H(x + h) = \frac{1 + h/2}{1 - h/2} H(x), \qquad H(0) = 1.$$

As before, we can easily determine the value of H at integer multiples of h by setting $x = 0, h, 2h$, etc., successively, in (3.21). We get

$$H(h) = \frac{1 + h/2}{1 - h/2}, \qquad H(2h) = \left(\frac{1 + h/2}{1 - h/2}\right)^2,$$

and for any integer n,

$$H(nh) = \left(\frac{1 + h/2}{1 - h/2}\right)^n.$$

Setting $h = 1/n$ gives

$$(3.22) \qquad H(1) = \left(\frac{1 + 1/2n}{1 - 1/2n}\right)^n = H_n.$$

We tabulate below values of H_n for various choices of n:

$$(3.23) \qquad \begin{aligned} H_4 &= 2.732\ldots \\ H_8 &= 2.721\ldots \\ H_{16} &= 2.719\ldots \\ H_{64} &= 2.7183\ldots \\ H_{256} &= 2.7182852\ldots \\ H_{1024} &= 2.7182820\ldots. \end{aligned}$$

The smaller h is, the closer $H(1)$ ought to be to $e = 2.71828\ldots$. That is, we expect the sequence $\{H_n\}$ to tend to e as n tends to ∞. This expectation is borne out by the numerical evidence of Table (3.23); note that $\{H_n\}$ tends to

e faster than either of the sequences $\{F_n\}$ or $\{G_n\}$, i.e., for the same value of n, H_n is closer to e than either F_n or G_n. This is to be expected since the quantity we neglected in approximating E by H was smaller than the quantities we neglected in approximating E by either F or G.

These methods can be used without alteration to compute e^s for any s. Instead of choosing the step size h to be $1/n$, we take it to be s/n. The analogues of (3.10), (3.16) and (3.22) are:

$$e^s = \lim_{n \to \infty} \begin{cases} \left(1 + \dfrac{s}{n}\right)^n \\[2ex] \left(1 + \dfrac{s}{n}\right)^{n+1} \\[2ex] \left(\dfrac{1 + s/2n}{1 - s/2n}\right)^n. \end{cases}$$

In Section 3.9 we have investigated the approximation of functions which have derivatives of all orders by Taylor polynomials. We recall from (9.9), Section 3.9, that the n-th Taylor polynomial of f at 0 is

$$t_n(x) = f(0) + f'(0)x + \frac{1}{2!}f''(0)x^2 + \cdots + \frac{1}{n!}f^{(n)}(0)x^n,$$

where $n!$ is an abbreviation for the product $1 \cdot 2 \cdot 3, \ldots, \cdot n$. Denote the n-th Taylor polynomial of the function E by E_n. Since all derivatives of E (E', E'', ..., $E^{(n)}$) are equal to E itself, we get that $E^{(n)}(0) = E(0) = 1$ for all n, so

(3.24)
$$E_n(x) = 1 + x + \frac{x^2}{2!} + \cdots + \frac{x^n}{n!}.$$

Below we list the value of

$$E_n(1) = 1 + \frac{1}{1!} + \frac{1}{2!} + \frac{1}{3!} + \cdots + \frac{1}{n!}$$

for various values of n:

n	$E_n(1)$	n	$E_n(1)$
1	2.0	10	2.718281801...
2	2.5	15	2.7182818284589
3	2.666666...	20	2.7182818284589
4	2.7083333...	30	2.7182818284589
5	2.7166666...		

Compare this list with the first ten digits of e listed below:

$$e = 2.718281828\ldots.$$

The sixth sample computer program at the end of this volume calculates the exponential function by adding up a finite number of terms of its Taylor series. The fifth sample computer program calculates $\log x$ by evaluating approximately, using Simpson's rule, the integral of $1/x$.

EXERCISES

3.1 (a) Consider the finite geometric series

$$(3.25) \qquad 1 - x + x^2 - x^3 + \cdots + (-x)^n = \sum_0^n (-x)^k.$$

Show that, for $0 \leq x < 1$, this sum converges, as n tends to ∞, to $1/(1 + x)$. Show that the difference between (3.25) and $1/(1 + x)$ is less than x^{n+1}.

(b) Let b lie in $0 \leq b < 1$; using the relation

$$\log(1 + b) = \int_0^b \frac{1}{1 + x}\, dx$$

and the approximation (3.25) to $1/(1 + x)$, show that

$$(3.26) \qquad b - \frac{b^2}{2} + \frac{b^3}{3} - \frac{b^4}{4} + \cdots + \frac{(-1)^n b^{n+1}}{n + 1}$$

tends to $\log(1 + b)$ as n tends to ∞.

(c) Show that the difference between (3.26) and $\log(1 + b)$ is less than $b^{n+2}/(n + 2)$.

(d) Choose n so large that setting $b = 0.1$ in (3.25) gives the value of $\log 1.1$ with an error $< 10^{-10}$, and calculate the first ten digits of $\log 1.1$.

(e) Show, using the formula for the n-th Taylor polynomial, that (3.26) is the $(n + 1)$-st Taylor polynomial for $\log(1 + b)$.

3.2 Make a list of the following powers of the number 1.1:

$$(3.27) \qquad 1.1, \quad 1.1^2, \quad 1.1^4, \quad 1.1^8, \quad 1.1^{16}, \quad 1.1^{32}, \quad 1.1^{64}.$$

Let c be any positive number $< 1.1^{64}$. We now describe a procedure for writing c as a product whose factors are among the seven numbers listed above, times a number of the form $1 + b$ where $b < 0.1$.

Find the largest power of 1.1 listed in (3.27) which is less than c, say $(1.1)^{m_1}$. Form the quotient

$$c_1 = \frac{c}{(1.1)^{m_1}}.$$

(a) Show that this quotient is less than 1.1^{m_1}.

(b) Show that if we repeat the procedure, we arrive at a number $c_k < 1.1$ in less than seven steps.

(c) Show that

$$c = (1.1)^{m_1 + m_2 + \cdots + m_k} c_k,$$

where k is the number of steps taken in (b), and therefore that

$$(3.28) \qquad \log c = (m_1 + m_2 + \cdots + m_k)\log 1.1 + \log c_k.$$

Formula (3.28) gives a practical method for calculating $\log c$ since we have already calculated $\log 1.1$ and can use the approximation (3.26) with $b = c_k - 1$ to approximate $\log c_k$.

(d) Write a computer program which performs the above steps to compute $\log c$ for $c < (1.1)^{64}$.

3.3 Integrate the differential equation

$$E' = E$$

satisfied by E over the interval $[0, 1]$. Using the fundamental theorem to evaluate the left side, we get

$$E(1) - E(0) = \int_0^1 E(x)dx.$$

Approximate the integral on the right by Simpson's rule; we get

(3.29) $\qquad E(1) - E(0) = \frac{1}{6}E(0) + \frac{2}{3}E(\frac{1}{2}) + \frac{1}{6}E(1) + \text{s.q.}$

where s.q. denotes the small quantity which is the difference between the integral and Simpson's approximation to it.

Denote $E(\frac{1}{2})$ by s; then $E(1) = E(\frac{1}{2})E(\frac{1}{2}) = s^2$. Substitute 1, s and s^2 for $E(0)$, $E(\frac{1}{2})$ and $E(1)$ into (3.29), and omit the small quantity; the resulting equation is quadratic in s. Find the solution of this quadratic equation. How good an approximation is s^2 to e?

3.4 Formula (3.3) describes Newton's method for finding a better approximation y_{new} to the solution of $\log y = 1$, given some approximation y_{old}. In the text we took $y_{old} = 2.6$ and found that $y_{new} = 2.7157$.

(a) Starting with $y_{old} = 2.7157$, find y_{new} and compare it to e. How many digits agree?

(b) Write a computer program which, for any y_{old}, generates a sequence of approximations to e using Newton's method.

3.5 The quantities F_n and G_n defined by (3.10) and (3.16), respectively, approximate e. We have shown that F_n lies below, G_n above, e; therefore it is a reasonable guess that their average is a better approximation to e than either F_n or G_n.

Make a list of

(3.30) $\qquad \dfrac{F_n + G_n}{2}$

for $n = 4, 8, 16, 64, 256, 1024$, and ascertain how many digits of (3.30) agree with those of e. Compare the approximation (3.30) with H_n defined by (3.22).

3.6 Denote by $E_n(x)$ the n-th degree Taylor polynomial approximation to $E(x)$, given by Formula (3.24). Let m be any integer; denote by $E_{n,m}$ the quantity

$$E_{n,m} = \left[E_n\left(\frac{1}{m}\right) \right]^m.$$

Write a computer program which tabulates the value of $E_{n,m}$ in the range $n = 1, 2, 3, 4, 5$ and $m = 8, 16, 64, 256, 1024$. Indicate in the table how close $E_{n,m}$ is to e, and record the number of multiplications needed to evaluate $E_{n,m}$.

3.7 (a) Find the equation of the tangent to the graph of $E(x)$ at the point $(a + h, E(a + h))$.
(b) Use the convexity of E and the fact that $E'(x) = E(x)$ to prove the inequality

(3.31) $$E(a) > (1 - h)E(a + h).$$

(c) By successively setting $a = 0, h, 2h, \ldots$, using the initial value $E(0) = 1$, and using (3.31), show that

(3.32) $$E((n + 1)h) < \frac{1}{(1 - h)^{n+1}}.$$

(d) Set $h = 1/(n + 1)$ in (3.32) and deduce that

$$G_n = \left(1 + \frac{1}{n}\right)^{n+1} > e.$$

3.8 Develop a computer program for solving the equation

$$F(y) = \log y - 1 = 0$$

by the method of bisection. When evaluating $\log y$, apply Simpson's rule to

$$\log y = \int_1^y \frac{dt}{t};$$

the choice of an appropriate subdivision is part of the problem.

6 Probability and its applications

The origins of calculus lie in Newtonian mechanics, of which a brief preview was given in Section 3.6. We saw in the examples of earth and moon motion that, once the force acting on a particle is ascertained and the initial position and velocity of the particle specified, the whole future course of the particle is predictable. Such a predictable motion is called *deterministic*. We will see in Section 8.1 that the motion of a particle under the combination of a restoring force and friction is likewise deterministic. In fact, any system of particles moved according to Newton's laws by ascertainable forces describes a predictable path. On the other hand, when the forces acting on a particle cannot be ascertained exactly, or even approximately, or when its initial position and velocity are not under our control or even our power to observe, then the path of the object is far from being predictable. Many—one is tempted to say almost all—motions observed in everyday life are of this kind; typical examples are the wafting of smoke, drifting clouds in the sky, dice thrown, cards shuffled and dealt. Such unpredictable motion is called *nondeterministic* or *random*.

Even though the outcome of a single throw of a die is unpredictable, the average outcome in the long run is quite predictable, at least if the die is the standard kind: each number will appear in about one sixth of a large number of throws. Similarly, if we repeatedly shuffle and deal out the top card of a deck of 52, each card will appear about 1/52 times the number of deals. With certain types of cloud formations experience may indicate rain in 3 out of 5 cases.

Probability is that branch of mathematics that deals with events whose individual outcome is unpredictable, but whose outcome on the average is predictable. In the next two sections we shall describe the rules of probability; in the next three we shall apply these rules to specific situations. As the reader will see, the notions and methods of calculus play an extremely important part in these applications; in particular, the logarithmic and exponential functions are ubiquitous. For these reasons this chapter has been included in this book.

6.1 Discrete probability

We shall consider experiments of which simple, almost simplistic examples are the tossing of a die, the shuffling of a deck and dealing the top card, the tossing of a coin. A more realistic example is the performance of a physical experiment. The two stages of an experiment are setting it up and observing its outcome. In many cases, such as meteorology, geology, oceanography, the setting up of the experiment is beyond our power; we can merely observe it being set up.

We shall be dealing with experiments that are *repeatable* and *nondeterministic*. Repeatable means that it can be set up repeatedly any number of times; nondeterministic means that any single performance of the experiment may result in a variety of *outcomes*. In the simple examples mentioned at the beginning of this section the possible outcomes are respectively: a whole number between 1 and 6, any one of 52 cards, head or tails. In this section we shall deal with experiments which, like the examples above, have *a finite number of possible outcomes*. We denote the number of possible outcomes by n, and shall number them from 1 to n.

Finally we assume that the outcome of the experiment, unpredictable in any individual instance, is *predictable on the average*. By this we mean: repeat the experiment infinitely many times. Denote by S_j the number of instances among the first N experimentations when the j-th outcome was observed to take place. Then the frequency S_j/N with which the j-th outcome has been observed to occur, tends to a limit as N tends to infinity. We call this limit the *probability of the j-th outcome*, and denote it by p_j:

(1.1)
$$p_j = \lim_{N \to \infty} \frac{S_j}{N}.$$

These probabilities have the following properties:

(i) Each probability p_j is a real number between 0 and 1:

(1.2)
$$0 \le p_j \le 1.$$

(ii) The sum of all probabilities equals 1:

(1.3)
$$p_1 + p_2 + \cdots + p_n = 1.$$

Both these properties follow from (1.1), for S_j/N lies between 0 and 1, and therefore so does its limit p_j; this proves (1.2). On the other hand, there are altogether n possible outcomes, so that each of the first N outcomes of the sequence of experiments performed falls into one of these n cases. Since S_j

is the number of instances among the first N when the j-th outcome was observed, it follows that

$$S_1 + S_2 + \cdots + S_n = N.$$

Dividing by N we get

(1.4)
$$\frac{S_1}{N} + \frac{S_2}{N} + \cdots + \frac{S_n}{N} = 1.$$

Now let N tend to infinity; by (1.1) the limit of S_1/N is p_1, of S_2/N is p_2, etc. We have seen in Section 1.4 that the limit of a sequence which is a sum of n sequences is the sum of the limits of the individual sequences. Applying this to (1.4) we obtain

$$\lim_{N \to \infty} \left\{ \frac{S_1}{N} + \frac{S_2}{N} + \cdots + \frac{S_n}{N} \right\} = \lim_{N \to \infty} \frac{S_1}{N} + \lim_{N \to \infty} \frac{S_2}{N} + \cdots + \lim_{N \to \infty} \frac{S_n}{N} = 1.$$

Therefore by (1.1):

$$p_1 + p_2 + \cdots + p_n = 1.$$

This is assertion (1.3).

Sometimes, in fact very often, we are not interested in all details of the outcome of an experiment, but merely in a particular aspect of it. For example, in drawing a card we may only be interested in the suit it belongs to; in throwing a die we may be interested only in whether the outcome is even or odd. An occurrence such as drawing a spade, or throwing an even number is called an *event*. In general *we define an event E as any collection of possible outcomes*. Thus drawing a spade is the collective name for the outcomes of drawing the deuce of spades, the three of spades, etc., all the way up to drawing the ace of spades. Similarly, an even throw of a die is the collective name for throwing a two, a four or a six.

We define the probability $p(E)$ of an event E similarly to the way we defined the probability of an outcome:

(1.5)
$$p(E) = \lim_{N \to \infty} \frac{S(E)}{N},$$

where $S(E)$ is the number of instances among the first N performances of the experiment when the event E took place. It is easy to show that this limit exists; in fact, it is easy to give a formula for $p(E)$. For, by definition, the event E takes place whenever the outcome belongs to the collection of those possible outcomes which make up the event E. Therefore $S(E)$, the number of instances in which E has occurred, is the sum of all S_j for those j which make up E:

(1.6)
$$S(E) = \sum_{j \text{ in } E} S_j.$$

Divide by N:

(1.7)
$$\frac{S(E)}{N} = \sum_{j \text{ in } E} \frac{S_j}{N}.$$

This relation says that the sequence $S(E)/N$ is the sum of the sequences $S_j/N, j$ in E. According to the fact used already in deriving (1.3), the limit of a sequence which is the sum of other sequences is the sum of the limits of the individual sequences. Since by (1.1)

$$\lim_{N \to \infty} \frac{S_j}{N} = p_j,$$

we deduce from (1.7) that, in the limit as N tends to ∞,

(1.8)
$$p(E) = \sum_{j \text{ in } E} p_j.$$

Two events E_1 and E_2 are called *disjoint* if both cannot take place simultaneously. That is, the set of outcomes which constitute the event E_1 and the set of outcomes which constitute the event E_2 have nothing in common. Here are some examples of disjoint events:

Example 1. $E_1 = $ drawing a spade; $E_2 = $ drawing a heart.

Example 2. $E_1 = $ throwing an even number; $E_2 = $ throwing a 3.

We define *the union of two events E_1 and E_2*, denoted by $E_1 + E_2$, as the event of either E_1 or E_2 taking place. That is, the outcomes that constitute $E_1 + E_2$ are the outcomes that constitute either E_1 or E_2.

The following observation is as important as it is simple: *The probability of the union of two disjoint events is the sum of the probabilities of each event*:

(1.9)
$$p(E_1 + E_2) = p(E_1) + p(E_2).$$

This is called the *addition rule for disjoint events*. This result follows from Formula (1.8) for the probability of an event; for, by definition of union,

$$p(E_1 + E_2) = \sum_{\substack{j \text{ in } E_1 \\ \text{or } E_2}} p_j.$$

On the other hand, disjointness means that any j may belong either to E_1 or E_2 but not to both. Therefore

$$p(E_1 + E_2) = \sum_{\substack{j \text{ in } E_1 \\ \text{or } E_2}} p_j = \sum_{j \text{ in } E_1} p_j + \sum_{j \text{ in } E_2} p_j$$
$$= p(E_1) + p(E_2),$$

as asserted in (1.9).

261

Next we turn to another important idea in probability, the *independence* of two experiments. Take two experiments such as throwing a die, and shuffling a deck and dealing the top card. Our common sense plus everything we know about the laws of nature tells us that these experiments are totally independent of each other in the sense that the outcome of one cannot possibly influence the other, nor is the outcome of both under the influence of a common cause. We state now, precisely in the language of probability theory, an important consequence of independence.

Given any two experiments, we can compound them into a single, *combined experiment* simply by performing them simultaneously. Let E be any event in the framework of one of the experiments, F any event in the framework of the other. The combined event of both E and F taking place will be denoted by EF. For instance, if E is the event of an even throw and F the event of drawing a spade, EF is the event of an even throw *and* drawing a spade. We claim that *if the experiments are independent, then the probability of the combined event EF is the product of the separate probabilities of the events E and F*:

(1.10) $$p(EF) = p(E)p(F).$$

We shall refer to this relation as the *product rule* for independent experiments.

We now show how to deduce the product rule. Imagine the combined experiment repeated infinitely often. We look at the first N experiments of this sequence. Among the first N, count the number of times E has occurred, F has occurred, and EF has occurred. We denote these numbers by $S(E)$, $S(F)$ and $S(EF)$. By definition (1.5) of the probability of an event

(1.11)$_E$ $$p(E) = \lim_{N \to \infty} \frac{S(E)}{N}$$

(1.11)$_F$ $$p(F) = \lim_{N \to \infty} \frac{S(F)}{N}$$

(1.12) $$p(EF) = \lim_{N \to \infty} \frac{S(EF)}{N}.$$

Suppose that we single out from the sequence of combined experiments the *subsequence* of those where E occurred. The frequency of occurrence of F in this subsequence is $S(EF)/S(E)$. If the two events E and F are truly independent, the frequency with which F occurs in this subsequence should be the same as the frequency with which F occurs in the original sequence. Therefore

(1.13) $$\lim_{N \to \infty} \frac{S(EF)}{S(E)} = \lim_{N \to \infty} \frac{S(F)}{N} = p(F).$$

262

Now we write the sequence $S(EF)/N$ as the product

$$\frac{S(EF)}{N} = \frac{S(EF)}{S(E)} \frac{S(E)}{N}.$$

According to the arithmetic of convergent sequences explained in Section 1.4 the limit of the product of two sequences is the product of their limits:

(1.14)
$$\lim_{N \to \infty} \frac{S(EF)}{N} = \lim_{N \to \infty} \frac{S(EF)}{S(E)} \cdot \lim_{N \to \infty} \frac{S(E)}{N}.$$

By (1.12) the quantity on the left is $p(EF)$, while according to (1.13) and $(1.11)_E$ the product on the right is $p(F)p(E)$; thus the product rule (1.10) follows from (1.14).

Suppose that one experiment has m possible outcomes, the other n possible outcomes. Denote their respective probabilities by $p_1, \ldots, p_m$ and $q_1, \ldots, q_n$. The combined experiment then has mn possible outcomes, namely all pairs of outcomes (j, k). If the experiments are independent, then the product rule (1.10) tells us that the outcome (j, k) of the combined experiment has probability

(1.15)
$$p_j q_k.$$

This formula plays a very important role in probability; we now give an illustration of its use.

Suppose the two experiments which we have been discussing are both the tossing of a die. Then the combined experiment is the tossing of a pair of dice. Each experiment has six possible outcomes, with probability $\frac{1}{6}$. According to Formula (1.10), there are 36 combined outcomes, each with probability $\frac{1}{36}$. We now ask the question: What is the probability of the event of tossing 7? Clearly, there are six ways of tossing 7: $(1, 6), (2, 5), (3, 4), (4, 3), (5, 2), (6, 1)$. According to Formula (1.8), the probability of tossing 7 is the sum of the probabilities of these six outcomes that constitute the event. That sum is

$$6 \times \tfrac{1}{36} = \tfrac{1}{6}.$$

Similarly we can calculate the probability of tossing any number between 2 and 12. We ask the reader to go through the calculations of determining the probability that the numbers $2, 3, \ldots, 12$ will be thrown. Here the result is tabulated:

(1.16)

Throw	2	3	4	5	6	7	8	9	10	11	12
Probability	$\frac{1}{36}$	$\frac{1}{18}$	$\frac{1}{12}$	$\frac{1}{9}$	$\frac{5}{36}$	$\frac{1}{6}$	$\frac{5}{36}$	$\frac{1}{9}$	$\frac{1}{12}$	$\frac{1}{18}$	$\frac{1}{36}$

We now turn to another important concept of probability, the *numerical outcome* of an experiment. In physical experiments designed to measure the value of a single physical quantity, the numerical outcome is simply the measured value of the quantity in question. For the simple example of throwing a pair of dice, the numerical outcome might be the sum of the face values of each die. For the experiment of dealing a bridge hand, the numerical outcome might be the point count of the bridge hand. In general, the *numerical outcome of an experiment means the assignment of a real number x_j to each of the possible outcomes, $j = 1, 2, \ldots, n$.* As the above examples show, the actual assignment of these numerical values is up to the experimenter and is dictated by the use he wishes to make of the experiment and his theories connected with it.

If the experiment is nondeterministic, its outcome is not predictable in any *individual* performance of the experiment. However, imagine performing the experiment infinitely many times, and list the outcomes, in the order they occur:

$$j_1, j_2, \ldots, j_N, \ldots \ .$$

Each j_N is an integer between 1 and n. List also the numerical outcomes:

(1.17) $$a_1, a_2, \ldots, a_N, \ldots \ ;$$

a_N, the N-th numerical outcome, is of course

(1.18) $$a_N = x_{j_N} .$$

In this section we have considered experiments that, although individually unpredictable, are predictable on the average. We shall now show that in this case the sequence of numerical outcomes (1.17) has an average value $\bar{x}$ defined to be the limit as N tends to infinity of the *arithmetic mean* of the first N numerical outcomes:

(1.19) $$\bar{x} = \lim_{N \to \infty} \frac{a_1 + a_2 + \cdots + a_N}{N}.$$

Furthermore we shall show that the average numerical outcome $\bar{x}$ is related to the numerical outcomes x_j and to the probabilities p_j of these outcomes by the following simple formula:

(1.20) $$\bar{x} = p_1 x_1 + \cdots + p_n x_n.$$

$\bar{x}$ is called the *weighted average* of the numbers $x_1, x_2, \ldots, x_n$; the $p_1, p_2, \ldots, p_n$ are called the weights.

Formula (1.20) is easy to prove. Denote by S_j the number of times the j-th outcome has occurred among the first N. Then, according to Formula (1.18), among $a_1, a_2, \ldots, a_N$ there are S_1 which are equal x_1, S_2 which are equal $x_2, \ldots, S_n$ which are equal x_n. Therefore

$$a_1 + a_2 + \cdots + a_N = S_1 x_1 + S_2 x_2 + \cdots + S_n x_n.$$

Dividing by N we get

(1.21) $$\frac{a_1 + \cdots + a_N}{N} = \frac{S_1}{N} x_1 + \cdots + \frac{S_n}{N} x_n.$$

Again we appeal to the rule concerning the sum of convergent sequences explained in Section 1.4: the limit of a sum of convergent sequences is the sum of their limits. We apply this to the sum of the sequences

$$\frac{S_1}{N} x_1, \; \frac{S_2}{N} x_2, \ldots, \; \frac{S_n}{N} x_n;$$

their limits are, by (1.1), $p_1 x_1, p_2 x_2, \ldots, p_n x_n$. Therefore the limit of their sum is $p_1 x_1 + \cdots + p_n x_n$. In view of (1.21) we obtain (1.20).

We now give an example of Formula (1.20) for the average numerical outcome. Take the experiment of throwing a pair of dice; we classify the outcomes as throwing a 2, 3, ... up to 12. We take these numbers to be the numerical outcomes of the experiment. The probability of each outcome is given in Table (1. 6). Using this table and Formula (1.20), we get the following value for the average numerical outcome of a throw with a pair of dice:

$$\bar{x} = \frac{1}{36} 2 + \frac{1}{18} 3 + \frac{1}{12} 4 + \frac{1}{9} 5 + \frac{5}{36} 6 + \frac{1}{6} 7$$

$$+ \frac{5}{36} 8 + \frac{1}{9} 9 + \frac{1}{12} 10 + \frac{1}{18} 11 + \frac{1}{36} 12 = 7.$$

EXERCISES

1.1 Let E be an event consisting of a certain collection of outcomes of an experiment. We may call these outcomes *favorable* from the point of view of the event that interests us. The collection of all unfavorable outcomes, i.e., those that do not belong to E, is called the event *complementary* to E, and is often denoted as $\bar{E}$.
   Prove that

$$p(E) + p(\bar{E}) = 1.$$

1.2 According to Formula (1.15), the probability of the outcome (j, k) of the combination of two independent experiments is

$$p_j q_k.$$

Show that the sum of all these probabilities is 1.

1.3 *An event E is included in the event F if, whenever E takes place, F also takes place.* Another way of expressing this relationship is to say that the outcomes that constitute E form a *subset* of the outcomes that make up F. The assertion "the event E is included in the event F" is expressed in symbols by $E \subset F$. For example, the event E of drawing a spade is included in the event F of drawing a black card.
   Show that if $E \subset F$, then

(1.22) $$p(E) \leq p(F).$$

1.4 Let $E_1, E_2, \ldots, E_m$ be a collection of m events which are *disjoint* in the sense that no outcome can belong to more than one event. Denote the union of the events E_j by E:

$$E = E_1 + \cdots + E_m.$$

Show that the *additive* rule holds:

$$p(E) = p(E_1) + \cdots + p(E_m).$$

1.5 Use a pseudo random number generator to simulate a sequence of outcomes of the roll of a pair of dice. See how close the average of the first thousand rolls comes to 7. [*Hint*: A pseudo random number generator is usually available as a function on most computers. Assume p is a random number uniformly distributed in the interval

$$0 \leq p \leq 1.$$

Subdivide this interval into six equal parts. Then assign the value $x_1 = 1$ to any number in the first interval, $x_2 = 2$ to any number in the second, ..., $x_6 = 6$ to any number in the last. Generate a *pair* of random numbers (p_j, p_k) and assign to the outcome of that experiment the sum of the values $x_j + x_k$.]

6.2 Information theory or how interesting is interesting

It is a universal human experience that some information is dull, some interesting. Man bites dog is news, dog bites man isn't. In this section we describe a way of assigning a quantitative measure to the value of a piece of information.

"Information," in this discussion, shall mean being informed that a certain event E, whose occurrence is subject to chance, has occurred. An event, as defined in Section 6.1, is a collection of possible outcomes of an experiment; the frequency with which the event E occurs in a large number of performances of the experiment is its probability $p(E)$. We assume in this theory that the information gained upon learning that an event has occurred depends only on the probability p of the event. We denote by $f(p)$ the information thus gained; in other words, we could think of $f(p)$ as a measure of the element of surprise generated by the event that has occurred.

What properties does this function f have? We claim that the following three are mandatory:

(2.1)
    (i) $f(p)$ increases as p decreases.
    (ii) $f(1) = 0$.
    (iii) $f(p)$ tends to ∞ as p tends to 0.

Property (i) expresses the fact that the occurrence of a less probable event is more surprising than the occurrence of a more probable one and therefore carries more information. Property (ii) says that the occurrence of an event that is a near certainty imparts no new information, while property (iii) says

266

that the occurrence of a rare event is of great interest and furnishes a lot of new information.

Next we deduce a crucial property of the function f. Suppose that two events E and F are *independent;* since such events are totally unrelated, being informed that both of them have occurred conveys no more information than learning that each has occurred separately, i.e., the information gained upon learning that both have occurred is the *sum* of the information gained by learning of each occurrence separately. Denote by p and q the probabilities of events E and F, respectively; according to the product rule (1.10), the probability of the combined event EF is the product pq. Therefore the rule stated above is expressed by the equation

$$(2.2) \qquad\qquad f(pq) = f(p) + f(q).$$

We saw in Section 5.2 that the logarithmic function satisfies the above functional equation, and conversely, the only function which satisfies the functional equation (2.2) is a constant multiple, k, of $\log p$. So we conclude that

$$(2.3) \qquad\qquad f(p) = k \log p.$$

What is the value of this constant? According to property (i) of (2.1), $f(p)$ increases with decreasing p. Since $\log p$ increases with increasing p, we conclude that the constant must be *negative*. What about its magnitude? There is no way of deciding that without first adopting an arbitrary unit of information. For convenience, we choose the constant to be -1, and so

$$(2.4) \qquad\qquad f(p) = -\log p.$$

What about property (ii) of (2.1), and property (iii)? We claim that they are satisfied by f as defined by (2.4); we leave the verification to the reader as an exercise.

Now consider an experiment with n possible outcomes having probabilities $p_1, p_2, \ldots, p_n$. If in a single performance of the experiment the j-th outcome occurs, we have, according to Formula (2.4), gained information in the amount $-\log p_j$. We now ask the following question: If we perform the experiment repeatedly many times, what is the *average information gain*? The answer to this question is contained in Formula (1.20) of Section 6.1 concerning the average numerical outcome of a series of experiments. According to that formula, if the j-th numerical outcome is x_j, the average numerical outcome is $p_1 x_1 + \cdots + p_n x_n$. In our case we take the numerical outcome to be the information gained in the j-th outcome,

$$x_j = -\log p_j.$$

So the average information gain I is

$$(2.5) \qquad I = -(p_1 \log p_1 + p_2 \log p_2 + \cdots + p_n \log p_n).$$

267

To indicate the dependence of I on the probabilities, we shall write

(2.6) $$I = I(p_1, \ldots, p_n).$$

Formula (2.5) is due to Claude Shannon.*

Let's look at the simplest case when there are only two possible outcomes, with probabilities p_1 and p_2. Since, according to (1.3), the sum of all probabilities is 1, $p_1 + p_2 = 1$, so we can express p_2 as

$$p_2 = 1 - p_1.$$

Using this, and dropping off the subscript, we can write the Formula (2.5) for information gain as follows:

(2.7)
$$\begin{aligned} I &= -(p_1 \log p_1 + p_2 \log p_2) \\ &= -p \log p + (p - 1)\log(1 - p). \end{aligned}$$

How does I depend on p? To study how I changes with p we use the methods of calculus: we differentiate I with respect to p and get

$$\frac{dI}{dp} = -\log p - 1 + \log(1 - p) + 1$$

$$= -\log p + \log(1 - p).$$

Using the functional equation of the log function, we can rewrite this as

(2.8) $$\frac{dI}{dp} = \log\left(\frac{1 - p}{p}\right).$$

Now we have seen in Section 5.2 that $\log x$ is positive for $x > 1$, negative for $x < 1$. Clearly

$$\frac{(1 - p)}{p} \quad \begin{cases} > 1 & \text{for} \quad 0 < p < \tfrac{1}{2} \\ < 1 & \text{for} \quad \tfrac{1}{2} < p < 1. \end{cases}$$

Therefore we conclude from (2.8) that

(2.9) $$\frac{dI}{dp} \quad \begin{cases} > 0 & \text{for } 0 < p < \tfrac{1}{2} \\ < 0 & \text{for } \tfrac{1}{2} < p < 1. \end{cases}$$

It follows from (2.9) that $I(p)$ is an increasing function of p from 0 to $\tfrac{1}{2}$, and a decreasing function as p goes from $\tfrac{1}{2}$ to 1. Clearly, *the largest value of I occurs when* $p = \tfrac{1}{2}$. In words: the most information that can be gained on the average from an experiment with two possible outcomes occurs when the probabilities of the two outcomes are equal.

* A good introduction to Shannon's information theory is C E. Shannon's and W. Weaver's little book *The Mathematical Theory of Communication*. Urbana: University of Illinois Press, 1949.

We now extend this result to experiments with n possible outcomes. Denote by $I(p_1, \ldots, p_n)$ the function defined in (2.5); the probabilities $p_1, \ldots, p_n$ are numbers between 0 and 1, satisfying (1.3):

$$(2.10) \qquad\qquad p_1 + \cdots + p_n = 1.$$

Theorem 2.1. *The function I, given by (2.5), is largest when*

$$p_1 = p_2 = \cdots = p_n = \frac{1}{n}.$$

PROOF. We have to show that

$$(2.11) \qquad\qquad I(p_1, \ldots, p_n) < I\left(\frac{1}{n}, \ldots, \frac{1}{n}\right)$$

unless all the p_j are equal to $1/n$.

In order to apply the methods of calculus to proving inequality (2.11), we consider the following functions $r_j(s)$:

$$(2.12) \qquad\qquad r_j(s) = sp_j + (1 - s)\frac{1}{n}, \qquad j = 1, \ldots, n.$$

These functions are so designed that at $s = 0$ the value of each r_j is $1/n$, and at $s = 1$ the value of r_j is p_j:

$$(2.13) \qquad\qquad r_j(0) = \frac{1}{n}, \qquad r_j(1) = p_j, \qquad j = 1, \ldots, n.$$

So, if we look at the function

$$J(s) = I(r_1(s), \ldots, r_n(s)),$$

(2.13) tells us that

$$J(0) = I\left(\frac{1}{n}, \ldots, \frac{1}{n}\right), \qquad J(1) = I(p_1, \ldots, p_n).$$

Therefore inequality (2.11) can be expressed simply as

$$J(1) < J(0).$$

We shall prove this by showing that $J(s)$ is a decreasing function of s. We shall use the monotonicity criterion to demonstrate the decreasing character of $J(s)$, by verifying that the derivative of J is negative. To calculate the derivative of $J(s)$, we need to know the derivative of each r_j with respect to s. This is easily calculated from (2.12):

$$(2.14) \qquad\qquad \frac{dr_j(s)}{ds} = p_j - \frac{1}{n}.$$

269

Note that the derivative of each r_j is constant. This is not surprising, since each r_j is a linear function of s.

Using the definition (2.5) of I,

(2.15) $$J(s) = -[r_1 \log r_1 + \cdots + r_n \log r_n].$$

We calculate the derivative of J using the chain rule:

(2.16) $$\frac{dJ}{ds} = -\left[(1 + \log r_1)\frac{dr_1}{ds} + \cdots + (1 + \log r_n)\frac{dr_n}{ds}\right];$$

using the value of dr_j/ds given by (2.14) in (2.16), we get the following expression:

(2.17) $$\frac{dJ}{ds} = -\left[(1 + \log r_1)\left(p_1 - \frac{1}{n}\right) + \cdots + (1 + \log r_n)\left(p_n - \frac{1}{n}\right)\right].$$

Set $s = 0$ in (2.17); using the fact, see (2.13), that at $s = 0$ the value of $r_j = 1/n$, we obtain

$$\frac{dJ}{ds}(0) = -\left(1 + \log\frac{1}{n}\right)\left(p_1 - \frac{1}{n} + \cdots + p_n - \frac{1}{n}\right).$$

According to (2.10), $p_1 + \cdots + p_n = 1$; therefore we conclude from the above relation that

(2.18) $$J'(0) = 0;$$

here we have switched to the notation J' for the derivative dJ/ds. We claim that, for all positive values of s,

(2.19) $$J'(s) < 0;$$

if we can show this, our proof that J is a decreasing function is complete. To verify (2.19) we shall show that $J'(s)$ itself is a decreasing function of s; since by (2.18), J' is zero at $s = 0$, J' would have to be negative for all positive s.

To show that J' is decreasing, we apply the monotonicity criterion once more, this time to J', and show that J'' is negative. We compute J'' by differentiating (2.17); using (2.14) we get

$$J'' = -\frac{1}{r_1}\left(p_1 - \frac{1}{n}\right)r_1' - \cdots - \frac{1}{r_n}\left(p_n - \frac{1}{n}\right)r_n'$$

$$= -\frac{1}{r_1}\left(p_1 - \frac{1}{n}\right)^2 - \cdots - \frac{1}{r_n}\left(p_n - \frac{1}{n}\right)^2.$$

Each term in the above sum is negative or zero; since not all p_j are equal to $1/n$, at least some terms are negative. This proves $J'' < 0$, and completes the proof of our theorem. ☐

EXERCISES

2.1 Let $p_1, \ldots, p_n$ be the probabilities of the n possible outcomes of an experiment, and $q_1, \ldots, q_m$ the probabilities of the outcomes of another experiment. Suppose that the experiments are *independent*, i.e., if we combine the two experiments, the probability of the first experiment having the j-th outcome and the second experiment having the k-th outcome is the product

$$r_{jk} = p_j q_k.$$

Show that, in this case, the average information gain from the combined experiment is the *sum* of the average information gains in the performances of each experiment separately:

$$I(r_{11}, \ldots, r_{mn}) = I(p_1, \ldots, p_n) + I(q_1, \ldots, q_m).$$

2.2 Suppose that an experiment has *three* possible outcomes, with probabilities p, q and r; naturally

$$p + q + r = 1.$$

Suppose that we simplify the description of our experiment by lumping the last two cases together, i.e., we look upon the experiment as having two possible outcomes, one with probability p, the other with probability $1 - p$. The average information gain when looking at the full description of the experiment is

$$-p \log p - q \log q - r \log r.$$

When looking at the simplified description the average information gain is

$$-p \log p - (1 - p)\log(1 - p).$$

Prove that the average information gain from the full experiment is greater than from its simplified description. The result is to be expected: if we lump data together, we lose information.

2.3 Suppose an experiment can have n possible outcomes, each with probability p_j, $j = 1, \ldots, n$. The information gained from this experiment is on the average

$$-p_1 \log p_1 - \cdots - p_n \log p_n.$$

Suppose we simplify the description of the experiment by lumping the last $n - 1$ outcomes together as failures of the first case. The average information gain from this description is

$$-p_1 \log p_1 - (1 - p_1)\log(1 - p_1).$$

Prove that we gain on the average more information from the full description than from the simplified description.

6.3 Continuous probability

The probability theory developed in Section 6.1 deals with experiments which have finitely many possible outcomes. This is a good model for experiments such as tossing a coin, throwing a die, drawing a card; but is artificial for

271

experiments such as making a physical measurement with an apparatus subject to random disturbances that can be reduced, but not totally eliminated. *Any real number is a possible outcome* of such an experiment; this section is devoted to developing a probability theory for such situations. The experiments we study are, just like the ones in Section 6.1, repeatable, nondeterministic but predictable on the average.

By "predictable on the average" we mean: Repeat the experiment infinitely many times and denote by $S(x)$ the number of instances among the first N performances when the numerical outcome was less than x. Then the frequency $S(x)/N$ with which this event occurs tends to a limit as N tends to infinity. This limit is called the *probability that the outcome is less than x*, and is denoted by $P(x)$:

$$(3.1) \qquad P(x) = \lim_{N \to \infty} \frac{S(x)}{N}.$$

Since the number of instances among the first N of an outcome less than x is $S(x)$, the number of outcomes greater than or equal to x is $N - S(x)$, and their frequency is $[N - S(x)]/N$. It follows from (3.1) that the limit of this frequency as N tends to infinity exists and is

$$(3.2) \qquad \lim_{N \to \infty} \frac{N - S(x)}{N} = \lim_{N \to \infty} \left(1 - \frac{S(x)}{N}\right) = 1 - P(x).$$

This shows that the probability of an outcome $\geq x$ is $1 - P(x)$.

The probability $P(x)$ has the following properties.

(i) Each probability lies between 0 and 1:

$$(3.3) \qquad 0 \leq P(x) \leq 1.$$

(ii) $P(x)$ is an increasing function of x.

Properties (i) and (ii) are consequences of the definition (3.1): for, the number $S(x)$ lies between 0 and N, so that the ratio $S(x)/N$ lies between 0 and 1; but then so does the limit $P(x)$, as asserted in (3.3). Secondly, $S(x)$ is an increasing function of x, so that the ratio $S(x)/N$ is an increasing function of x; but then so is the limit $P(x)$, as asserted in (ii).

We shall assume two further properties of $P(x)$:

(iii) $P(x)$ tends to 0 as x tends to $-\infty$.
(iv) $P(x)$ tends to 1 as x tends to $+\infty$.

Property (iii) says that the probability of a very large negative outcome is very small. Property (iv) says that $1 - P(x)$ is small when x is large positive; since by (3.2) $1 - P(x)$ is the probability that the outcome is $\geq x$, (iv) asserts that very large positive outcomes are very improbable.

As in Section 6.1, we shall be interested in collections of outcomes; we call these *events*. Examples of events are:

Example 1. The outcome is less than x.

Example 2. The outcome lies in the interval I.

Example 3. The outcome lies in a given collection of intervals.

The probability of an event E, which we denote by $P(E)$, is defined as in Section 6.1, as the limit of the frequencies:

$$\lim_{N \to \infty} \frac{S(E)}{N} = P(E),$$

$S(E)$ the number of times the event E took place among the first N of an infinite sequence of performances of an experiment.

The simple argument presented in Section 6.1 can be used in the present context to show the additive rules for disjoint events: Suppose E and F are two events which have probabilities $P(E)$ and $P(F)$ and suppose that they are *disjoint* in the sense that one event precludes the other; that is, no outcome can belong to both E and F. In this case the *union* $E + F$ of the events, consisting of all outcomes either in E or in F, also has a probability which is the sum of the probabilities of E and F:

(3.4) $$P(E + F) = P(E) + P(F).$$

We apply this to the events

$$E: \text{the outcome } x \text{ is } < a$$

and

$$F: \text{the outcome } x \text{ lies in the interval } a \le x < b.$$

The union of these two is

$$E + F: \text{the outcome } x \text{ is } < b.$$

Using the notation introduced in (3.1),

$$P(E) = P(a), \quad P(E + F) = P(b).$$

We conclude from (3.4) that

(3.5) $$P(F) = P(b) - P(a)$$

is the probability of an outcome less than b, but greater than or equal to a.

We now make the following assumption:

(v) $P(x)$ is a differentiable function.

This assumption holds in many important cases and allows us to use the methods of calculus. We denote the derivative of P by p:

(3.6)
$$\frac{dP(x)}{dx} = p(x).$$

The function $p(x)$ is called the *probability density*. According to the mean value theorem of Section 3.8, for any a and b there is a number c lying between a and b such that

(3.7)
$$P(b) - P(a) = p(c)(b - a).$$

According to the fundamental theorem of calculus, P can be expressed as the integral of its derivative:

(3.8)
$$P(b) - P(a) = \int_a^b p(x)dx.$$

Let a tend to $-\infty$; since by assumption (iii) $P(a)$ tends to 0, we conclude from (3.8) that

(3.9)
$$P(b) = \int_{-\infty}^b p(x)dx.$$

Let b tend to $+\infty$; since by assumption (iv) $P(b)$ tends to 1, we conclude from (3.9) that

(3.10)
$$1 = \int_{-\infty}^\infty p(x)dx.$$

According to property (ii), $P(x)$ is an increasing function of x. Since the derivative of an increasing function is nowhere negative, we conclude that $p(x)$ is nonnegative for all x:

(3.11)
$$0 \le p(x).$$

We now define the *average numerical outcome* $\bar{x}$ of an experiment analogously to the discrete case discussed in Section 6.1. Imagine the experiment performed infinitely often, and denote the sequence of outcomes by

$$a_1, a_2, \ldots, a_N, \ldots .$$

Theorem 3.1. *If an experiment is predictable on the average, and if the outcomes are restricted to lie in a finite interval, then*

(3.12)
$$\bar{x} = \lim_{N \to \infty} \frac{a_1 + \cdots + a_N}{N}$$

exists and is equal to

(3.13)
$$\bar{x} = \int_{-\infty}^\infty xp(x)dx.$$

The assumption that the outcomes lie on a finite interval is a realistic one if one thinks of the experiment as a measurement. After all, every measuring apparatus has a finite range. However there are probability densities of great theoretical interest, such as the ones we shall discuss in Sections 6.4 and 6.5 which are positive for all real x. Theorem 3.1 remains true for these experiments, too, under some additional assumptions, of which the most important is that the improper integral (3.13) defining $\bar{x}$ should exist.

PROOF. Divide the interval I in which all outcomes lie into n subintervals $I_1, \ldots, I_n$. The probability P_j of an outcome lying in the interval I_j is, by (3.5), the difference of the values of P at the endpoints of I_j. According to Formula (3.7), this difference is equal to

(3.14)
$$P_j = p(x_j)|I_j|$$

where x_j is a point in I_j, and $|I_j|$ denotes the length of I_j. We now simplify the original experiment by recording merely the intervals I_j in which the outcome falls, and calling the numerical outcome in this case x_j, that point in I_j which appears in Formula (3.14). The actual outcome of the full experiment and the numerical outcome of the simplified experiment always belong to the same subinterval of the partition we have taken; therefore *these two outcomes differ at most by w_n, the length of largest of the subintervals I_j,* i.e., $w_n = \max_{1 \le j \le n} |I_j|$.

Now consider the sequence of outcomes $a_1, a_2, \ldots, a_N, \ldots$ of the original experiment. The corresponding outcomes of the simplified experiment are $b_1, b_2, \ldots, b_N, \ldots$. As explained above, b_k is obtained by determining the interval I_j to which a_k belongs, and then setting b_k equal to x_j (see Figure 6.1).

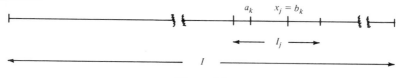

Figure 6.1

The simplified experiment has a finite number of outcomes. For such discrete experiments we have shown in Section 6.1 that the average of the numerical outcomes tends to a limit, called the average numerical outcome. We denote it by $\bar{x}_n$:

(3.15)
$$\bar{x}_n = \lim_{N \to \infty} \frac{b_1 + \cdots + b_N}{N},$$

where n is the number of subintervals of I. We have furthermore shown how to express $\bar{x}_n$ in terms of the probabilities of the numerical outcomes. According to Formula (1.20)

(3.16)
$$\bar{x}_n = P_1 x_1 + \cdots + P_n x_n.$$

We have seen that each exact outcome a_k differs from the corresponding simplified outcome at most by w_n; it follows that the same is true of their average:

(3.17) $$\frac{a_1 + \cdots + a_N}{N} \quad \text{and} \quad \frac{b_1 + \cdots + b_N}{N}$$

differ at most by w_n. According to (3.15), by choosing N large enough the averages of $b_1, \ldots, b_N$ can be made very close to $\bar{x}_n$, certainly within w_n of $\bar{x}_n$. It follows from this and the comparison statement (3.17) that for such N

(3.17)' $$\frac{a_1 + \cdots + a_n}{N} \quad \text{and} \quad \bar{x}_n$$

differ at most by $2w_n$.

We substitute the expressions (3.14) for P_j into Formula (3.16) for $\bar{x}_n$, and obtain

$$\bar{x}_n = p(x_1)x_1|I_1| + \cdots + p(x_n)x_n|I_n|.$$

We recognize this as an approximating sum for the integral of $xp(x)$ over I. If the partition is fine enough, *the approximating sum $\bar{x}_n$ differs very little from the value of the integral*

(3.18) $$\int_I xp(x)dx.$$

But if the partition is fine enough, then w_n, the length of the longest subinterval in the partition, is very small. So it follows from the comparison (3.17)' that, for N large enough, $(a_1 + \cdots + a_N)/N$ differs from $\bar{x}_n$ by as little as we choose. This demonstrates the existence of the limit (3.12). Since $\bar{x}_n$ in turn differs by very little from the integral (3.18), we conclude that for N large enough, $(a_1 + \cdots + a_N)/N$ differs from the value of the integral (3.18) by a very small amount. This shows that $(a_1 + \cdots + a_N)/N$ tends to (3.18) as N tends to ∞.

Since we have assumed that all observations fall within I, it follows that the probability density $p(x)$ is zero for x outside I. This shows that the integral (3.18) has the same value as the integral (3.13) and completes the proof of Theorem 3.1. $\square$

We now give some examples of Formula (3.13).

Example 1. Let A be positive, and define $p(x)$ by

$$p(x) = \begin{cases} 0 & \text{for } x < 0 \\ 1/A & \text{for } 0 \leq x < A \\ 0 & \text{for } A \leq x; \end{cases}$$

see Figure 6.2. This choice of p satisfies condition (3.10), i.e.,

$$\int_{-\infty}^{\infty} p(x)dx = \int_0^A \frac{1}{A} dx = 1.$$

Figure 6.2

We now compute $\bar{x}$:

$$\bar{x} = \int_{-\infty}^{\infty} xp(x)dx = \int_0^A \frac{x}{A}dx$$

$$= \frac{x^2}{2A}\Big|_0^A = \frac{A}{2}.$$

The symbol $f(x)|_a^b$ means $f(b) - f(a)$.

Example 2. Let A be a positive number; set

$$p(x) = \begin{cases} 0 & \text{for } x < 0 \\ Ae^{-Ax} & \text{for } 0 \leq x; \end{cases}$$

see Figure 6.3. Again we can verify that condition (3.10) is satisfied. Using the fundamental theorem of calculus we have

$$\int_{-\infty}^{\infty} p(x)dx = \int_0^{\infty} Ae^{-Ax}\,dx = -e^{-Ax}\Big|_0^{\infty} = 1.$$

We now compute $\bar{x}$; using integration by parts and then the fundamental theorem, we have

$$\bar{x} = \int_{-\infty}^{\infty} xp(x)dx = \int_0^{\infty} xAe^{-Ax}\,dx = \int_0^{\infty} e^{-Ax}\,dx$$

$$= \frac{-e^{-Ax}}{A}\Big|_0^{\infty} = \frac{1}{A}.$$

Figure 6.3

Example 3. Assume that $p(x)$ is an even function. Then $xp(x)$ is an odd function and so

$$\bar{x} = \int_{-\infty}^{\infty} xp(x)dx = 0.$$

Let $f(x)$ be any function of x. We define the *average value of f, with respect to the probability density $p(x)$*, denoted by $\bar{f}$, to be

(3.19) $$\bar{f} = \int_{-\infty}^{\infty} f(x)p(x)dx.$$

One can show, analogously to the foregoing discussion, that if $a_1, \ldots, a_N, \ldots$ is a sequence of outcomes, then

(3.20) $$\lim_{N \to \infty} \frac{f(a_1) + \cdots + f(a_N)}{N} = \bar{f}.$$

We now turn to the important concept of *independence*. The intuitive notion is the same as in the discrete models discussed in Section 6.1: Two experiments are independent if the outcome of either has no influence on the other, nor are they both influenced by a common cause. We analyse the consequences of independence the same way we did in Section 6.1, by constructing a *combined experiment* consisting of performing *both* experiments.

We first analyse the case when the outcome of the first experiment may be any real number, but the second experiment can have only a finite number of outcomes. As before, we denote by $P(a)$ the probability that the numerical outcome of the first experiment is less than a. The second experiment has n possible numerical outcomes $a_1, \ldots, a_n$; they occur with probabilities $Q_1, Q_2, \ldots, Q_n$. The same reasoning that led to the product formula (1.10), in Section 6.1, now leads to the following: If the two experiments are independent, the probability that in the combined experiment the numerical outcome of the first experiment is less than a *and* in the second experiment the numerical outcome is a_j is the *product* of the probabilities of the separate events:

(3.21) $$Q_j P(a)$$

We define the *numerical outcome of the combined experiment to be the sum of the separate numerical outcomes of the two experiments which constitute it.*

We now derive a useful and important formula for the probability that the numerical outcome of the combined experiment is less than x. We denote this event by $E(x)$, and denote its probability by $U(x)$. We shall show that

(3.22) $$U(x) = Q_1 P(x - a_1) + \cdots + Q_n P(x - a_n).$$

PROOF. The numerical outcome of the second experiment is one of the n numbers a_j. The numerical outcome of the combined experiment is then less than x if, and only if, the outcome of the first experiment is less than $x - a_j$.

278

We denote this event by $E_j(x)$. Thus the event $E(x)$ is the union of the events $E_j(x)$:

(3.23) $$E(x) = E_1(x) + \cdots + E_n(x).$$

The events $E_j(x)$ are clearly disjoint, that is, an outcome cannot belong to two distinct events $E_j(x)$ and $E_k(x)$. It follows then from the *addition rule for disjoint events* that the probability of their *union* $E(x)$ is the *sum* of the probabilities of the events $E_j(x)$.

Since the two experiments are independent, the probability of $E_j(x)$ is given by Formula (3.21), with $x - a_j$ in place of a, i.e., by

$$Q_j P(x - a_j).$$

The sum of the probabilities of the $E_j(x)$ is $U(x)$, the probability of $E(x)$; this completes the proof of Formula (3.22). $\square$

Assume, as we have throughout this section, that $P(x)$ is differentiable; denote its derivative by $p(x)$, *the probability density of the first experiment.* It follows then, from Formula (3.22), that $U(x)$ is also differentiable. Its derivative $u(x)$, the probability density of $E(x)$, can be determined by differentiating (3.22):

(3.24) $$u(x) = Q_1 p(x - a_1) + \cdots + Q_n p(x - a_n).$$

We now turn to the situation when both experiments can have any real number as outcome. We denote by $P(a)$ and $Q(a)$ the probability that the outcome is less than a in each of the two experiments, respectively. For any interval I, we denote the probability that the outcome of the first experiment lies in I by $P(I)$, and the probability that the outcome of the second experiment lies in an interval J by $Q(J)$. Denote by IJ the event in the combined experiment when the outcome of the first experiment lies in I and that of the second in J; denote by $C(IJ)$ the probability of this event. The same argument used in Section 6.1 to derive (1.10) and appealed to in this section for (3.21) shows that, if the two experiments are independent, then

(3.25) $$C(IJ) = P(I)Q(J).$$

We define *the numerical outcome of the combined experiment as the sum of the outcomes of the two experiments which constitute it.*

We shall prove the following analogue of Formula (3.22): Suppose that $Q(x)$ is a differentiable function; denote its derivative by $q(x)$. Then $U(x)$, the probability that the outcome of the combined experiment is less than x is given by the following formula:

(3.26) $$U(x) = \int_{-\infty}^{\infty} q(a)P(x - a)\,da.$$

The proof deduces (3.26) from (3.22). First of all, to simplify the argument we assume that the outcome of the second experiment always lies in some finite interval I. We partition I into a finite number n of subintervals I_j. Let's denote the probability $Q(I_j)$ by Q_j. According to Formula (3.7), $Q(I_j)$ can be expressed in terms of the derivative q of Q as follows:

$$(3.27) \qquad\qquad Q_j = Q(I_j) = q(a_j)|I_j|,$$

where a_j is some point in I_j, and $|I_j|$ denotes the length of I_j.

We *discretize* the second experiment by lumping together all outcomes that lie in the same interval I_j and *redefine* the numerical outcome in that case to be a_j, the number appearing in Formula (3.27). The probability of the outcome a_j is Q_j.

Denote by w_n the width of the longest of the n subintervals I_j. Since a_j lies in I_j, it follows that the numerical outcome of the discretized version of the experiment differs from the numerical outcome of the full version of the experiment at most by w_n. Consider the combination of the first experiment with a discretized version of the second experiment. Since the numerical outcome of a combined experiment is defined as the sum of the outcomes of each experiment, it follows that the outcome of the discretized combination differs from the outcome of the exact combination at most by w_n. Denote by $U_n(x)$ the probability that the outcome of the discretized combination is less than x. Consider an infinite sequence of repetitions of the combined experiment. Denote by $S(x)$ the number of instances among the first N when the outcome of the combined experiment was less than x. Now lump the outcome of the second experiment in the manner described above, and denote by $S_n(x)$ the number of instances among the first N when the sum of the outcomes of the first experiment and the discretized second experiment is less than x.

We saw that the outcome of the combined experiment differs from the outcome of the discretized combined experiment at most by w_n. Therefore in each case when the outcome of the combined experiment is less than x, the outcome of the discretized combination is less than $x + w_n$. Similarly when the outcome of the discretized combination is less than $x - w_n$, the outcome of the combined experiment is less than x. These statements are expressed by the inequalities

$$S_n(x - w_n) \le S(x) \le S_n(x + w_n).$$

Divide all members by N:

$$(3.28) \qquad \frac{S_n(x - w_n)}{N} \le \frac{S(x)}{N} \le \frac{S_n(x + w_n)}{N}.$$

By definition of the probability $U_n(x)$, as N tends to ∞, the left side tends to $U_n(x - w_n)$ and the right side to $U_n(x + w_n)$. According to (3.22),

$$U_n(x) = Q_1 P(x - a_1) + \cdots + Q_n P(x - a_n).$$

Substitute for each Q_j the expressions given in (3.27); we get that

$$(3.29) \qquad U_n(x) = q(a_1)P(x - a_1)|I_1| + \cdots + q(a_n)P(x - a_n)|I_n|.$$

The experienced reader instantly recognizes the sum on the right as an approximating sum for the integral

$$(3.30) \qquad \int_{-\infty}^{\infty} q(a)P(x - a)da.$$

This function was denoted as $U(x)$ in Formula (3.26). Since approximating sums tend to the integral as the partition is made finer and finer, we conclude that for every x, $U_n(x)$ tends to $U(x)$ as the partition becomes refined.

We claim that $U_n(x - w_n)$ and $U_n(x + w_n)$ also tend to $U(x)$ as the partition becomes more refined. To see this assume that $|I| = 1$ and that all subintervals of the n-th partition have equal length $1/n$. Thus $w_n = 1/n$; replacing x by $x - 1/n$ in (3.29) gives

$$U_n(x - w_n) = \left\{ q(a_1)P\left(x - \frac{1}{n} - a_1\right) + \cdots + q(a_n)P\left(x - \frac{1}{n} - a_n\right) \right\} \frac{1}{n}.$$

Similarly,

$$U_n(x + w_n) = \left\{ q(a_1)P\left(x + \frac{1}{n} - a_1\right) + \cdots + q(a_n)P\left(x + \frac{1}{n} - a_n\right) \right\} \frac{1}{n}.$$

These quantities are approximating sums to the integral (3.30), and therefore, for n large enough, both $U_n(x - w_n)$ and $U_n(x + w_n)$ are very close to $U(x)$. Since, for N large enough, $S_n(x - w_n)/N$ and $S_n(x + w_n)/N$ are very close to $U_n(x - w_n)$ and $U_n(x + w_n)$, it follows from inequality (3.28) that, for N large enough, $S(x)/N$ is very close to $U(x)$. But this is what we mean when we say that $S(x)/N$ converges to $U(x)$ as N tends to ∞. Thus

$$(3.31) \qquad \lim_{N \to \infty} \frac{S(x)}{N} = U(x) = \int_{-\infty}^{\infty} q(a)P(x - a)da.$$

This is precisely what we set out to prove. $\qquad\square$

Now suppose that $P(x)$ is differentiable; denote its derivative by $p(x)$. It follows from (2.24) in Section 4.2 that $U(x)$, as defined by (3.26), is differentiable, and its derivative, which we denote by $u(x)$, can be obtained by differentiating the integrand on the right in (3.26) with respect to x:

$$(3.32) \qquad u(x) = \int_{-\infty}^{\infty} q(a)p(x - a)da.$$

We summarize what we have proved:

Theorem 3.2. *Consider two experiments whose outcomes lie in some finite interval and have probability densities p and q, respectively. Suppose the experiments are independent. Consider the combined experiment consisting of performing both experiments; define the outcome of the combined experiment to be the sum of the outcomes of the individual experiments. Then the combined experiment has probability density $u(x)$ given by Formula (3.32).*

The restriction of the outcomes of the experiments to a finite interval is too confining for many important applications. Fortunately the theorem, although not our proof, holds under more general conditions. The function u defined by Formula (3.32) is called the *convolution* of the functions q and p. This relation is denoted by

(3.33)
$$u = q * p.$$

Convolution is an important operation among functions, with many uses; one of them will make its appearance in Section 6.5. We now state and prove some of its basic properties.

Theorem 3.3

(i) *Convolution is distributive, i.e.,*

$$(q_1 + q_2) * p = q_1 * p + q_2 * p.$$

(ii) *Let k be any constant; then*

$$(kq) * p = k(q * p).$$

(iii) *Convolution is commutative, i.e.,*

$$q * p = p * q.$$

PROOF. The first result, (i), follows from the *additivity* of integrals:

$$(q_1 + q_2) * p = \int_{-\infty}^{\infty} [q_1(a) + q_2(a)]p(x - a)da$$

$$= \int_{-\infty}^{\infty} q_1(a)p(x - a)da + \int_{-\infty}^{\infty} q_2(a)p(x - a)da$$

$$= q_1 * p + q_2 * p.$$

The second result, (ii), follows from another basic property of the integral. If you multiply the integrand by a constant k, the value of the integral is multiplied by k:

$$(kq) * p = \int_{-\infty}^{\infty} kq(a)p(x - a)da = k\int_{-\infty}^{\infty} q(a)p(x - a)da$$

$$= k(q * p).$$

The third result, (iii), follows if we make the change of variable $b = x - a$; then

$$q * p = \int_{-\infty}^{\infty} q(a)p(x - a)da$$

$$= \int_{-\infty}^{\infty} q(x - b)p(b)db$$

$$= p * q.$$

This completes the proof of Theorem 3.3. □

282

Since p and q are probability densities, they each satisfy relation (3.10):

(3.34) $$\int_{-\infty}^{\infty} p(x)dx = 1, \qquad \int_{-\infty}^{\infty} q(a)da = 1.$$

Since $u(x)$ too is a probability density, it follows that

(3.35) $$\int_{-\infty}^{\infty} u(x)dx = 1.$$

We now give a mild but useful generalization of this result.

Theorem 3.4. *Suppose p and q are a pair of functions, both zero outside some finite interval. Denote their convolution by u:*

$$u = p * q.$$

Then the integral of the convolution satisfies

(3.36) $$\int_{-\infty}^{\infty} u(x)dx = \int_{-\infty}^{\infty} p(x)dx \int_{-\infty}^{\infty} q(a)da.$$

PROOF. Suppose p and q are non-negative and satisfy conditions (3.34). Then they can be regarded as probability densities, and so u satisfies (3.35). This is precisely the assertion (3.36) of Theorem 3.4 in this special case.

Next we show how to use Theorem 3.3 to reduce the general case to this special case. First we give up the requirement that p and q satisfy (3.34), then the requirement that they be nonnegative. Suppose p and q are two non-negative functions; denote their integrals over the whole real axis by P and Q:

(3.37) $$\int_{-\infty}^{\infty} p(x)dx = P, \qquad \int_{-\infty}^{\infty} q(a)da = Q.$$

We define p_1 and q_1 by

(3.38) $$p_1(x) = \frac{1}{P} p(x), \qquad q_1(a) = \frac{1}{Q} q(a).$$

Using rule (ii) of Theorem 3.3

$$u = p * q = Pp_1 * Qq_1 = PQp_1 * q_1$$

so

(3.39) $$\int_{-\infty}^{\infty} u(x)dx = PQ \int_{-\infty}^{\infty} p_1 * q_1 \, dx$$

It follows from (3.37) and (3.38) that p_1 and q_1 satisfy (3.34), because

$$\int_{-\infty}^{\infty} p_1(x)dx = \frac{1}{P} \int_{-\infty}^{\infty} p(x)dx = \frac{P}{P} = 1$$

283

and similarly for q_1. Therefore, by our earlier result,

$$\int_{-\infty}^{\infty} p_1 * q_1 \, dx = 1.$$

Substituting this into (3.39), we get

$$\int_{-\infty}^{\infty} u(x)dx = PQ.$$

In view of the definition of P and Q in (3.37), this is precisely relation (3.36).

Now consider any p and q, not necessarily non-negative. We write p and q as the *difference* of non-negative functions:

(3.40) $$p = p_+ - p_-, \qquad q = q_+ - q_-.$$

To see that every function can be decomposed in this fashion, set

$$p_+(x) = \max\{0, p(x)\}$$
$$p_-(x) = \max\{0, -p(x)\}.$$

Obviously, both p_+ and p_- are non-negative; see Figure 6.4. We claim that p_+ and p_- give a decomposition of p described in (3.40). For if $p(x)$ is positive, $p_+(x) = p(x)$, $p_-(x) = 0$; if $p(x)$ is negative, $p_+(x) = 0$, $p_-(x) = -p(x)$. If $p(x) = 0$, $p_+(x) = 0$, $p_-(x) = 0$. In all three cases p is the difference of p_+ and p_-.

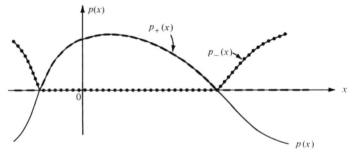

Figure 6.4

Using the decomposition (3.40) and the distributive law for convolutions, we can rewrite $p * q$ as

(3.41)
$$p * q = (p_+ - p_-) * (q_+ - q_-)$$
$$= p_+ * q_+ - p_+ * q_- - p_- * q_+ + p_- * q_-.$$

To avoid writing the integral sign too often we revert temporarily to the notation used for the integral at the beginning of Chapter 4:

$$\int_{-\infty}^{\infty} u(x)dx = I(u).$$

Integrating (3.41) from $-\infty$ to ∞ gives

(3.42) $\quad I(p * q) = I(p_+ * q_+) - I(p_+ * q_-) - I(p_- * q_+) + I(p_- * q_-).$

For non-negative functions, we have already proved relation (3.36); so, since $p_\pm$ and $q_\pm$ are non-negative,

$$I(p_\pm * q_\pm) = I(p_\pm)I(q_\pm).$$

Substituting this into (3.42) we get

(3.43) $\quad I(p * q) = I(p_+)I(q_+) - I(p_+)I(q_-) - I(p_-)I(q_+) + I(p_-)I(q_-).$

The right side can be written as the product

(3.44) $$[I(p_+) - I(p_-)][I(q_+) - I(q_-)].$$

Integrating (3.40) from $-\infty$ to ∞ gives

$$I(p) = I(p_+) - I(p_-), \qquad I(q) = I(q_+) - I(q_-);$$

substituting this into (3.44) gives the value

$$I(p)I(q)$$

for (3.44). Since (3.44) is equal to the right side of (3.43) we deduce that

$$I(p * q) = I(p)I(q).$$

This is precisely relation (3.36) and completes the proof of Theorem 3.4. $\quad\square$

Remark 1. The basic step of the proof was the observation that (3.36) holds for non-negative functions p and q whose integral equals 1, because the convolution of two probability densities is again a probability density. It is possible to prove Theorem 3.4 without referring to probabilities at all; such a proof is outlined in Exercise 3.4.

Remark 2. In our derivation of the convolution formula, we have assumed that at least one of the functions p or q is zero outside a finite interval. For that reason we have imposed the same condition on p and q in Theorem 3.4. This condition is unnecessary and can be replaced by requiring that p_+, p_-, q_+ and q_- all be integrable from $-\infty$ to ∞.

The numerical outcome of the combination of two experiments was defined as the *sum* of the numerical outcomes of its two constituents. We now give some realistic examples to illustrate why this definition is of interest.

Suppose the outcomes of the two experiments represent *income* from two entirely different sources. Their sum is then the total income; its probability distribution is of considerable interest.

Here is another example: Suppose the two outcomes represent amounts of water entering a reservoir in a given period from two different sources. Their sum represents the total inflow, again an object of considerable interest.

Consider the following example of evaluating the convolution of two functions:

$$(3.45) \qquad p(t) = \begin{cases} 0 & \text{for } t < 0 \\ e^{-At} & \text{for } 0 \le t \end{cases}, \qquad q(t) = \begin{cases} 0 & \text{for } t < 0 \\ e^{-Bt} & \text{for } 0 \le t \end{cases},$$

A and B positive numbers. Substitute these definitions of p and q into the definition of the convolution

$$(3.46) \qquad u(x) = (p * q)(x) = \int_{-\infty}^{\infty} p(a)q(x - a)da.$$

Both $p(t)$ and $q(t)$ were defined to be zero for $t < 0$. It follows from this that if $x < 0$, one of the factors in the integrand on the right in (3.46) is zero, no matter what a is. This shows that $u(x) = 0$ for $x < 0$. For $x > 0$ the same analysis shows that the integrand is nonzero only in the range $0 \le a \le x$. So, for $x > 0$,

$$u(x) = \int_{0}^{x} e^{-Aa - B(x-a)} da = e^{-Bx} \int_{0}^{x} e^{(B-A)a} da$$

$$(3.47)$$

$$= e^{-Bx} \frac{e^{(B-A)a}}{B - A}\Big|_{0}^{x} = \frac{1}{B - A}(e^{-Ax} - e^{-Bx}).$$

EXERCISES

3.1 Define p by

$$p(x) = \begin{cases} 0 & \text{for } x < 0 \\ \dfrac{2}{A}\left(1 - \dfrac{x}{A}\right) & \text{for } 0 \le x \le A \\ 0 & \text{for } A < x. \end{cases}$$

(a) Show that

$$\int_{-\infty}^{\infty} p(x)dx = 1.$$

(b) Calculate the average value of x with respect to this probability density, i.e., find

$$\bar{x} = \int_{-\infty}^{\infty} xp(x)dx.$$

(c) Calculate the average value of x^2:

$$\overline{x^2} = \int_{-\infty}^{\infty} x^2 p(x)dx.$$

3.2 Define p by

$$p(x) = k|x|e^{-kx^2}, \quad k > 0.$$

(a) Show that

$$\int_{-\infty}^{\infty} p(x)dx = 1.$$

(b) Calculate

$$\bar{x} = \int_{-\infty}^{\infty} xp(x)dx.$$

3.3 Let A and B be two positive numbers; define p and q by

$$p(t) = \begin{cases} 0 & \text{for } t < 0 \\ 1/A & \text{for } 0 \le t \le A, \\ 0 & \text{for } A < t \end{cases} \qquad q(t) = \begin{cases} 0 & \text{for } t < 0 \\ 1/B & \text{for } 0 \le t \le B. \\ 0 & \text{for } B < t \end{cases}$$

(a) Show that p and q are probability densities, i.e., that they satisfy

$$\int_{-\infty}^{\infty} p(t)dt = 1, \qquad \int_{-\infty}^{\infty} q(t)dt = 1.$$

(b) Let u denote the convolution of p and q, defined by (3.46). Show that $u(x) = 0$ for $x < 0$ and for $x > A + B$.

(c) Assume $A < B$; show that, for $0 < x < A + B$, the integrand in (3.46) equals $1/AB$ in the intervals indicated below, and is zero outside these intervals:

For $0 < x < A$ and $0 < a < x$
For $A < x < B$ and $0 < a < A$
For $B < x < A + B$ and $x - B < a < A$:

Using this information show that

$$u(x) = \begin{cases} 0 & \text{for } x < 0 \\[2mm] \dfrac{x}{AB} & \text{for } 0 \le x < A \\[2mm] \dfrac{1}{B} & \text{for } A \le x < B \\[2mm] \dfrac{A + B - x}{AB} & \text{for } B \le x < A + B \\[2mm] 0 & \text{for } A + B \le x. \end{cases}$$

(d) Verify that u, as defined above, is a probability density which satisfies

$$\int_{-\infty}^{\infty} u(x)dx = 1.$$

(e) Determine u when $A > B$.

3.4 The purpose of this exercise is to give an alternative proof of Theorem 3.4. Let p and q be a pair of functions, both zero outside some interval J. Let u be the convolution of p and q, defined by Formula (3.46).

(a) Let h be a small number. Show that the sum

$$(3.48) \qquad\qquad \sum_i p(ih)q(x - ih)h$$

is an approximating sum to the integral defining $u(x)$.

(b) Show that

$$(3.49) \qquad\qquad \sum_j u(jh)h$$

is an approximating sum to the integral

$$\int_{-\infty}^{\infty} u(x)dx.$$

(c) Substitute the approximations (3.48) for $u(x)$ into (3.49) with $x = jh$; show that the result is the *double sum*

$$\sum_{i,j} p(ih)q((j - i)h)h^2.$$

(d) Denote $j - i$ by l and rewrite the above double sum as

$$\sum_{i,l} p(ih)q(lh)h^2.$$

(e) Show that this double sum can be written as the product of two single sums:

$$\left(\sum_i p(ih)h\right)\left(\sum_l q(lh)h\right).$$

(f) Show that the single sums are approximations to the integrals

$$\int p(x)dx \qquad \text{and} \qquad \int q(x)dx.$$

(g) Show that as h tends to zero, you obtain identity (3.36) of Theorem 3.4.

3.5 Let p and q be a pair of probability distributions; denote their convolution by

$$u = p * q.$$

Denote the average values of x with respect to the probability densities p, q and u by $\bar{x}_p$, $\bar{x}_q$ and $\bar{x}_u$:

$$\bar{x}_p = \int_{-\infty}^{\infty} xp(x)dx, \qquad \bar{x}_q = \int_{-\infty}^{\infty} xq(x)dx, \qquad \bar{x}_u = \int_{-\infty}^{\infty} xu(x)dx.$$

(a) Show that

$$(3.50) \qquad\qquad \bar{x}_u = \bar{x}_p + \bar{x}_q.$$

[*Hint*: Apply Theorem 3.2 to the pair $xp(x)$, q and the pair p, $xq(x)$.]

(b) Give an intuitive probabilistic interpretation for (3.50).

3.6 (a) Deduce the commutative law for convolutions of probability densities from the interpretation of convolution given in (5.11) of Section 6.5.

(b) Formulate an associative law for convolution, and deduce it for probability densities by using the interpretation of convolution given in (5.11).

(c) Prove the associative law for the convolution of arbitrary functions by using decompositions of the form (3.40), and the distributive law.

3.7 Compute $p * q$ for

(a) $p = \cos t, \quad q = p, \quad t > 0$
(b) $p = \cos t, \quad q = t, \quad t > 0$
(c) $p = \sin t, \quad q = t, \quad t > 0$
(d) $p = \sin at, \quad q = \sin bt, \quad a \neq b, \quad t > 0.$

6.4 The law of errors

In this section we shall analyse a particular experiment of the kind described in Section 6.3. The experiment consists of dropping pellets from a fixed point at a certain height onto a horizontal plane. If the hand that releases the pellet were perfectly still and if there were no air currents diverting the pellet on its downward path, then we could predict with certainty that the pellet will end up directly below the point where it was released. But even the steadiest hand west of the Pecos trembles a little, and even on the stillest day minute air currents buffet the pellet in its downward flight, in a random fashion. These effects become magnified and very noticeable if the pellets are dropped from a great height, say the 10th floor. Under such circumstances the experiment appears to be nondeterministic, i.e., it is impossible to predict where each pellet is going to land.*

Although it is impossible to predict where any particular pellet would fall, the outcome can be predicted very well on the average, in the sense explained in Section 6.3. That is, let G be any region such as a square, rectangle, triangle, circle, etc. Denote by $S(G)$ the number of those instances among the first N in a sequence of experiments when the pellet has landed in G. Then the frequencies $S(G)/N$ tend to a limit, called the probability of landing in G and denoted by $C(G)$

(4.1)
$$\lim_{N \to \infty} \frac{S(G)}{N} = C(G).$$

In this section we shall investigate the nature of this probability.

---

* G. I. Taylor (1886–1975), a famous British applied mathematician, described the following experience during the first World War: Taylor was working on a project to develop aerial darts; his task was to record the patterns created when a large number of darts were dropped from an airplane. This he did by putting a piece of paper under each dart where it had fallen in the field. He had just finished this tedious task when a cavalry officer came by on horseback and demanded what Taylor was doing. Taylor explained the dart project, whereupon the officer exclaimed: "And you chaps managed to hit all those bits of paper? Good show!"

Suppose that the region G is a very small one; then we expect the probability of landing in G to be nearly proportional to the area $A(G)$ of G. We can express this surmise more precisely as follows: Let g be any point in the plane; then there is a number $c = c(g)$, called the *probability density* at g, such that for any region G containing g

(4.2)
$$C(G) = [c(g) + \text{small}]A(G),$$

where "small" means a quantity that tends to zero as G shrinks so that its boundary approaches g.

What can we say about the probability density $c(g)$? It depends on how close g is to the bull's eye, i.e., the point directly underneath where the pellet is released. The closer g is, the greater the probability of a hit near g; in particular, the maximum value of c is achieved when g is the bull's eye. We now adopt the following two hypotheses about the way in which the uncontrolled tremors of the hand and the unpredictable gusts of wind influence the distribution of hits and misses:

(i) $c(g)$ depends only on the distance of g from the bull's eye, and not on the direction in which g lies.
(ii) Let x and y be perpendicular directions; displacement of pellets in the x direction is independent of their displacement in the y-direction.

To express these hypotheses in a mathematical form, we introduce a Cartesian coordinate system with the origin, naturally, at the bull's eye. We denote by (a, d) the coordinates of the point g; see Figure 6.5. We denote by $P(a)$ the probability that the pellet falls in the halfplane

$$x < a.$$

As explained in Section 6.3, the probability that the pellet falls in the strip $a \le x < b$ is given by Formula (3.5) as

(4.3)
$$P(b) - P(a).$$

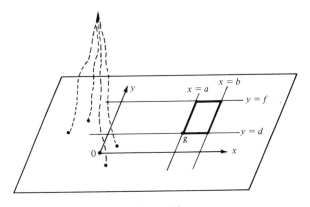

Figure 6.5

As in Section 6.3, we assume that P is a differentiable function; we denote its derivative by $p(x)$. According to the mean value theorem, the difference (4.3) can be expressed as

$$(4.4) \qquad p(\bar{a})(b - a),$$

$\bar{a}$ some value between a and b.

It follows from hypothesis (i) that the probability of a pellet falling in the strip $d \leq y < f$ parallel to the x-axis is of the same form as the probability of falling in the strip parallel to the y-axis, i.e., by Formulas (4.3) and (4.4):

$$(4.5) \qquad P(f) - P(d) = p(\bar{d})(f - d),$$

$\bar{d}$ between d and f.

What is the probability that the pellet falls in the rectangle

$$(4.6) \qquad a \leq x < b, \quad d \leq y < f\,?$$

This event occurs when the pellet falls in the strip $a \leq x < b$ and the strip $d \leq y < f$. According to hypothesis (ii) these two events are now independent, and therefore, according to the *product rule*, the probability of the combined event is the product of the probabilities of the two separate events whose simultaneous occurrence constitutes the combined event. Thus the probability of a pellet falling in the rectangle (4.6) is

$$(4.7) \qquad p(\bar{a})(b - a)p(\bar{d})(f - d).$$

Since the product $(b - a)(f - d)$ is the area A of the rectangle (4.6), we can rewrite (4.7) as

$$(4.8) \qquad p(\bar{a})p(\bar{d})A.$$

Now consider a sequence of rectangles which tend to the point $g = (a, d)$ by letting b tend to a, f tend to d. Since $\bar{a}$ lies between a and b, $\bar{d}$ between d and f, and since p is a continuous function, it follows that $p(\bar{a})$ tends to $p(a)$ and $p(\bar{d})$ tends to $p(d)$. Thus, in this case, we can rewrite (4.8) as

$$(p(a)p(d) + \text{small})A.$$

Comparing this with (4.2), we conclude that the probability density c at the point $g = (a, d)$ is $p(a)p(d)$:

$$(4.9) \qquad c(g) = p(a)p(d).$$

Next we exploit the symmetry of the experimental setup around the bull's eye by introducing another coordinate system whose origin is still the bull's eye but where one of the coordinate axes is deliberately chosen to go through the point g whose coordinates in the old system were (a, d). Clearly, the coordinates of g in the new system, see Figure 6.6, are

$$(0, \sqrt{a^2 + d^2}).$$

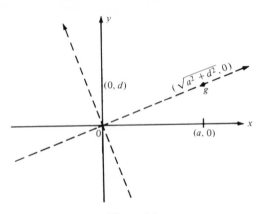

Figure 6.6

According to hypothesis (i), we can apply relation (4.9) in any coordinate system. Then c, in the new coordinate system is

(4.10) $$c(g) = p(0)p(\sqrt{a^2 + d^2}).$$

Comparing (4.9) and (4.10), we conclude that

(4.11) $$p(a)p(d) = p(0)p(\sqrt{a^2 + d^2}).$$

This is a functional equation for $p(x)$ which can be solved by the trick of writing the function $p(x)$ in terms of another function f related to p as follows:

(4.12) $$f(x) = p(\sqrt{x}).$$

Set $x = a^2$ and $y = d^2$ in (4.11). Using (4.12) we can rewrite (4.11) as

(4.13) $$f(x)f(y) = f(0)f(x + y).$$

This, at last, is the familiar functional equation satisfied by exponential functions and only by them, as explained in Section 5.1. So we conclude that

(4.14) $$f(x) = f(0)e^{Kx}.$$

According to relation (4.12), $p(a) = f(a^2)$; we deduce from this and (4.14) that

(4.15) $$p(a) = p(0)e^{Ka^2}.$$

We claim that the constant K is negative; for, as a tends to ∞, the probability density $p(a)$ must tend to zero, and this is the case only if K is negative. To put this into evidence we rename K as $-k$, and rewrite (4.15) as

(4.15)' $$p(x) = p(0)e^{-kx^2}.$$

Every probability density must satisfy condition (3.10):

$$\int_{-\infty}^{\infty} p(x)dx = 1.$$

Substituting (4.15)' into this relation gives

$$p(0) \int_{-\infty}^{\infty} e^{-kx^2} \, dx = 1.$$

Introduce

$$y = \sqrt{k} x$$

as a new variable of integration; we get

$$\int_{-\infty}^{\infty} e^{-kx^2} \, dx = \frac{1}{\sqrt{k}} \int_{-\infty}^{\infty} e^{-y^2} \, dy.$$

We shall show numerically in Exercise 4.1 and probabilistically in Exercise 4.2 that the value of the integral

$$\int_{-\infty}^{\infty} e^{-y^2} \, dy$$

is $\sqrt{\pi}$! Anticipating this result yields $p(0) = \sqrt{k/\pi}$. Substituting this into (4.15)' gives

(4.16)
$$p(x) = \sqrt{\frac{k}{\pi}} e^{-kx^2}.$$

Substituting (4.16) into (4.9), we deduce

(4.17)
$$c(x, y) = \frac{k}{\pi} e^{-k(x^2 + y^2)}.$$

That the probabilities governing such random events as the dropping of pellets are of the form (4.16), (4.17) is called Gauss' *law of errors*, after its discoverer, the mathematician Carl Friedrich Gauss (1777–1855). *Functions of the form* (4.16) *and* (4.17) *are called* gaussian.

The derivation of the law of errors presented above is due to the physicist James Clark Maxwell (1831–1879), who made profound investigations of the significance of probability densities of form (4.16) and (4.17) in physics; for this reason such *densities* in physics are called *maxwellian*. In Section 6.5 we shall study another instance of a gaussian law.

In Figure 6.7 we see the shape of the gaussian function $p(x)$ for three different values of k: $k = 0.5$, $k = 1$, $k = 2$. These graphs indicate that the larger the value of k, the greater the concentration of the probability near the bull's eye. We make this intuitive observation more precise by proving the following proposition about p as defined by (4.16).

(i) The probability density is an increasing function of k at the origin.
(ii) For any fixed a, the probability of a hit between $x = -a$ and $x = a$ is an increasing function of k.
(iii) The average value of $|x|$ is a decreasing function of k, i.e., the larger k, the closer the outcome is to the bull's eye on the average.

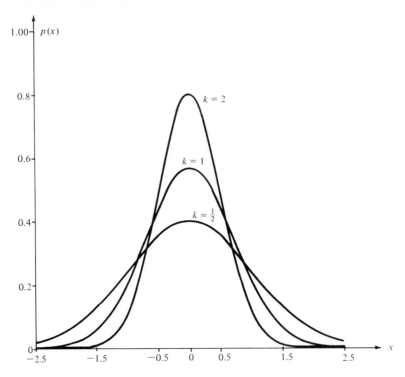

Figure 6.7 Graphs of gaussian probability densities $p(x)$. For clarity, the vertical axis has been shifted to $x = -2.5$.

Proposition (i) follows from Formula (4.16) for $x = 0$:

$$(4.18) \qquad p(0) = \sqrt{\frac{k}{\pi}}.$$

To prove proposition (ii), we have to study the integral formula (3.9) for the probability of a hit between $-a$ and a:

$$\int_{-a}^{a} p(x)dx = \sqrt{\frac{k}{\pi}} \int_{-a}^{a} e^{-kx^2} \, dx.$$

Introducing $y = \sqrt{k}x$ as new variable of integration, we can rewrite the above integral as

$$(4.19) \qquad \sqrt{\frac{k}{\pi}} \int_{-\sqrt{k}a}^{\sqrt{k}a} e^{-y^2} \frac{dy}{\sqrt{k}} = \frac{1}{\sqrt{\pi}} \int_{-\sqrt{k}a}^{\sqrt{k}a} e^{-y^2} \, dy.$$

In this form the integrand e^{-y^2} is independent of k, whereas the length of the interval of integration $(-\sqrt{k}a, \sqrt{k}a)$ is an increasing function of k. Since the integrand is positive, the larger k, the larger the value of the integral.

294

To prove (iii) we compute the average value of $|x|$. By Formula (3.19), with $f(x) = |x|$, this is

$$\overline{|x|} = \int_{-\infty}^{\infty} |x| p(x) dx = \sqrt{\frac{k}{\pi}} \int_{-\infty}^{\infty} |x| e^{-kx^2} dx.$$

Since the integrand is even, this can be written as

$$\overline{|x|} = \sqrt{\frac{k}{\pi}} \int_0^{\infty} 2x e^{-kx^2} dx.$$

The integrand is the derivative of

$$\frac{-1}{k} e^{-kx^2};$$

so the integral can be evaluated by appealing to the fundamental theorem of calculus:

(4.20)
$$\overline{|x|} = \sqrt{\frac{k}{\pi}} \left(\frac{-e^{-kx^2}}{k} \right) \Big|_0^{\infty} = \frac{1}{\sqrt{\pi k}}.$$

This is clearly a decreasing function of k.

EXERCISES

4.1 The purpose of this exercise is to evaluate numerically the integral

(4.21)
$$\int_{-\infty}^{\infty} e^{-y^2} dy.$$

We shall use the rectangle rule, i.e., the approximating sums

(4.22)
$$h \sum_n e^{-(nh)^2}.$$

The integrand in (4.21) is an even function, so

$$\int_{-\infty}^{\infty} e^{-y^2} dy = 2 \int_0^{\infty} e^{-y^2} dy$$

and

(4.22)′
$$h \sum_{-\infty}^{\infty} e^{-(nh)^2} = h + 2h \sum_1^{\infty} e^{-(nh)^2}.$$

Since the interval of integration is infinite, we truncate it by considering, for large B,

(4.21)$_t$
$$\int_{-B}^{B} e^{-y^2} dy = 2 \int_0^{B} e^{-y^2} dy$$

instead of (4.21), and the corresponding approximating sums

(4.22)$_t$
$$h \sum_{n=-N}^{N} e^{-(nh)^2} = h + 2h \sum_{n=1}^{N} e^{-(nh)^2}.$$

(a) Prove the following inequalities:

(4.23)
$$\sum_{n=N}^{\infty} e^{-(nh)^2} \le \sum_{n=N}^{\infty} e^{-Nnh^2} = \frac{e^{-N^2h^2}}{1 - e^{-Nh^2}}.$$

(b) Using the truncated version (4.22)$_t$, evaluate the approximating sum numerically for $h = 0.5, 0.3, 0.1$, and use (4.23) to estimate the error of omitting the rest of the terms.

(c) Show, using (4.23), that for $h = 1$, the contribution of all terms in (4.22) with $n > 3$ is less than 10^{-6}. Using a five place table for the exponential function, we obtain

k	0	1	4	9
e^{-k}	1	0.36788	0.01832	0.00012

So

$$\sum_{-3}^{3} e^{-n^2} = 1 + 2(0.36788 + 0.01832 + 0.00012)$$

$$= 1.77264.$$

As remarked earlier, the value of the integral (4.21) is $\sqrt{\pi}$. The tabulated value of $\sqrt{\pi}$ is

$$\sqrt{\pi} = 1.77245\ldots.$$

Thus we see that the rectangle rule (4.22) gives an astonishingly good approximation to the value of the integral (4.21), even when we divide the interval of integration into subintervals of length $h = 1$, a rather crude partition. There is a good reason for this, but its explanation goes beyond elementary calculus.

4.2 The purpose of this exercise is to give a probabilistic proof that

(4.24)
$$\int_{-\infty}^{\infty} e^{-y^2} \, dy = \sqrt{\pi}.$$

The basis of the argument is relation (4.10), which asserts that the probability density at the point (x, y) is

$$c(x, y) = p(0)p(\sqrt{x^2 + y^2}).$$

Combined with expression (4.15)′ for $p(x)$, in which we have, for simplicity, taken $k = 1$, this relation can be written as

(4.25)
$$c(x, y) = p^2(0)e^{-(x^2 + y^2)}.$$

Let h be some small positive quantity. We divide the x, y-plane into ring shaped regions R_n, $n = 1, 2, \ldots$ of thickness $\sqrt{h}$; see Figure 6.8. The point (x, y) falls into the ring R_n if

(4.26)
$$(n - 1)\sqrt{h} \le \sqrt{x^2 + y^2} < n\sqrt{h}.$$

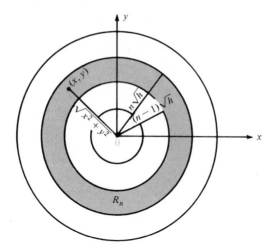

Figure 6.8

(a) Show that the area of R_n is

(4.27) $$A(R_n) = \pi(2n - 1)h.$$

(b) Show, using (4.25), that the probability $C(R_n)$ of a hit in R_n lies between the following limits:

(4.28) $$p^2(0)A(R_n)\exp\{-n^2h\} \le C(R_n) \le p^2(0)A(R_n)\exp\{-(n - 1)^2h\}.$$

(c) Show by summing (4.28) and making use of (4.27) that

$$\sum_1^\infty C(R_n)$$

lies between

(4.29) $$\pi p^2(0) \sum_1^\infty e^{-n^2h}(2n - 1)h$$

and

(4.29)' $$\pi p^2(0) \sum_1^\infty e^{-(n-1)^2h}(2n - 1)h.$$

(d) Show that both (4.29) and (4.29)' are approximating sums for the integral

$$\pi p^2(0) \int_0^\infty 2re^{-r^2}\, dr,$$

where $r^2 = x^2 + y^2$. Show that the value of the above integral is

$$\pi p^2(0).$$

(e) On probabilistic grounds

$$\sum_{1}^{x} C(R_n) = 1.$$

Using this fact and (c) and (d) above, show that

$$p^2(0) = \frac{1}{\pi}.$$

(f) Using relation (4.15)'

$$p(x) = p(0)e^{-x^2}$$

and the relation

$$\int_{-\infty}^{\infty} p(x)dx = 1,$$

deduce (4.24).

6.5 Diffusion

When a small drop of ink is dropped into a glass of water, the ink begins to spread out, becoming more and more uniformly distributed as time goes on. This process is called *diffusion*. In this section we shall investigate its mathematical theory.

Diffusion is a manifestation of the molecular nature of matter. Water, which appears as a continuous fluid, is actually composed of a very large number of molecules. These molecules move with a great deal of irregularity, but this irregular motion takes place on such a small scale that it is not observable by the naked eye. What we see directly is the *average motion* of very many molecules; this averaging wipes out the irregularities and creates the illusion of a smoothly flowing liquid. Although irregular molecular motion cannot be seen directly, its effect on a small floating particle can be observed indirectly through a microscope. The impact of molecules impinging on a sufficiently small particle does *not* average out during a time long enough to cause a significant displacement of the particle. The ensuing irregular motion of the particle, see Figure 6.9, is called *brownian*, after the botanist Brown (1773–1858), who first observed this phenomenon.

Brownian motion is nondeterministic: the motion of an individual particle cannot be predicted, but the motion is predictable on the average in the sense explained in Section 6.3. Suppose we observe an "infinite" sequence of identical small particles, each placed at the same point in a glass of water. Denote by $S(G)$ the number of these particles among the first N which, after a specified time t, have ended up in the same region G; the sequence $S(G)/N$ tends to a limit which we denote by $C(G)$:

(5.1) $$\lim_{N \to \infty} \frac{S(G)}{N} = C(G).$$

$C(G)$ is the probability that a particle ends up in G.

Suppose that instead of releasing N particles in sequence we release them all at once at one point; if the N particles have no effect on each other so that each moves as if the others were not there, then the proportion of those which at the end of the experiment happen to be in G is $S(G)/N$. For large N this is, according to (5.1), very nearly $C(G)$; thus $C(G)$ can be interpreted as that fraction of the mass, released at the beginning of the experiment, which at the end of the experiment is in G.

Ink is a suspension of tiny solid particles, so small that when immersed in water each undergoes brownian motion. If we assume—and this is a very good approximation to the truth—that each ink particle moves independently of all the others, then according to the foregoing analysis, the fraction of ink contained in any region G at time t is equal to $C(G)$, the probability that a single particle released at the specified point ends up in G after undergoing brownian motion for time t.

We shall study this function C; to make the study as easy as possible we make the following simplifying assumptions:

(i) Diffusion takes place in 2 dimensions.
(ii) The diffusion takes place in the whole plane.

Assumption (i) can be simulated by carrying out the diffusion experiment in a flat dish containing water. Assumption (ii) can be simulated by taking a rather large dish, so that the edge of the dish has, at least for a limited time, little influence on the process of diffusion. In Exercise 5.5 we indicate how to remove assumption (i), and in Exercise 5.4 how to remove (ii).

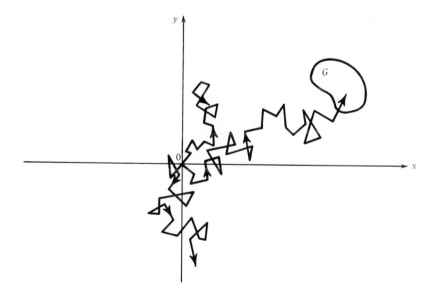

Figure 6.9

Suppose G is a small region containing some point g; then we expect $C(G)$ to be very nearly proportional to the area A of G, i.e., to be of the form

(5.2)
$$C(G) = [c(g) + \text{small}]A(G).$$

If we interpret $C(G)$ as probability, $c(g)$ would be called the *probability density* at g. As we have shown above, $C(G)$ can also be interpreted as the fraction of ink contained in G. Suppose the ink drop contains a unit mass of ink; then $C(G)$ is the actual amount of ink in G. It follows from (5.2) that $c(g)$ is the *density* of ink at g. Thus *the probability density is equal to the ink density.*

To deduce what kind of a function c is of g, we note that at any instant a particle undergoing brownian motion is as likely to move in one direction as in another. Furthermore, the way the particle moved in the past has no influence on the way it will move in the future. These facts have the following two consequences:

(i) The probability density $c(g)$ is the same for all points g which have the same distance from the origin.
(ii) Let x and y be perpendicular axes; then displacement in the x direction is independent of displacement in the y direction.

These suppositions are identical to the ones we have assumed valid in Section 6.4 for the probability distribution governing the random outcome of dropping pellets from a height. We have shown there that these assumptions almost completely determine the probability distribution. Specifically, the probability density c at any point is given by Formula (4.17):

(5.3)
$$c(x, y) = \frac{k}{\pi} e^{-k(x^2 + y^2)}.$$

Let us denote by $P(a)$ the probability that a particle starting at $(0,0)$ and undergoing brownian motion for time t ends up to the left of $x = a$. Denote by p the derivative of P; we call $p(a)$ the *density along the line* $x = a$. According to Formula (4.16)

(5.4)
$$p(x) = \sqrt{\frac{k}{\pi}} e^{-kx^2}.$$

Formulas (5.3) and (5.4) contain the number k about which our analysis in Section 6.4 says nothing except that it is positive. The value of k depends on the circumstances of the experiment, such as the size and weight of ink particles, and the temperature of the water which affects the average speed of the water molecules. Most conspicuously, k depends on the *duration* of the experiment. Clearly, the longer the ink diffuses, the farther it will spread out from the point where it was originally dropped into the water. Therefore the larger t, the smaller the concentration at the origin. According to (5.3) the concentration at the origin is k/π; so we conclude that $k(t)$ is a decreasing function of t. The rest of this section is devoted to determining very precisely what kind of function k is of t.

We have assumed that ink particles diffuse independently of each other, and that diffusion takes place in the whole plane. These assumptions have the following consequences:

1. If M units of ink are released at time $t = 0$ anywhere along the line $x = 0$, the resulting ink density at time t along any line $x = b$ is

    $$Mp(b).$$

2. If M units of ink are released at time $t = 0$ anywhere along the line $x = a$, the resulting ink density at time t along any line $x = b$ is

    $$Mp(b - a).$$

3. If n blobs of ink of $M_1, M_2, \ldots, M_n$ units, respectively, are released at time $t = 0$ along n lines $x = a_1$, $x = a_2$, $\ldots$, $x = a_n$, the resulting ink density at time t along any line $x = b$ is the sum of the ink densities due to each individual blob and is therefore given by

    (5.6) $\qquad M_1 p(b - a_1) + M_2 p(b - a_2) + \cdots + M_n p(b - a_n).$

Suppose now that ink is distributed initially not on n distinct lines but continuously over the whole plane. We denote the total amount of mass contained initially in the halfplane $x < a$ by $M(a)$. We assume that M is a differentiable function, and denote its derivative by $m(a)$. We call $m(a)$ the *initial mass density* along $x = a$. The total mass contained initially in the strip $a \leq x < b$ is $M(b) - M(a)$. According to the fundamental theorem of calculus, this can be expressed as

(5.7) $$\int_a^b m(x)dx.$$

We compute the ink density which results after this initial deposit of ink has diffused for time t. For the sake of simplicity we assume at first that there is no mass at points whose x-coordinate lies outside some finite interval I. We partition I into n subintervals $I_j, j = 1, 2, \ldots, n$. We denote by M_j the mass contained in the j-th strip consisting of points whose x-coordinate lies in I_j:

$$M_j = \int_{I_j} m(x)dx.$$

According to the mean value theorem of integral calculus (see Section 4.2) there is a value a_j in I_j such that

(5.8) $$M_j = \int_{I_j} m(x)dx = m(a_j)|I_j|,$$

$|I_j|$ the length of I_j. We now *lump* all mass contained in the j-th strip on the line $x = a_j$. In this process of lumping, each ink particle is moved at most the width of one of the strips. Therefore, if the partition is fine enough, in the sense that each interval I_j is very short, the lumped distribution of ink is very

301

close to the original initial ink distribution. It is plausible to conclude then that the ink density at time t resulting from the lumped initial distribution differs little from the ink density resulting from the continuous initial ink distribution.

The ink density at time t on the line $x = b$ resulting from the lumped initial ink distribution is given by Formula (5.6); setting the expression (5.8) for M_j into the formula, we get

$$(5.9) \qquad m(a_1)p(b - a_1)|I_1| + \cdots + m(a_n)p(b - a_n)|I_n|.$$

We recognize this sum as an approximating sum to the integral

$$(5.10) \qquad \int_{-\infty}^{\infty} m(a)p(b - a)da;$$

as the partition gets more refined, (5.9) tends to (5.10).

Therefore we draw the plausible conclusion: *Denote by $m(a)$ the density on the line $x = a$ of some initial ink distribution. Then after diffusing for time t, the ink density on the line $x = b$ is given by Formula (5.10). The function p in this formula is given by (5.4).* Comparing (5.10) with (3.32) we recognize that (5.10) is the convolution of the functions m and p, denoted in (3.33) by

$$(5.10)' \qquad\qquad m * p.$$

To indicate explicitly the dependence of p on t, we shall write it as $p[t]$. The value of $p[t]$ on the line x will be denoted as $p(x, t)$. Let s be an arbitrary time interval. We shall choose the initial ink distribution $m(a)$ to be $p(a, s)$. After diffusing for time t, the resulting ink distribution is, according to (5.10)', the convolution

$$p[s] * p[t].$$

Since $p[s]$ itself is the result of a single drop of ink diffusing for time s, the final ink distribution can be thought of as the distribution resulting when a drop of ink diffuses for time $s + t$. That distribution is denoted by $p[s + t]$. It follows therefore that

$$(5.11) \qquad\qquad p[s + t] = p[s] * p[t].$$

We now present an alternative derivation of (5.11). In Section 6.3 we have shown [see (3.32), (3.33)] that if two experiments are independent, and if we define the numerical outcome of the combination of the two experiments as the sum of the individual outcomes, then the probability density for the combined experiment is the convolution of the probability densities for the individual experiments. We now take for the first experiment diffusion for time s; this is followed by the second experiment, diffusion for time t. Clearly, since ink particles have no memory, these two experiments are independent of one another. If we define the numerical outcome of each experiment as the displacement of ink particles in the direction of the x-axis, then clearly the numerical outcome of the combined experiment, which is diffusion lasting time $s + t$, is the sum of the outcomes of each individual experiment. This shows that (5.11) is a corollary of (3.32), (3.33).

Next we need the following result about convolution:

Theorem 5.1. *Suppose p and q are two functions which satisfy the conditions*

$$\text{(5.12)} \qquad \int_{-\infty}^{\infty} p(x)dx = 1, \qquad \int_{-\infty}^{\infty} q(x)dx = 1,$$

and

$$\text{(5.13)} \qquad \int_{-\infty}^{\infty} xp(x)dx = 0, \qquad \int_{-\infty}^{\infty} xq(x)dx = 0.$$

Denote the convolution of p and q by

$$u = p * q,$$

and denote by S_p, S_q, S_u the quantities

$$\text{(5.14)} \quad S_p = \int_{-\infty}^{\infty} x^2 p(x)dx, \qquad S_q = \int_{-\infty}^{\infty} x^2 q(x)dx, \qquad S_u = \int_{-\infty}^{\infty} x^2 u(x)dx.$$

Then

$$\text{(5.15)} \qquad S_u = S_p + S_q.$$

PROOF. We define three functions u_1, u_2 and u_3 as follows:

$$u_1(x) = (x^2 p) * q = \int_{-\infty}^{\infty} a^2 p(a)q(x - a)da$$

$$\text{(5.16)} \qquad u_2(x) = (2xp) * xq = 2\int_{-\infty}^{\infty} ap(a)(x - a)q(x - a)da$$

$$u_3(x) = p * (x^2 q) = \int_{-\infty}^{\infty} p(a)(x - a)^2 q(x - a)da$$

We now apply identity (3.36) of Theorem 3.4 to u_1, u_2, u_3 and obtain

$$\int_{-\infty}^{\infty} u_1(x)dx = \int_{-\infty}^{\infty} a^2 p(a)da \int_{-\infty}^{\infty} q(x)dx,$$

$$\int_{-\infty}^{\infty} u_2(x)dx = 2\int_{-\infty}^{\infty} ap(a)da \int_{-\infty}^{\infty} xq(x)dx,$$

$$\int_{-\infty}^{\infty} u_3(x)dx = \int_{-\infty}^{\infty} p(a)da \int_{-\infty}^{\infty} x^2 q(x)dx.$$

We can use the assumed properties (5.12) and (5.13) and the notation (5.14) to simplify these relations:

$$\int_{-\infty}^{\infty} u_1(x)dx = S_p,$$

$$\int_{-\infty}^{\infty} u_2(x)dx = 0,$$

$$\int_{-\infty}^{\infty} u_3(x)dx = S_q.$$

Adding these relations we get

(5.17) $$\int_{-\infty}^{\infty} (u_1 + u_2 + u_3)dx = S_p + S_q.$$

On the other hand, adding the three relations (5.16) we get, after a little algebra and using the definition of $u = p * q$, that

$$u_1(x) + u_2(x) + u_3(x) = \int_{-\infty}^{\infty} [a^2 + 2a(x - a) + (x - a)^2]p(a)q(x - a)da$$

$$= \int_{-\infty}^{\infty} x^2 p(a)q(x - a)da = x^2 u(x).$$

Integrating this with respect to x over $(-\infty, \infty)$ we get, in the notation (5.14)

$$\int_{-\infty}^{\infty} (u_1 + u_2 + u_3)dx = \int_{-\infty}^{\infty} x^2 u(x)dx = S_u.$$

Comparing this with (5.17) leads inescapably to the conclusion (5.15) of Theorem 5.1. □

We shall apply Theorem 5.1 to $p = p[s]$, $q = p[t]$ as given by Formula (5.4). Since $p[s]$, $p[t]$ are probability densities, the assumptions (5.12) are satisfied. Since, as shown by Formula (5.4), $p(x, s)$ is an *even* function of x, it follows that (5.13) are fulfilled. Therefore we can draw the conclusion (5.15); we state this conclusion by substituting for the notation (5.14) the following more appropriate one:

(5.18) $$\int_{-\infty}^{\infty} x^2 p(x, t)dx = S(t).$$

Since $u = p * q = p[s] * p[t]$, using relation (5.11) we conclude that

$$u = p[s + t].$$

By means of the notation (5.18) we can restate (5.15) as

(5.19) $$S(s + t) = S(s) + S(t).$$

This is a very simple functional equation; we now show how to find all solutions of it which are differentiable. As first step, we set $s = 0$, $t = 0$ in (5.19); we get

$$S(0) = 2S(0).$$

We deduce from this that

(5.20) $$S(0) = 0.$$

Next, subtract $S(t)$ from both sides of (5.19) and divide by s; the resulting equation is

$$\frac{S(s + t) - S(t)}{s} = \frac{S(s)}{s}.$$

We let s tend to zero; on the left we obtain the derivative of S at t; on the right the derivative of S at 0. So we conclude that the derivative of S is constant; call it

(5.21) $$\frac{dS(t)}{dt} = D.$$

According to the fundamental theorem of calculus

$$S(s) - S(0) = \int_0^s \frac{dS(t)}{dt}\, dt.$$

Using (5.20) and (5.21) we conclude that

(5.22) $$S(s) = Ds.$$

As remarked earlier, $p(x, t)$ is given by Formula (5.4):

(5.23) $$p(x, t) = \sqrt{\frac{k}{\pi}}\, e^{-kx^2},$$

$k = k(t)$ some function of t.

To determine what kind of function k is of t, we relate it to S by substituting p as expressed by (5.23) into the definition (5.18) of S. We obtain

(5.24) $$\sqrt{\frac{k}{\pi}} \int_{-\infty}^{\infty} x^2 e^{-kx^2}\, dx = S(t).$$

We rewrite the integral on the left as

$$\sqrt{\frac{k}{\pi}} \int_{-\infty}^{\infty} x^2 e^{-kx^2}\, dx = \sqrt{\frac{k}{\pi}} \int_{-\infty}^{\infty} \frac{x}{2k} (2kx e^{-kx^2})\, dx.$$

We note that $2kx \exp\{-kx^2\}$ is the derivative of $-\exp\{-kx^2\}$; integrate by parts and obtain

(5.24)′ $$\sqrt{\frac{k}{\pi}} \frac{1}{2k} \int_{-\infty}^{\infty} e^{-kx^2}\, dx = S(t).$$

305

Since

$$\int_{-\infty}^{\infty} p(x, t)dx = \sqrt{\frac{k}{\pi}} \int_{-\infty}^{\infty} e^{-kx^2}\, dx = 1,$$

we conclude from (5.24)′ that

(5.24)″ $$\frac{1}{2k} = S(t).$$

We can determine k as

(5.25) $$k(t) = \frac{1}{2S(t)}.$$

According to (5.22), $S(t) = Dt$, so by (5.25)

(5.26) $$k(t) = \frac{1}{2Dt}.$$

Substituting this into (5.23), we get

(5.27) $$p(x, t) = \sqrt{\frac{1}{2\pi Dt}}\exp\left\{-\frac{x^2}{2Dt}\right\}.$$

Since k is positive, we conclude that the constant D is positive. D is called the *diffusion constant* and depends on physical properties of ink and water, such as the size of the ink particles and the temperature of water.

   Substituting Formula (5.26) for k into Formula (5.3) for ink density, we get

(5.28) $$c(x, y, t) = \frac{1}{2\pi Dt}\exp\left\{-\frac{x^2 + y^2}{2Dt}\right\}.$$

If we set both x and y equal to zero we get

(5.29) $$c(0, 0, t) = \frac{1}{2\pi Dt}.$$

This formula shows how fast the ink density decreases at the spot where the ink has dropped into the water. An interesting application of Formula (5.29) is to predict how fast diffusion diminishes the concentration of some pollutant dumped into the ocean.

   Gaussian probability densities, i.e., densities of the form

(5.30) $$p(x) = \sqrt{\frac{k}{\pi}}\exp\{-kx^2\},$$

have figured prominently in both Sections 6.4 and 6.5. We now show that gaussian densities have an *extremal* property that is both interesting and important. We have seen in (5.24)″ that, for p given by (5.30),

(5.31) $$\int_{-\infty}^{\infty} x^2 p(x)dx = \frac{1}{2k}.$$

Theorem 5.2. *Among all probability densities $q(x)$ that satisfy*

(5.32)
$$\int_{-\infty}^{\infty} x^2 q(x) dx = \frac{1}{2k},$$

the quantity

(5.33)
$$E(q) = -\int_{-\infty}^{\infty} q(x)\log q(x) dx$$

is largest for the gaussian, i.e., *when $q = p$ given by (5.30).*

Remark. This result is a continuous analogue of Theorem 2.1 in Section 6.2, which asserts that, among all probability distributions for n events,

(5.34)
$$-\sum_{1}^{n} p_j \log p_j$$

is largest when all the p_j are equal. Our proof of Theorem 5.2 is similar to the proof given in Section 6.2 in the discrete case.

PROOF. Given any q satisfying (5.32), we construct the following one-parameter family of probability densities $r(s)$ in the interval $0 \leq s \leq 1$:

(5.35)
$$r(s) = sq + (1 - s)p,$$

p given by (5.30). This function is so designed that

(5.36)
$$r(0) = p, \qquad r(1) = q.$$

To show that

(5.37)
$$E(p) \geq E(q),$$

it suffices to verify that $E(r(s))$, which we abbreviate as $F(s)$, is a decreasing function of s. According to the monotonicity criterion, the decreasing character of $F(s)$ can be shown by verifying that the derivative of $F(s)$ is negative. To this end we calculate the derivative of

(5.38)
$$F(s) = -\int_{-\infty}^{\infty} r(s)\log r(s) dx.$$

We shall appeal to the differentiation theorem for integrals in Section 4.2 which says: If $f[s]$ is a one-parameter family of functions of x which depends differentiably on the parameter s, then

(5.39)
$$F(s) = \int_{a}^{b} f(s) dx$$

depends differentiably on s, and its derivative with respect to s is

(5.40)
$$\frac{dF(s)}{ds} = \int_{a}^{b} \frac{d}{ds} f(s) dx.$$

307

We apply this to

$$f(s) = -r(s)\log r(s).$$

Differentiation with respect to s gives

(5.41) $$\frac{d}{ds} f(s) = -[1 + \log r(s)]\frac{dr}{ds}.$$

From the definition (5.35) of r we get

(5.42) $$\frac{dr}{ds} = (q - p);$$

substituting this into (5.41) gives the integrand

(5.41)′ $$\frac{d}{ds} f(s) = [1 + \log r(s)](p - q)$$

of (5.40), which now reads

(5.43) $$\frac{dF(s)}{ds} = \int_{-\infty}^{\infty} [1 + \log r(s)][p - q]dx.$$

According to (5.36)

$$r(0) = p;$$

and by definition (5.30) of p,

$$\log p = \log\sqrt{\frac{k}{\pi}} - kx^2.$$

So, if we set $s = 0$ in (5.43), we get

(5.44)
$$\frac{dF}{ds}(0) = \int_{-\infty}^{\infty} \left[1 + \log\sqrt{\frac{k}{\pi}} - kx^2\right](p - q)dx$$

$$= \left(1 + \log\sqrt{\frac{k}{\pi}}\right)\int_{-\infty}^{\infty} (p - q)dx - k\int_{-\infty}^{\infty} x^2(p - q)dx.$$

Since both p and q are probability densities,

$$\int_{-\infty}^{\infty} p\, dx = \int_{-\infty}^{\infty} q\, dx = 1.$$

Furthermore, according to (5.31) and (5.32),

$$\int_{-\infty}^{\infty} x^2 p(x)dx = \int_{-\infty}^{\infty} x^2 q(x)dx = \frac{1}{2k}.$$

It follows from these relations that the two integrals on the second line in (5.44) are both zero. Therefore

(5.45)
$$\frac{dF}{ds}(0) = 0.$$

To show that

(5.46)
$$\frac{dF}{ds}(s) \leq 0$$

for all s between 0 and 1, we shall verify that

$$\frac{d^2F}{ds^2} \leq 0.$$

According to the monotonicity criterion this implies that dF/ds is a decreasing function of s. Since by (5.45), dF/ds is zero at $s = 0$, (5.46) follows.

We now calculate the second derivative of F by again applying the differentiation theorem for integrals of Chapter 4 to the function dF/ds as defined by the integral (5.43). We get that

$$\frac{d^2F(s)}{ds^2} = \int_{-\infty}^{\infty} (p - q)\frac{1}{r(s)}\frac{dr}{ds}dx.$$

Using Formula (5.42) for dr/ds, we get

(5.47)
$$\frac{d^2F(s)}{ds^2} = \int_{-\infty}^{\infty} -\frac{(p - q)^2}{r(s)}dx.$$

The integrand in (5.47) is nonpositive, hence so is the integral:

$$\frac{d^2F(s)}{ds^2} \leq 0.$$

This completes the proof of the decreasing character of $F(s)$ and thus proves Theorem 5.2. $\qquad\qquad\square$

Remark 1. In our proof we have applied the differentiation theorem for integrals of Chapter 4 to improper integrals over the infinite interval $(-\infty, \infty)$, whereas this differentiation theorem was proved only for proper integrals. To get around this difficulty we assume that $q(x)$ equals $p(x)$ outside a sufficiently large interval (a, b) and derive the inequality

$$E(q) \leq E(p)$$

for this subclass of q. From this we can deduce the inequality for any q by approximating any q by a sequence of q's belonging to the subclass. We omit the details of this step in the proof.

309

Remark 2. We have shown in Section 6.2 that the function (5.34) has an information theoretic interpretation. The function $E(q)$ defined by (5.33) has an interpretation in statistical mechanics as the *entropy* of the probability density q. Therein lies the importance of Theorem 5.2.

EXERCISES

5.1 Let m and p be the functions

$$m(a) = \sqrt{\frac{j}{\pi}} e^{-ja^2},$$

(5.48)

$$p(x) = \sqrt{\frac{h}{\pi}} e^{-hx^2}.$$

Determine their convolution

$$u = m * p,$$

given by Formula (5.10), by carrying out the following steps:
(a) Substitute (5.48) for m and p into (5.10); show that the resulting integral can be written as

(5.49)
$$u(x) = \frac{\sqrt{jh}}{\pi} \int_{-\infty}^{\infty} e^{B(a, x)} da$$

where

$$B(a, x) = -ja^2 - h(x - a)^2$$

(b) Show that B can be rewritten as

(5.50)
$$B = -(j + h)b^2 - \frac{jh}{j + h} x^2$$

where

(5.51)
$$b = a - \frac{h}{j + h} x.$$

(c) Substitute (5.50) for B into the integral on the right in (5.49) and introduce b as new variable of integration. Show that the value of the resulting integral is

(5.52)
$$u(x) = \sqrt{\frac{jh}{(j + h)\pi}} \exp\left\{ -\frac{jh}{j + h} x^2 \right\}.$$

(d) Identify h as $k(t)$, j as $k(s)$ and $jh/(j + h)$ as $k(s + t)$, and show that this identification is consistent with (5.26).

5.2 Using Formula (5.27) for $p(x, t)$

    (a) determine the derivative of p with respect to x;

    (b) determine the second derivative of p with respect to x;

    (c) determine the derivative of p with respect to t;

    (d) verify that

(5.53)
$$p'_t = \frac{D}{2} p''_{xx},$$

where p'_t abbreviates the t derivative of p and p''_{xx} the second x derivative of p.

5.3 Let m be any function which is zero outside some finite interval. Consider the function $u(x, t)$ defined by Formula (5.10):

(5.10)ₜ
$$u(x, t) = \int_{-\infty}^{\infty} m(a)p(x - a, t)da.$$

    (a) Show that the derivatives of u with respect to x and t are given by the formulas

$$u'_t = \int_{-\infty}^{\infty} m(a)p'_t(x - a, t)da,$$

$$u'_x = \int_{-\infty}^{\infty} m(a)p'_x(x - a, t)da,$$

$$u''_{xx} = \int_{-\infty}^{\infty} m(a)p''_{xx}(x - a, t)da.$$

[*Hint*: Use the rules for differentiating a function defined as an integral given in Section 4.2.]

    (b) Show, using (5.53), that u satisfies the differential equation

(5.54)
$$u'_t = \frac{D}{2} u''_{xx}.$$

5.4 (a) Suppose that we place a barrier which turns back ink particles along the line $x = 0$. At time $t = 0$, a unit mass of ink is placed at $(a, 0)$. Show by a plausible argument that the density of ink at time t at the point (x, y), $x > 0$, is

$$\frac{1}{2\pi Dt} \left[\exp\left\{ -\frac{(x - a)^2 + y^2}{2Dt} \right\} + \exp\left\{ -\frac{(x + a)^2 + y^2}{2Dt} \right\} \right].$$

    (b) Suppose that the barrier placed at $x = 0$ is made of material that *absorbs* whatever ink particle reaches it. Give a plausible argument to show that, in this case, the density at the point (x, y), $x > 0$ and at time t is

$$\frac{1}{2\pi Dt} \left[\exp\left\{ -\frac{(x - a)^2 + y^2}{2Dt} \right\} - \exp\left\{ -\frac{(x + a)^2 + y^2}{2Dt} \right\} \right].$$

5.5 (a) Suppose that all of 3-dimensional space is filled with water. Release a drop of ink containing a unit mass at the origin and let it diffuse. Show by an argument analogous to the one employed in this section, that the ink density at the point (x, y, z) and at time t is

$$[2\pi Dt]^{-3/2} \exp\left\{ -\frac{x^2 + y^2 + z^2}{2Dt} \right\}.$$

311

(b) Suppose that the halfspace $z \leq 0$ is filled with water, and that any ink particle reaching the surface is reflected back in the water. Show that if a unit mass of ink is released at the origin and allowed to diffuse, the ink density at the point x, y, z, at time t is

$$2(2\pi Dt)^{-3/2} \exp\left\{-\frac{x^2 + y^2 + z^2}{2Dt}\right\}.$$

5.6 Define $p(x)$ to be

(5.55) $$p(x) = \sqrt{k/\pi} \exp\{-k(x-a)^2\}.$$

(a) Show that

$$\int_{-\infty}^{\infty} p(x)dx = 1,$$

$$\int_{-\infty}^{\infty} xp(x)dx = a,$$

$$\int_{-\infty}^{\infty} x^2 p(x)dx = \frac{1}{2k} + a^2.$$

(b) Show that among all probability densities that satisfy

$$\int_{-\infty}^{\infty} xq(x)dx = a, \qquad \int_{-\infty}^{\infty} x^2 q(x)dx = \frac{1}{2k} + a^2.$$

the value of

(5.56) $$E(q) = -\int_{-\infty}^{\infty} q(x)\log q(x)dx$$

is largest for $q = p$ given by (5.55).

5.7 Define $p(x)$ to be

(5.57) $$p(x) = \begin{cases} \dfrac{1}{k}e^{-kx} & \text{for } x \geq 0 \\ 0 & \text{for } x < 0 \end{cases}$$

(a) Show that

$$\int_0^{\infty} xp(x)dx = \frac{1}{k}.$$

(b) Show that among all probability densities $q(x)$ which vanish for $x < 0$ and which satisfy

$$\int_0^{\infty} xq(x)dx = \frac{1}{k},$$

the value of $E(q)$ defined by (5.56) is largest for $q = p$ given by (5.57).

5.8 (a) In the diffusion equation (5.54) replace the derivatives by the difference approximations

$$u'_t \approx \frac{u(x, t + k) - u(x, t)}{k}$$

and

$$u''_{xx} \approx \frac{u(x + h, t) - 2u(x, t) + u(x - h, t)}{h^2}.$$

Set the diffusion coefficient $D = 2$. Show that the resulting equation is

(5.58) $u(x, t + k) = u(x, t) + \dfrac{k}{h^2} [u(x + h, t) - 2u(x, t) + u(x - h, t)].$

(b) Show that if h and k are related by

(5.59) $\dfrac{k}{h^2} = \dfrac{1}{2},$

then (5.58) takes the form

(5.58)′ $u(x, t + k) = \tfrac{1}{2}u(x - h, t) + \tfrac{1}{2}u(x + h, t).$

(c) Deduce from (5.58)′ that

$$u(x, 2k) = \tfrac{1}{4}u(x - 2h, 0) + \tfrac{1}{2}u(x, 0) + \tfrac{1}{4}u(x + 2h, 0)$$

and

$$u(x, 3k) = \tfrac{1}{8}u(x - 3h, 0) + \tfrac{3}{8}u(x - h, 0) + \tfrac{3}{8}u(x + h, 0) + \tfrac{1}{8}u(x + 3h, 0).$$

(d) Show, for every integer n, that if u satisfies (5.58)′, then

(5.60) $u(x, nk) = \dfrac{1}{2^n} \displaystyle\sum_{j=0}^{n} \binom{n}{j} u(x + (2j - n)h, 0),$

where the $\binom{n}{j}$ are the *binomial coefficients* defined by Equation (9.17), Section 3.9.
(e) Recall that the $\binom{n}{j}$ satisfy the binomial formula

(5.61) $\displaystyle\sum_{j=0}^{n} \binom{n}{j} s^j = (1 + s)^n.$

Take the initial values $u(x, 0)$ of the ink density to be

(5.62) $u(x, 0) = e^{px}$

p any real or complex number. Show, by setting (5.62) into (5.60) and using
(5.61), that

(5.63) $u(x, nk) = e^{px} \left(\dfrac{e^{ph} + e^{-ph}}{2} \right)^n.$

(f) Show that if h and k tend to zero so that (5.59) is satisfied, and n tends to ∞ so
that nk tends to t, then $u(x, nk)$ given by (5.63) tends to

(5.64) $v(x, t) = e^{px} e^{p^2 t}.$

[*Hint*: Show, using the Taylor series for the exponential function, that
$(\exp\{ph\} + \exp\{-ph\})/2 = 1 + (p^2h^2/2)f$, where f tends to 1 as h tends to
zero. Then use (5.59) and $nk = t$.]

(g) Show that the function v, defined in (5.64), satisfies the diffusion equation

$$v'_t = v''_{xx}.$$

(h) In Formula $(5.10)_t$, set $D = 2$ and replace the integral on the right by an approximating sum based on the subdivision of the real line into the subintervals

$$(x + (2l - 1)h, \quad x + (2l + 1)h), \qquad l = 0, 1, 2, \ldots .$$

The length of each interval is $2h$ and the midpoint of the l-th interval is $x + 2lh$. Therefore verify that the midpoint formula gives the following approximation to $u(x, t)$:

(5.65)
$$\frac{h}{\sqrt{\pi t}} \sum_l m(x + 2lh) e^{-l^2 h^2/t}.$$

(i) In Formula (5.60) take n to be even and set $l = j - n/2$ as new variable of summation; set $u(x, 0) = m(x)$ and get

(5.60)'
$$u(x, nk) = \frac{1}{2^n} \sum_{l=-n/2}^{n/2} m(x + 2lh) \binom{n}{l + n/2}.$$

(j) Both (5.65) and (5.60)' furnish an approximation to the density of ink whose initial density is $m(x)$ and which has been diffusing for time t. So we expect the *weights* occurring in these formulas to be approximately equal:

$$\frac{1}{2^n} \binom{n}{l + n/2} \approx \frac{h}{\sqrt{\pi t}} e^{-l^2 h^2/t}.$$

On the right set $t = nk$ and $2k = h^2$. On the left use Formula (9.17), Section (3.9) for the binomial coefficient and obtain

(5.66)
$$\frac{1}{2^n} \frac{n(n - 1) \cdots (n - (l + n/2) + 1)}{(l + n/2)!} \approx \sqrt{\frac{2}{\pi n}} e^{-2l^2/n}.$$

(k) Let's test the validity of (5.66) in the simplest case $l = 0$:

$$\frac{1}{2^n} \frac{n(n - 1) \cdots (n/2 + 1)}{1 \cdot 2 \cdots n/2} \approx \sqrt{\frac{2}{\pi n}}.$$

Multiply both numerator and denominator on the left by $(n/2)!$, and multiply both sides by $\sqrt{n}$, to get that

$$\frac{1}{2^n} \frac{n!}{[(n/2)!]^2} \sqrt{n} \approx \sqrt{\frac{2}{\pi}}.$$

Now write $n!$ as a product of even and odd factors; divide each of the $n/2$ even factors by 2 and obtain

$$\frac{1}{2^{n/2}} \frac{(n - 1)(n - 3) \cdots}{(n/2)!} \sqrt{n} \approx \sqrt{\frac{2}{\pi}}$$

or

(5.67)
$$\frac{(n - 1)(n - 3) \cdots}{n(n - 2) \cdots} \sqrt{n} \approx \sqrt{\frac{2}{\pi}}.$$

Use a hand calculator to compute the value of the left side for $n = 10, 20$, and 40 and compare the result to $\sqrt{2/\pi}$. For a proof, study Exercise 2.3 in Section 7.2.

Rotation and the trigonometric functions 7

It is often asserted in trigonometry texts, especially the older ones, that the importance of trigonometry lies in its usefulness for surveying and navigation. Since the proportion of our population engaged in these pursuits is rather small, one wonders what kind of stranglehold surveyors and navigators have over professional education to be able to enforce the universal teaching of this abstruse subject. Or is it merely inertia? The answer, of course, is that the importance of trigonometry lies elsewhere: in the description of *rotation* and *vibration*. It is an astonishing fact of mathematical physics that the vibration of as diverse a collection of objects as:

springs

strings

airplane wings

steel beams   light beams

and water streams

building sways

and ocean waves . . .

and many others are described in terms of trigonometric functions. That such diverse phenomena can be treated with a common tool is one of the most striking successes of calculus. Some simple and some not so simple examples will be discussed in the next chapters.

In trigonometry texts one learns that there are six trigonometric functions:

sine, cosine, tangent, cotangent

secant, and cosecant.

This turns out to be a slight exaggeration. There are only two basic functions, sine and cosine; all the others can be defined in terms of them, when

315

necessary. Furthermore sine and cosine are so closely related that each can be expressed in terms of the other; so one can say that there is really only one trigonometric function.

7.1 Rotation

We start with the measurement of angles. Consider an angle A in the Cartesian plane, swept out as the positive x-axis is rotated counterclockwise about the origin until it coincides with some given ray R; see Figure 7.1a. Draw a circle of unit radius about the origin.

Definition. *The radian measure of angle A is the length of that arc of the unit circle which lies inside the angle A* (indicated by a heavy line in Figure 7.1b).

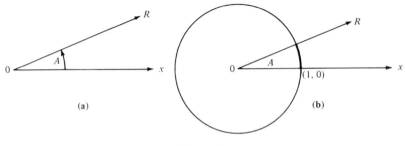

Figure 7.1

This is a fine definition except for one thing: we have not yet learned how to calculate the length of an arc of a circle. So our definition of radian measure is circular in more than one sense. In Volume II we shall describe a way to calculate the length of arcs of arbitrary curves; but for the present we shall get around this difficulty by relating arc length to area.

The length s of an arc of the unit circle is proportional to the area A_s of the sector bounded by the arc and the radii connecting its endpoints to the center. Denoting the constant of proportionality by k, we may write

$$s = kA_s.$$

This is in accordance with our geometric intuition of arc length and area. We shall now determine the constant of proportionality by looking at a very slender sector; see Figure 7.2. A slender sector looks very much like a slender triangle, except that one of its sides is slightly curved instead of being straight; so to a very good approximation the area of the sector is $\frac{1}{2}$ base times altitude.

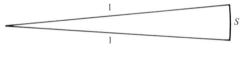

Figure 7.2

Choose as base the curved side; the altitude, being the distance from the opposite vertex to the base, is equal to 1, the radius of the circle. So approximately, the area of the slender sector is $\frac{1}{2}$ the length of the arc: $A_s = \frac{1}{2}s$. We infer from this intuitive argument that the constant of proportionality k is 2; that is,

$$s = 2A_s.$$

The length of an arc of the unit circle is twice the area of the sector bounded by the arc and the radii connecting its endpoints to the center. Since the area of a sector can be defined as an integral, this is an adequate definition of the length of circular arcs. The area of the unit circle is denoted by π; calculations in Section 4.6 give the value of π as

$$\pi = 3.14159\ldots\,.$$

So the circumference of the unit circle is

$$2\pi = 6.28318\ldots\,.$$

Consider as before an angle A in the Cartesian plane, formed by the positive x-axis and a ray R issuing from the origin. Denote the radian measure of A by α. Let (x, y) be the point on R whose distance from the origin is 1; see Figure 7.3. We have defined α to be the length of the arc from $(1, 0)$ to (x, y). We now define x to be the cosine of α, and y to be the sine of α:

(1.1) $$\boxed{x = \cos \alpha, \qquad y = \sin \alpha}.$$

Sine and cosine as defined above are *functions* of α; we shall now study the basic properties of these functions.

By definition of unit circle, the distance of the point (x, y) from the origin is 1 and can be expressed, by the Pythagorean theorem, as $\sqrt{x^2 + y^2}$. So we deduce that

(1.2) $$\cos^2 \alpha + \sin^2 \alpha = 1$$

for all angles α.

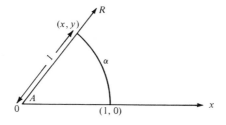

Figure 7.3

Next we compute the value of sin α and cos α for some simple special values of α. For this purpose we restate the definition of sine and cosine. *Starting at the point* $(1, 0)$ *and proceeding in the counterclockwise direction, measure off, along the circumference of the unit circle, an arc of length* α. *The x and y coordinates of the endpoint of this arc are the cosine and sine of* α. Clearly, for $\alpha = 0$, we are constructing an arc of zero length; its endpoint is the same as the beginning point, so we conclude that $(x, y) = (1, 0)$, i.e.,

$$\cos 0 = 1, \qquad \sin 0 = 0.$$

Since the circumference of the unit circle is 2π, an arc of length π takes us half way around the circle. The endpoints of such a semicircle are diametrically opposite each other relative to the center, so that in this case, $(x, y) = (-1, 0)$, that is,

$$\cos \pi = -1, \qquad \sin \pi = 0.$$

An arc of length $\pi/2$ takes us one quarter of the way around the circle; see Figure 7.4a. If one endpoint of such an arc lies on the positive x-axis, then the other lies on the positive y-axis, at $(x, y) = (0, 1)$, so

$$\cos \frac{\pi}{2} = 0, \qquad \sin \frac{\pi}{2} = 1.$$

An arc of length $\pi/4$ takes us $(1/8)$-th the distance about the unit circle; see Figure 7.4b. The endpoint (x, y) of such an arc lies halfway between the positive x axis and the positive y axis, on the line $y = x$. Using the relation (1.2), we conclude that $x^2 + y^2 = 2x^2 = 1$. Thus

$$\cos \frac{\pi}{4} = \frac{\sqrt{2}}{2}, \qquad \sin \frac{\pi}{4} = \frac{\sqrt{2}}{2}.$$

An arc of length $\pi/3$ covers $(1/6)$-th of the unit circle. Let (x, y) be the co-ordinates of the endpoint of this arc; see Figure 7.5. Since the angle at 0 is $\pi/3$ radians, since the triangle with vertices $(0, 0)$, $(1, 0)$ and (x, y) is isosceles, and since the sum of its angles is π, we see that the triangle is equilateral.

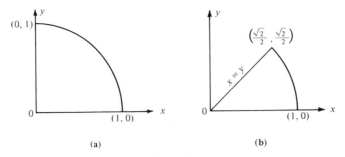

(a)   (b)

Figure 7.4

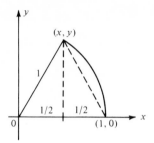

Figure 7.5

In particular the vertex (x, y) on top lies over the midpoint of the base, i.e., $x = \frac{1}{2}$. Using the Pythagorean relation, we get $y = \sqrt{1 - x^2} = \sqrt{1 - 1/4} = \sqrt{3/4} = \sqrt{3}/2$, so

$$\cos \frac{\pi}{3} = \frac{1}{2}, \qquad \sin \frac{\pi}{3} = \frac{\sqrt{3}}{2}.$$

An arc of length $\pi/6$ covers $(\frac{1}{12})$-th of the unit circle. A glance at Figure 7.6 reveals that the points $(\cos(\pi/6), \sin(\pi/6))$ and $(\cos(\pi/3), \sin(\pi/3))$ lie symmetrically with respect to the line $y = x$, so that $\cos(\pi/6) = \sin(\pi/3)$ and $\sin(\pi/6) = \cos(\pi/3) = \frac{1}{2}$. Thus

$$\cos \frac{\pi}{6} = \frac{\sqrt{3}}{2}, \qquad \sin \frac{\pi}{6} = \frac{1}{2}.$$

In a similar fashion we can determine the sine and cosine of all angles which are integer multiples of $\pi/4$ and $\pi/6$; we list these values in Table (1.3). As the length α of the arc increases, the coordinates x, y of its endpoint vary continuously, so $\cos \alpha$, $\sin \alpha$ are continuous functions of α. Using the continuity of these functions and their values given in Table (1.3), we present in

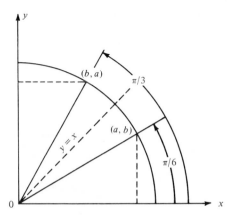

Figure 7.6

Table 1.3

α	0	$\dfrac{\pi}{6}$	$\dfrac{\pi}{4}$	$\dfrac{\pi}{3}$	$\dfrac{\pi}{2}$	$\dfrac{2\pi}{3}$	$\dfrac{3\pi}{4}$	$\dfrac{5\pi}{6}$	π
$\cos\alpha$	1	$\dfrac{\sqrt{3}}{2}$	$\dfrac{\sqrt{2}}{2}$	$\dfrac{1}{2}$	0	$-\dfrac{1}{2}$	$-\dfrac{\sqrt{2}}{2}$	$-\dfrac{\sqrt{3}}{2}$	-1
$\sin\alpha$	0	$\dfrac{1}{2}$	$\dfrac{\sqrt{2}}{2}$	$\dfrac{\sqrt{3}}{2}$	1	$\dfrac{\sqrt{3}}{2}$	$\dfrac{\sqrt{2}}{2}$	$\dfrac{1}{2}$	0

α	$\dfrac{7\pi}{6}$	$\dfrac{5\pi}{4}$	$\dfrac{4\pi}{3}$	$\dfrac{3\pi}{2}$	$\dfrac{5\pi}{3}$	$\dfrac{7\pi}{4}$	$\dfrac{11\pi}{6}$	2π
$\cos\alpha$	$-\dfrac{\sqrt{3}}{2}$	$-\dfrac{\sqrt{2}}{2}$	$-\dfrac{1}{2}$	0	$\dfrac{1}{2}$	$\dfrac{\sqrt{2}}{2}$	$\dfrac{\sqrt{3}}{2}$	1
$\sin\alpha$	$-\dfrac{1}{2}$	$-\dfrac{\sqrt{2}}{2}$	$-\dfrac{\sqrt{3}}{2}$	-1	$-\dfrac{\sqrt{3}}{2}$	$-\dfrac{\sqrt{2}}{2}$	$-\dfrac{1}{2}$	0

Figure 7.7, a sketchy graph of the sine and cosine functions in the range $0 \le \alpha \le 2\pi$.

In the above table and graphs we considered angles between 0 and 2π; this restriction is dictated by the observation that the length of a *proper* arc of a circle lies between 0 and the total circumference. However, there is a perfectly natural interpretation of cos α and sin α for angles exceeding 2π. If we wrap a string of length greater than 2π around the unit circle, we pass the starting point as shown in Figure 7.8. Clearly, the endpoint of an arc of length $\alpha + 2\pi$ is the same as the endpoint of an arc of length α. Since sine

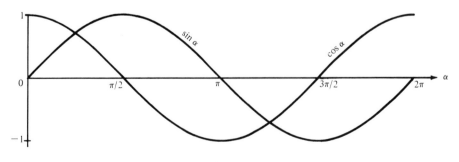

Figure 7.7

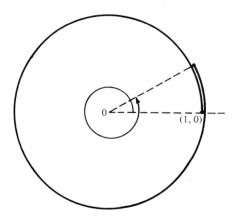

Figure 7.8

and cosine are defined in terms of the endpoint, it follows that for every α,

(1.4) $$\cos(\alpha + 2\pi) = \cos \alpha, \qquad \sin(\alpha + 2\pi) = \sin \alpha.$$

This property of sine and cosine is called *periodicity*.

The next property of sine and cosine has to do with their values for small angles α. (The term "angle α" abbreviates "angle of radian measure α.") The sector of angle α *contains* the triangle formed by the 3 vertices $(0, 0)$, $(1, 0)$ and $(\cos \alpha, \sin \alpha)$; see Figure 7.9a. The area of the triangle $= \frac{1}{2}$ base $\times$ altitude $= \frac{1}{2} 1 \cdot \sin \alpha = \frac{1}{2} \sin \alpha$. At the beginning of this section, we saw that

$$\text{Area of sector} = \tfrac{1}{2}\alpha.$$

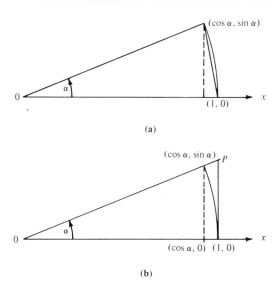

Figure 7.9

321

Therefore we conclude that

(1.5)
$$\tfrac{1}{2} \sin \alpha \le \tfrac{1}{2}\alpha.$$

On the other hand, the sector is *contained* in the right triangle shown in the Figure 7.9b with vertices $(0, 0)$, $(1, 0)$, and P. This triangle is similar to the triangle formed by $(0, 0)$, $(\cos \alpha, 0)$, and $(\cos \alpha, \sin \alpha)$. Therefore the third vertex P is at the point

$$P: \left(1, \frac{\sin \alpha}{\cos \alpha}\right).$$

The area of this triangle is $\tfrac{1}{2}$ base $\times$ altitude $= \tfrac{1}{2} \times 1 \times \sin \alpha/\cos \alpha$. Since the sector is contained in the triangle, its area is less than that of the triangle; thus

(1.6)
$$\frac{1}{2} \alpha \le \frac{1}{2} \frac{\sin \alpha}{\cos \alpha}.$$

Combining this inequality with (1.5), we get

(1.7)
$$\sin \alpha \le \alpha \le \frac{\sin \alpha}{\cos \alpha}.$$

It follows from the left inequality that, as α tends to zero, so does $\sin \alpha$. Therefore it follows from relation (1.2) that $\cos \alpha$ tends to 1. Now divide (1.7) by $\sin \alpha$ and obtain the inequalities

$$1 \le \frac{\alpha}{\sin \alpha} \le \frac{1}{\cos \alpha}.$$

As α tends to zero, $1/\cos \alpha$ tends to 1, and so we conclude that $\alpha/\sin \alpha$, squeezed between 1 and $1/\cos \alpha$, also tends to 1:

(1.8)
$$\lim_{\alpha \to 0} \frac{\alpha}{\sin \alpha} = 1.$$

This relation describes the behavior of the sine function for small angles.

By means of the relation (1.2) the cosine function can be expressed in terms of the sine function as

(1.9)
$$\cos \alpha = \sqrt{1 - \sin^2 \alpha}.$$

Here we take the positive square root because for small arcs α, the abscissa $x = \cos \alpha$ of the endpoint is in the right half plane. Since we already know how the sine function behaves for small inputs, (1.9) enables us to study the behavior of the cosine function. The derivative of the sine function will be helpful to this end.

At $\alpha = 0$, the derivative of the sine function is defined as

$$\lim_{\alpha \to 0} \frac{\sin \alpha - \sin 0}{\alpha} = \lim_{\alpha \to 0} \frac{\sin \alpha}{\alpha};$$

so by (1.8),

$$\frac{d \sin \alpha}{d\alpha} = 1 \qquad \text{at } \alpha = 0.$$

The chain rule enables us to express the derivative of the cosine in terms of the derivative of the sine:

$$\frac{d \cos \alpha}{d\alpha} = \frac{d\sqrt{1 - \sin^2 \alpha}}{d\alpha} = -\frac{\sin \alpha}{\sqrt{1 - \sin^2 \alpha}} \cdot \frac{d \sin \alpha}{d\alpha}.$$

Setting $\alpha = 0$ and using the fact that $\sin 0 = 0$, we find

$$(1.10) \qquad \frac{d \cos \alpha}{d\alpha} = 0 \qquad \text{at } \alpha = 0.$$

Recalling the definition of the derivative, we can restate (1.10) as

$$(1.11) \qquad \lim_{\alpha \to 0} \frac{\cos \alpha - \cos 0}{\alpha} = \lim_{\alpha \to 0} \frac{\cos \alpha - 1}{\alpha} = 0.$$

Next, we use the identities from trigonometry

$$(1.12) \qquad \begin{aligned} \cos(\alpha + \beta) &= \cos \alpha \cos \beta - \sin \alpha \sin \beta \\ \sin(\alpha + \beta) &= \sin \alpha \cos \beta + \cos \alpha \sin \beta. \end{aligned}$$

In Section 7.5 we shall derive these from the rules of multiplication for complex numbers.

The functional equations (1.12) for the cosine and sine express another important property of these functions. Set $\beta = \pi/2$ in the second of these formulas, obtaining

$$(1.13) \qquad \sin\left(\alpha + \frac{\pi}{2}\right) = \sin \alpha \cos \frac{\pi}{2} + \cos \alpha \sin \frac{\pi}{2}.$$

Previously, we observed that

$$\cos \frac{\pi}{2} = 0, \qquad \sin \frac{\pi}{2} = 1,$$

so that (1.13) yields

$$(1.14) \qquad \sin\left(\alpha + \frac{\pi}{2}\right) = \cos \alpha.$$

This relation says that a shift by $\pi/2$ in the argument (input) of the sine yields the cosine function. Geometrically this means that the graph of the cosine function is obtained by shifting the graph of the sine function to the left by the amount $\pi/2$. Computationally this means that $\cos\alpha$ can be obtained from any algorithm for $\sin\alpha$ by using $\alpha + \pi/2$ as input instead of α. It was this relation between cosine and sine that prompted us to say at the beginning of this section that they can hardly be considered two different functions.

We come now to the main point of this section, the differentiation of cosine and sine. By subtracting from Equations (1.12) $\cos\alpha$ and $\sin\alpha$, respectively, and dividing by β, we can write the difference quotients as follows:

(1.15)
$$\frac{\cos(\alpha + \beta) - \cos\alpha}{\beta} = \cos\alpha\,\frac{\cos\beta - 1}{\beta} - \sin\alpha\,\frac{\sin\beta}{\beta}$$
$$\frac{\sin(\alpha + \beta) - \sin\alpha}{\beta} = \sin\alpha\,\frac{\cos\beta - 1}{\beta} + \cos\alpha\,\frac{\sin\beta}{\beta}.$$

Let β tend to 0; according to (1.11), $(\cos\beta - 1)/\beta$ tends to 0, and according to (1.8), $\sin\beta/\beta$ tends to 1. Therefore it follows from (1.15) that

$$\lim_{\beta\to 0}\frac{\cos(\alpha + \beta) - \cos\alpha}{\beta} = -\sin\alpha$$

$$\lim_{\beta\to 0}\frac{\sin(\alpha + \beta) - \sin\alpha}{\beta} = \cos\alpha.$$

That is, the derivatives of cosine and sine are

(1.16)
$$\frac{d}{d\alpha}\cos\alpha = -\sin\alpha$$
$$\frac{d}{d\alpha}\sin\alpha = \cos\alpha.$$

In the next section we shall show that *all* properties of cosine and sine are contained in the differential equations (1.16) and in the specifications

$$\cos 0 = 1, \qquad \sin 0 = 0$$

of their values at 0.

EXERCISES

1.1 (a) By setting $\beta = \alpha$ in (1.12), show that

(1.17)
$$\cos 2\alpha = \cos^2\alpha - \sin^2\alpha,$$
$$\sin 2\alpha = 2\sin\alpha\cos\alpha.$$

(b) Using relation (1.2), show that

(1.18)
$$\cos 2\alpha = 2\cos^2\alpha - 1,$$
$$\cos 2\alpha = 1 - 2\sin^2\alpha.$$

(c) Deduce from (1.18) that

$$\cos \alpha = \pm \sqrt{\frac{1 + \cos 2\alpha}{2}},$$

(1.19)

$$\sin \alpha = \pm \sqrt{\frac{1 - \cos 2\alpha}{2}}.$$

(d) Using (1.19) and that $\cos \pi = -1$, determine the values of $\cos(\pi/2)$, $\cos(\pi/4)$, $\cos(\pi/8)$, $\cos(3\pi/2)$, and $\sin(\pi/2)$, $\sin(\pi/4)$, $\sin(\pi/8)$ and $\sin(3\pi/2)$.

(e) Setting $\beta = 2\alpha$ in (1.12), and using (1.17), show that

(1.20) $$\cos 3\alpha = 4 \cos^3 \alpha - 3 \cos \alpha.$$

Using (1.20) and that $\cos \pi = -1$, determine $\cos(\pi/3)$. Using (1.20) and that $\cos(\pi/2) = 0$, determine $\cos(\pi/6)$.

(f) Using (1.18) repeatedly, show that

$$\cos 4\alpha = 8 \cos^4 \alpha - 8 \cos^2 \alpha + 1.$$

1.2 The tangent function, denoted as $\tan \alpha$, is defined by

(1.21) $$\tan \alpha = \frac{\sin \alpha}{\cos \alpha}.$$

(a) Draw a crude sketch of the graph of the tangent function for $0 \le \alpha \le 2\pi$; at what α is $\tan \alpha$ undefined?

(b) Show that

(1.22) $$\frac{d}{d\alpha} \tan \alpha = \frac{1}{\cos^2 \alpha}.$$

(c) Give a geometric interpretation of $\tan \alpha$.

(d) Using the addition formulas (1.12), prove that

(1.23) $$\tan(\alpha + \beta) = \frac{\tan \alpha + \tan \beta}{1 - \tan \alpha \tan \beta}.$$

1.3 Differentiate the following functions:

(a) $x^2 \sin \dfrac{1}{x}$ (b) $\cos \sqrt{1 + x^2}$ (c) $\cos nx$ (d) $\cos^n x$ (e) $e^{\cos x}$

(f) $\log \sin x$ (g) $\dfrac{\sin x}{1 + \cos^2 x}$ (h) $\dfrac{1}{\sin x}$ (i) $\dfrac{1}{\cos x}$ (j) $\dfrac{1}{\tan x}$.

7.2 Properties of cosine, sine, arcsine, and arctan

In this section we shall show that all properties of the cosine and sine functions, abbreviated here as c and s, are contained in the differential equations

(2.1) $$c' = -s, \qquad s' = c,$$

satisfied by c and s, and in their initial values

(2.2) $$c(0) = 1, \qquad s(0) = 0.$$

Let us start by deducing property (1.2):

(2.3) $$c^2 + s^2 = 1.$$

When we differentiate $c^2 + s^2$ and use (2.1), we get

$$(c^2 + s^2)' = 2cc' + 2ss' = 2c(-s) + 2sc = 0.$$

A function which has derivative 0 is constant. The constant value of $c^2 + s^2$ can be determined at the point 0 from (2.2):

$$c^2(0) + s^2(0) = 1^2 + 0^2 = 1.$$

This completes the proof of (2.3).

Next we show that the functions c and s are uniquely determined by the differential equation (2.1) and by the initial values (2.2). For, suppose that there were two sets of solutions, c_1 and s_1, as well as c_2 and s_2, each satisfying (2.1), i.e.,

(2.1)′ $$\begin{aligned} c_1' &= -s_1, & s_1' &= c_1 \\ c_2' &= -s_2, & s_2' &= c_2, \end{aligned}$$

and each having initial values (2.2):

(2.2)′ $$\begin{aligned} c_1(0) &= 1 = c_2(0) \\ s_1(0) &= 0 = s_2(0). \end{aligned}$$

Subtract corresponding equations in (2.1)′ to obtain

(2.4) $$(c_1 - c_2)' = -(s_1 - s_2), \qquad (s_1 - s_2)' = c_1 - c_2.$$

If we introduce the abbreviations

(2.5) $$c_1 - c_2 = C, \qquad s_1 - s_2 = S;$$

then Equations (2.4) can be written as

$$C' = -S, \qquad S' = C.$$

Thus the pair C, S of differences, (2.5), satisfies the *same* system of differential equations as the original quantities c, s. It follows therefore, as before, that $C^2 + S^2$ is a constant, whose value can be determined at the point 0. The initial values (2.2)′ tell us that

$$C(0) = c_1(0) - c_2(0) = 0, \quad \text{and} \quad S(0) = s_1(0) - s_2(0) = 0.$$

Therefore

$$C^2(0) + S^2(0) = 0,$$

and consequently

$$C^2 + S^2 = 0 \quad \text{for all values;}$$

but this can be only if each C and S is zero. Recalling the definition (2.5) of C and S, we conclude that c_1 and c_2 as well as s_1 and s_2 are *identical*. This concludes the proof of the assertion that cosine and sine are *uniquely* determined by (2.1) and (2.2).

We now give another proof of the uniqueness of solutions of (2.1) subject to (2.2), not because we lack confidence in the above proof, but because the proposition is important enough to warrant a second proof, which is intimately connected with the functional equations for sine and cosine.

As before, denote by c_1, s_1 and c_2, s_2 two allegedly different pairs of solutions. Let a be any number, and consider the function

(2.6) $$c_1(x)s_2(a - x) + s_1(x)c_2(a - x).$$

The derivative of this function is

$$c_1'(x)s_2(a - x) - c_1(x)s_2'(a - x) + s_1'(x)c_2(a - x) - s_1(x)c_2'(a - x).$$

Using the differential equations (2.1)′ we can rewrite the above function as

$$-s_1(x)s_2(a - x) - c_1(x)c_2(a - x) + c_1(x)c_2(a - x) + s_1(x)s_2(a - x),$$

which is clearly zero, since the terms cancel pairwise. So we conclude that the derivative of the function (2.6) is 0, which makes the function (2.6) constant. In particular its values at $x = 0$ and $x = a$ are the same;

$$c_1(0)s_2(a) + s_1(0)c_2(a) = c_1(a)s_2(0) + s_1(a)c_2(0).$$

According to (2.2)′, the above relation expresses simply that

$$s_2(a) = s_1(a).$$

Since a is an arbitrary number, it follows that the functions s_1 and s_2 are identical, whereupon it follows from (2.1)′ that c_1 and c_2 also are identical.

Having shown that $s_1 = s_2$ and $c_1 = c_2$, we return to the function (2.6), dropping all subscripts:

(2.7) $$c(x)s(a - x) + s(x)c(a - x).$$

As noted before, this function is constant, and its value at $x = 0$ is $s(a)$. So for any x,

$$c(x)s(a - x) + s(x)c(a - x) = s(a).$$

Let us denote $a - x$ by y; then $a = x + y$ and the above relation can be rewritten as

(2.8) $$c(x)s(y) + s(x)c(y) = s(x + y).$$

This is the second of the functional equations (1.12) for cosine and sine. The first,

$$(2.9) \qquad c(x)c(y) - s(x)s(y) = c(x + y),$$

can be derived by differentiating (2.8) with respect to x. *Thus we have succeeded in deducing the functional equations of sine and cosine from the differential equations they satisfy.*

Next we shall determine, still using only the differential equations and initial values of sine and cosine, the shape of their graphs. From (2.2) and (2.1) we deduce that the sine function, at $x = 0$, satisfies

$$s(0) = 0, \qquad s'(0) = 1$$

i.e., that the graph of $s(x)$ starts at the origin and increases with positive x. As seen in Figure 7.10a, $s(x)$ is positive for small positive x. It follows then, from $c' = -s$, that $c(x)$, which has the value 1 when $x = 0$, is a decreasing function for small positive x.

What happens to $s(x)$ and $c(x)$ as x keeps on increasing? It follows from $s' = c$ that, as long as $c(x)$ is positive, $s(x)$ keeps on increasing; and, since $c' = -s$, as long as $s(x)$ keeps increasing $c(x)$ decreases faster and faster down through zero and into negative values. But at this point $s' = c$ dictates that s stop increasing, turn around and become a decreasing function of x. Call t the value of x where $s(x)$ turns around; it is also the point where $c(t) = 0$. In view of $c^2 + s^2 = 1$, we see that $s(t) = 1$. The graph of $s(x)$ and $c(x)$, according to our analysis thus far, is displayed in Figure 7.10b.

Using the same kind of reasoning as before, we can show that beyond the point $x = t$, the functions $s(x)$ and $c(x)$ both decrease until $s(x)$ hits zero; at that point $c(x)$ reaches -1. Beyond that point $c(x)$ starts increasing, and $s(x)$ keeps decreasing until $c(x)$ hits zero; at that point $s(x)$ hits bottom with the value -1. Then $s(x)$ turns around and starts increasing until it reaches 0;

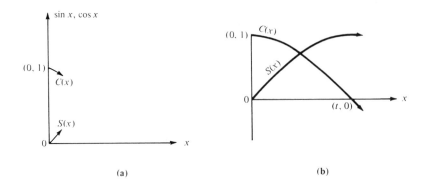

(a) (b)

Figure 7.10   (a) Behavior of sine and cosine for small positive arguments. (b) Graphs of sine and cosine for one quarter of their period.

328

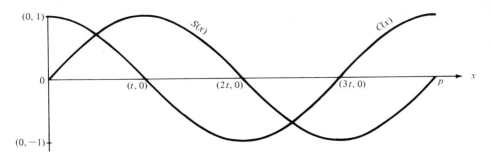

Figure 7.11

simultaneously $c(x)$ reaches 1; see Figure 7.11. At the point $x = p$ where $s(p) = 0$ and $c(p) = 1$, these functions are back at the values they had at the beginning, i.e., at $x = 0$. Now s and c start repeating their course.

The argument outlined above shows how to deduce qualitative informa-tion about the behavior of a pair of functions from the differential equations they satisfy. In the next section we shall show how to use differential equations to obtain quantitative information.

The function $s(x)$ is monotonic increasing around $x = 0$; from Section 7.1 we see that $s(x)$ is increasing between $-\pi/2$ and $\pi/2$. Therefore in the interval $[-\pi/2, \pi/2]$, $s(x)$ has an *inverse*, which we denote temporarily by $a(y)$:

$$s(x) = y, \qquad a(y) = x, \qquad -\frac{\pi}{2} \le x \le \frac{\pi}{2}.$$

In particular, for

$$s\left(-\frac{\pi}{2}\right) = -1, \qquad s(0) = 0, \quad \text{and} \quad s\left(\frac{\pi}{2}\right) = 1,$$

we have

$$a(-1) = -\frac{\pi}{2}, \qquad a(0) = 0, \quad \text{and} \quad a(1) = \frac{\pi}{2}.$$

According to the rule for differentiating the inverse of a function, the deriva-tive of the inverse is the reciprocal of the derivative of the original function:

$$a'(y) = \frac{1}{s'(x)}, \qquad x = a(y).$$

By (2.1), $s'(x) = c(x)$, and according to (2.3) we can express $c(x)$ as $\sqrt{1 - s^2(x)}$; the positive square root is to be taken, since cosine is positive in the part of the domain $-\pi/2 \le x \le \pi/2$ under consideration. Substituting this into the expression for $a'(y)$ above, and using the fact that $s(x) = y$, we get the chain of equalities

$$a'(y) = \frac{1}{c(x)} = \frac{1}{\sqrt{1 - s^2(x)}} = \frac{1}{\sqrt{1 - y^2}}.$$

329

The standard name for the inverse of the sine function is *arc sine*, "abbreviated" as *arcsin*; so we have shown that

(2.10)
$$\frac{d}{dy} \arcsin y = \frac{1}{\sqrt{1 - y^2}}.$$

By means of the fundamental theorem of calculus, we can express the arc sine as the integral of its derivative. Since arcsin $0 = 0$, we get

(2.11)
$$\arcsin y = \int_0^y \frac{1}{\sqrt{1 - z^2}} \, dz, \quad -1 < y < 1.$$

We could just as well reverse our steps; define $a(y)$ as the function whose derivative is $1/\sqrt{1 - y^2}$ and which is zero at $y = 0$. Define $s(x)$ as the inverse of $a(y)$, and $c(x)$ as $\sqrt{1 - s^2(x)}$. From the rule for differentiating the inverse of a function, we get

$$s'(x) = \frac{1}{a'(y)} = \sqrt{1 - y^2} = \sqrt{1 - s^2(x)} = c(x).$$

Differentiating

$$c(x) = \sqrt{1 - s^2(x)}$$

and using the above relation $s' = \sqrt{1 - s^2}$, we get

$$c' = -\frac{2ss'}{2\sqrt{1 - s^2}} = -s;$$

these are the differential equations (2.1) satisfied by the sine and cosine functions.

At this point the reader might reasonably object that this last calculation was unnecessary and downright circular. What is the point of rediscovering the derivative of sine and cosine from the derivative of arc sine, when the formula for the derivative of arc sine was deduced from that of sine? Our reason for doing the calculation in the reverse direction is similar to the argument used in Chapter 5 where we showed that the exponential function is inverse to the logarithmic function whose existence was guaranteed by the fundamental theorem of calculus. The fundamental theorem of calculus assures us that there is a function $a(y)$ whose derivative is $1/\sqrt{1 - y^2}$, whereas we have not yet *proved* that a pair of differential equations like (2.1) have a solution.

We now turn to the tangent function, defined in Exercise 1.2 as the ratio of sine to cosine:

$$\tan x = \frac{\sin x}{\cos x}.$$

Differentiating the quotient gives

$$\frac{d}{dx}\tan x = \frac{1}{\cos^2 x}.$$

The tangent is not defined at points where $\cos x$ vanishes, which are $\pm\pi/2$, $\pm 3\pi/2$, etc. Let us look at the tangent function on the interval $(-\pi/2, \pi/2)$, the longest interval containing the origin on which the function $\tan x$ is defined. Since $(\tan x)'$ is positive, $\tan x$ is increasing on this interval and so has an inverse there. Let us temporarily denote this inverse by $A(y)$:

$$\tan x = y, \quad A(y) = x, \qquad -\frac{\pi}{2} < x < \frac{\pi}{2}.$$

Since $\tan 0 = 0$ and $\tan \pi/4 = 1$, we have

$$A(0) = 0, \qquad A(1) = \frac{\pi}{4}.$$

As x approaches $\pi/2$, $\tan x$ approaches $+\infty$. Therefore as y approaches $+\infty$, $A(y)$ approaches $\pi/2$.

According to the rule for differentiating the inverse of a function,

$$A'(y) = \frac{1}{(\tan x)'} = \frac{1}{1/\cos^2 x} = \cos^2 x.$$

Now, the following relation holds between tangent and cosine:

$$\tan^2 x = \frac{\sin^2 x}{\cos^2 x} = \frac{1 - \cos^2 x}{\cos^2 x} = \frac{1}{\cos^2 x} - 1.$$

This shows that

$$\cos^2 x = \frac{1}{1 + \tan^2 x}.$$

Substituting this into the expression for A', and using the relation $\tan x = y$, we get

$$\frac{d}{dy} A(y) = \frac{1}{1 + y^2}.$$

The inverse A of the tangent function is called *arc tangent* which is abbreviated as *arctan*. So, we have succeeded in showing that

(2.12) $$\frac{d}{dy}\arctan y = \frac{1}{1 + y^2}.$$

Many integrals involving the sine and cosine can be evaluated exactly. For example, using the fundamental theorem we get

(2.13) $$\int_0^\pi \sin x \, dx = -\cos \pi + \cos 0 = 2.$$

We describe now a method that can be used to evaluate the integral of any rational function R of $\sin t$ and $\cos t$, i.e., one that is obtained from $\sin t$ and $\cos t$ by a sequence of arithmetic operations: additions, multiplications, divisions and multiplication by a constant. A horrible enough example of such a function is

$$R(t) = \frac{\sin^3 t - 3 \sin t \cos t + 5}{\sin^2 t \cos^3 t - \cos t + 8}.$$

The method is based on the *rational representation* of sine and cosine, as follows:

(2.14)
$$\sin t = \frac{2s}{1 + s^2}, \qquad \cos t = \frac{1 - s^2}{1 + s^2}.$$

The variable s has a geometrical interpretation:

(2.15)
$$s = \tan \frac{t}{2}$$

We· shall verify this relation by calculus. Differentiate the first equation in (2.14); using the chain rule we get

$$\cos t \, \frac{dt}{ds} = \frac{2}{1 + s^2} - \frac{4s^2}{(1 + s^2)^2} = \frac{1 - s^2}{1 + s^2} \frac{2}{1 + s^2}.$$

Dividing by $\cos t$ and using the second relation in (2.14) we get

(2.16)
$$\frac{dt}{ds} = \frac{2}{1 + s^2}.$$

It follows from (2.14) that $t = 0$ when $s = 0$; integrating (2.16) gives

$$t = 2 \arctan s,$$

from which (2.15) follows. With s as a new variable of integration, the integral

$$\int R(t) dt$$

is transformed into

$$\int R \frac{dt}{ds} \, ds = \int Q(s) ds$$

where

$$Q(s) = R(t(s)) \frac{2}{1 + s^2}.$$

Since R is a rational function of $\sin t$ and $\cos t$, and $\sin t$, $\cos t$ are rational functions of s, it follows that $Q(s)$ is a rational function of s.

We have shown in the appendix to Chapter 2 that rational functions can be decomposed into partial fractions. We show here how this decomposition can be used to integrate rational functions.

Example 1. Integrate

$$\int_0^{\pi/3} \frac{1}{\cos t} \, dt.$$

By the change of variables (2.14), (2.15), (2.16), this becomes

$$\int_0^{1/\sqrt{3}} \frac{1 + s^2}{1 - s^2} \frac{2}{1 + s^2} \, ds = \int_0^{1/\sqrt{3}} \frac{2}{1 - s^2} \, ds.$$

The integrand can be decomposed into partial fractions:

$$\frac{2}{1 - s^2} = \frac{1}{1 - s} + \frac{1}{1 + s},$$

so the integral equals

$$\log \frac{1 + s}{1 - s} \Big|_0^{1/\sqrt{3}} = \log\left(\frac{\sqrt{3} + 1}{\sqrt{3} - 1}\right)$$

A change of variables like (2.14) also works to evaluate the integral of any rational function R of x and $\sqrt{1 - x^2}$. A horrendous example of such a function is

$$\frac{x^2\sqrt{1 - x^2} + 3(1 - x^2)^{3/2}}{7x + \sqrt{1 - x^2}}$$

The exact form we use is

(2.17) $$x = \frac{2s}{1 + s^2}, \quad \sqrt{1 - x^2} = \frac{1 - s^2}{1 + s^2}.$$

From the first relation (2.17) we get

(2.18) $$\frac{dx}{ds} = \frac{2(1 - s^2)}{(1 + s^2)^2}.$$

Introducing s as variable of integration we get

$$\int R \, dx = \int R \frac{dx}{ds} \, ds = \int Q(s) \, ds.$$

Since R is a rational function of x and $\sqrt{1 - x^2}$, it follows from (2.17), (2.18) that Q is a rational function of s.

We now discuss a nonhorrible example:

Example 2. Integrate

$$\int \frac{\sqrt{1-x^2}}{x} \, dx.$$

The substitution (2.17), (2.18) changes this integral into

$$\int \frac{1-s^2}{2s} \frac{2(1-s^2)}{(1+s^2)^2} \, ds = \int \frac{(1-s^2)^2}{s(1+s^2)^2} \, ds.$$

The method of decomposition into partial fractions described in the appendix to Chapter 2 is no longer applicable, since there we have only treated the case where the denominator has distinct real roots. Here the denominator, $s(1+s^2)^2$, has two complex roots, $\pm i$, each of multiplicity 2. The method of partial fractions is easily extended to all rational functions including those whose denominator has complex and multiple roots. We shall not present the general theory but only an example. In the case at hand, we can decompose the integrand as follows

$$\frac{(1-s^2)^2}{s(1+s^2)^2} = \frac{1}{s} - \frac{4s}{(1+s^2)^2}.$$

Clearly the right side is the derivative of

$$\log s + \frac{2}{1+s^2},$$

so the fundamental theorem can be used to evaluate the integral in question.

Another way of tackling integrals of rational functions of x and $\sqrt{1-x^2}$ is to introduce a new variable of integration t, related to x by

(2.19) $$x = \sin t, \qquad \sqrt{1-x^2} = \cos t.$$

Then $dx/dt = \cos t$, so this substitution transforms the integral into one of a rational function of $\sin t$, $\cos t$. In some cases the resulting integration can be carried out by inspection.

Example 3. Evaluate the integral

$$\int_0^1 \sqrt{1-x^2} \, dx.$$

By (2.19)

$$\int_0^1 \sqrt{1-x^2} \, dx = \int_{\text{arc sin } 0}^{\text{arc sin } 1} \cos t \, \frac{dx}{dt} \, dt = \int_0^{\pi/2} \cos^2 t \, dt.$$

The integrand $\cos^2 t = \frac{1}{2}(1 + \cos 2t)$ is the derivative of $\frac{1}{2}t + \frac{1}{4}\sin 2t$, so according to the fundamental theorem,

$$\int_0^1 \sqrt{1 - x^2}\, dx = \frac{1}{2}t + \frac{1}{4}\sin 2t \Big|_0^{\pi/2} = \frac{\pi}{4}.$$

In other cases, this procedure may yield complicated rational functions of $\sin t$ and $\cos t$ as integrand; then one must resort to the rational substitution (2.17). But always try the trigonometric substitution first.

Many more examples will be given in the exercises.

EXERCISES

2.1 Differentiate the functions

(a) $\arcsin(\cos x)$

(b) $\arctan \sqrt{y}$

(c) $y \arctan y$

(d) $y^2 \arctan y$

(e) $\log \cos x$

(f) $e^{\cos x}$

(g) $\arctan(\sin x/\cos x)$

(h) $\log[\arctan(e^{\cos x})]$.

2.2 Find the integrals

(a) $\displaystyle\int_0^{\pi/4} \tan x\, dx$

(b) $\displaystyle\int_0^1 \arctan y\, dy$.

[*Hints*: (a) Recall the definition of $\tan x$ and observe it is of the form f'/f; (b) Try integrating by parts.]

2.3 (a) Write the integral $\int_0^\pi \sin^2 x\, dx$ in the form $\int_0^\pi f'(x)g(x)dx$, with $f'(x) = \sin x$, $g(x) = \sin x$. Integrate by parts and, by using the relation $\cos^2 x = 1 - \sin^2 x$, show that

$$\int_0^\pi \sin^2 x\, dx = \frac{\pi}{2}.$$

(b) Denote by S_n the value of the integral

$$S_n = \int_0^\pi \sin^n x\, dx.$$

Write S_n as $\int_0^\pi f'g\, dx$ with $f'(x) = \sin x$ and $g(x) = \sin^{n-1}x$. Integrate by parts, and use the relation $\cos^2 x = 1 - \sin^2 x$ to show that

$$S_n = \frac{n-1}{n} S_{n-2}.$$

(c) Show that, if n is an *odd integer*, then

$$S_n = \frac{(n-1)(n-3)\ldots 2}{n(n-2)\ldots 3}\, 2,$$

and that, if n is an *even integer*, then

$$S_n = \frac{(n-1)(n-3)\ldots 1}{n(n-2)\ldots 2}\, \pi.$$

(d) Show that the sequence S_n decreases as n increases.

(e) Show, using (b), that S_n/S_{n-2} tends to 1 as n tends to infinity. Then, using (d), show that S_n/S_{n-1} tends to 1 as n tends to infinity.

(f) Show that as m tends to infinity

$$\frac{2^2 \cdot 4^2 \cdot 6^2 \ldots (2m)^2}{3^2 \cdot 5^2 \cdot 7^2 \ldots (2m-1)^2(2m+1)}$$

tends to $\pi/2$. [*Hint*: Combine (c) and (e).] This result is called Wallis' product formula for $\pi/2$.

2.4 (a) Denote by $A_n(y)$ the sum

$$A_n(y) = 1 - y^2 + y^4 - y^6 + \cdots + (-1)^n y^{2n}.$$

Show that

$$A_n(y) = \frac{1 + y^{2n+2}(-1)^{n+1}}{1 + y^2}.$$

(b) Show that

$$\int_0^1 \frac{y^{2n+2}}{1 + y^2}\, dy < \int_0^1 y^{2n+2}\, dy$$

and conclude that, as n tends to ∞,

$$\int_0^1 \frac{y^{2n+2}}{1 + y^2}\, dy$$

tends to zero.

(c) Show that, as n tends to ∞, $\int_0^1 A_n(y)dy$ tends to $\int_0^1 dy/(1 + y^2)$.

(d) Show that, as n tends to infinity, the sum

$$1 - \frac{1}{3} + \frac{1}{5} - \frac{1}{7} + \cdots + \frac{(-1)^n}{2n+1}$$

tends to $\pi/4$.

The result under (d) is called the Leibniz series for $\pi/4$.

2.5 The purpose of this exercise is to obtain information about the tangent function, given the definition of its inverse,

$$\arctan y = \int_0^y \frac{1}{1 + s^2}\, ds.$$

(a) Graph the integrand $1/(1 + s^2)$ for $0 \le s \le 5$.

(b) Evaluate the above integral, using Simpson's rule with $n = 1, 10, 100, 1000$ subdivisions for $y = 0.1, 0.5, 1.0, 5.0$.

(c) Using the results of (b) for $n = 1000$, sketch a graph of $\arctan y$ for $0 \le y \le 5$.

(d) Show that $\arctan y$ is an odd function and extend your sketch of part (c) to $-5 \le y \le 0$.

(e) Show that $\arctan y$ increases monotonically. Therefore, it has an inverse, the tangent function. Sketch its graph by reflecting the graph of $\arctan y$ about the line through the origin with slope 1.

2.6 Define the functions $C(x)$ and $S(x)$ by

$$C(x) = \frac{e^x + e^{-x}}{2}, \qquad S(x) = \frac{e^x - e^{-x}}{2}.$$

(a) Show that

$$C' = S, \quad S' = C, \qquad C(0) = 1, \quad S(0) = 0.$$

(b) Show that $C^2 - S^2 = 1$.

(c) Show that $C(x)$ and $S(x)$ are positive and increasing for $x > 0$. Denote by A the inverse of S for $x > 0$:

$$A(y) = x, \quad \text{where } y = S(x).$$

Show that $A'(y) = 1/\sqrt{1 + y^2}$.

(d) Define $T(x)$ by

$$T(x) = \frac{S(x)}{C(x)}.$$

Show that $T'(x) = 1/C^2(x)$.

(e) Denote by R the inverse of T:

$$R(y) = x \quad \text{where } y = T(x).$$

Show that $R'(y) = 1/(1 - y^2)$.

(f) Show that

$$\left(\frac{1}{2} \log \frac{1 + y}{1 - y} \right)' = \frac{1}{1 - y^2}.$$

2.7 Suppose the functions $a(x)$, $b(x)$ satisfy the differential equations

$$a' = b^3, \qquad b' = -a^3,$$

with initial conditions

$$a(0) = 0, \qquad b(0) = 1.$$

(a) Prove that

$$a^4 + b^4 = 1.$$

(b) Prove that the life cycle of the pair of functions $a(x)$ and $b(x)$ is similar to that of sine and cosine, i.e., that starting with $a(0) = 0$ the function $a(x)$ increases as x increases until it reaches the value 1, then decreases until it reaches the value -1, then starts increasing again and reaches the value 0.

(c) a is increasing near 0, so has an inverse there; denote this inverse by A:

$$a(x) = y, \qquad A(y) = x.$$

Prove that $A'(y) = (1 - y^4)^{-3/4}$.

2.8 Find a function f, given that its derivative is

(a) $f' = \dfrac{1}{\cos t + \sin t}$

(b) $f' = \dfrac{1}{1 + \cos t}$

(c) $f' = \dfrac{1}{1 + \sin t + \cos t}$

(d) $f' = \dfrac{1}{2 \sin t \cos t}$.

2.9 Find a function g, given that its derivative is

(a) $g' = (1 - x^2)^{-1/2}$

(b) $g' = \dfrac{1}{x\sqrt{1 - x^2}}$

(c) $g' = \dfrac{1}{x^2(1 - x^2)^{1/2}}$.

2.10 (a) Evaluate

$$\int_0^t \sqrt{c^2 - x^2} \, dx.$$

[*Hint*: First introduce the new variable of integration $y = x/c$, then use the result of Example 3.]

(b) Show that the area of the ellipse

$$\frac{x^2}{a^2} + \frac{y^2}{b^2} = 1$$

is πab.

7.3 The computation of cosine, sine, and arctangent

We have seen that the arctangent function can be expressed as an integral,

$$(3.1) \qquad\qquad \arctan y = \int_0^y \frac{1}{1 + z^2} \, dz.$$

The integrand, $1/(1 + z^2)$, is a nice, smooth function; so for moderate values of y, the integral in (3.1) can be effectively evaluated to any desired degree of accuracy with the aid of Simpson's rule. We show in Table (3.1)′ such approximations to the integral (3.1), for $y = 1/\sqrt{3}$ and $y = 1$, employing subdivisions of the interval of integration into n equal subintervals, $n = 1, 5, 10$ and 100. The exact value of $\arctan 1/\sqrt{3}$ is $\pi/6$, and of $\arctan 1$, $\pi/4$; they are listed alongside the approximations, which are accurate to 12 decimal places.

Computing $\arctan y$ by evaluating the integral (3.1) is impractical for large values of y; here is a device for avoiding this difficulty.

Table 3.1′

n	approximation to $\arctan \dfrac{1}{\sqrt{3}}$	approximation to $\arctan 1$
1	0.523686	0.783333
5	0.52359895	0.78539815
10	0.523598787	0.7853981632
100	0.523598775599	0.785398163397
⋮	⋮	⋮
$\to \infty$	$0.523598775598 \cong \pi/6$	$0.785398163397 \cong \pi/4$

Referring to Figure 7.11 we see that, if we replace x by $x - \pi/2$ and use the oddness of the sine function and the evenness of cosine function, we obtain

$$\sin\left(x - \frac{\pi}{2}\right) = -\sin\left(\frac{\pi}{2} - x\right) = -\cos x,$$

$$\cos\left(x - \frac{\pi}{2}\right) = \cos\left(\frac{\pi}{2} - x\right) = \sin x;$$

therefore

$$\sin\left(\frac{\pi}{2} - x\right) = \cos x, \qquad \cos\left(\frac{\pi}{2} - x\right) = \sin x.$$

This result can also be obtained directly from the functional equations. Divide the first identity by the second, and obtain

$$\tan\left(\frac{\pi}{2} - x\right) = \frac{1}{\tan x};$$

so, by setting $y = \tan x$ and taking the arctan of both sides, we find that $\pi/2 - x = \arctan(1/y)$, so

(3.2) $$\arctan y = \frac{\pi}{2} - \arctan\frac{1}{y}.$$

For $y > 1$, one can use (3.2) to express the arctan of a number y greater than 1 in terms of arctan of $1/y$, where $1/y$ is less than 1.

Next we describe a more efficient scheme for approximating arctan y. It is based on approximating the integrand, $1/(1 + z^2)$, by the finite geometric series

(3.3) $$1 - z^2 + z^4 - z^6 + \cdots + (-1)^n z^{2n}.$$

As already noted in Exercise 2.4, the absolute value of the difference between $1/(1 + z^2)$ and the sum (3.3) is

$$\frac{z^{2n+2}}{1 + z^2}.$$

This quantity is less than z^{2n+2}, so we conclude that the integrals of $1/(1 + z^2)$ and of (3.3) differ by less than

(3.4) $$\int_0^y z^{2n+2} \, dz = \frac{y^{2n+3}}{2n + 3}.$$

For small y, this difference is small for moderate values of n; e.g., for $y \leq \frac{1}{10}$ and $n = 3$, the deviation (3.4) from the integral does not exceed $10^{-9}/9$. The

integral of (3.3) can be evaluated explicitly, yielding the following approximation:

(3.5)
$$\arctan y = y - \frac{y^3}{3} + \frac{y^5}{5} - \frac{y^7}{7} + \text{error},$$

where for $y < 10^{-1}$, the error is $< 10^{-9}/9$.

In light of (3.2), the problem of evaluating the arctan of any number can
be reduced to the problem of finding arctan of a number less than 1. Next
we show how to evaluate the arc tangent of any number less than 1 if we know
how to find the arc tangent of numbers less than $\frac{1}{10}$. We start with the addition
formula for the tangent, derived in Exercise 1.2:

$$\tan(x_1 + x) = \frac{\tan x_1 + \tan x}{1 - \tan x_1 \tan x}.$$

The tangent is an odd function, i.e., $\tan(-x) = -\tan x$. Replacing x by $-x$ in
the addition formula, we obtain the subtraction formula

(3.6)
$$\tan(x_1 - x) = \frac{\tan x_1 - \tan x}{1 + \tan x_1 \tan x}.$$

Set

$$y_1 = \tan x_1, \qquad y = \tan x,$$

then

(3.7)
$$\arctan y_1 = x_1, \qquad \arctan y = x.$$

Take the arctan of (3.6) to obtain

$$x_1 - x = \arctan \frac{y_1 - y}{1 + y_1 y}.$$

Using the relations (3.7) we write this as

(3.8)
$$\arctan y_1 = \arctan y + \arctan \frac{y_1 - y}{1 + y_1 y}.$$

Suppose y_1 is any number between 0.1 and 1.0. Choose y to be 0.1 and use
(3.8) to express $\arctan y_1$ as the sum

$$\arctan 0.1 + \arctan y_2 \qquad \text{where} \qquad y_2 = \frac{y_1 - 0.1}{1 + y_1(0.1)}.$$

The important fact to note is that y_2 is less than $y_1 - 0.1$. If y_2 turns out to be
less than 0.1, $\arctan y_2$ can be evaluated using Formula (3.5). If not, we can,
repeating the previous procedure, write $\arctan y_2$ as the sum

$$\arctan 0.1 + \arctan y_3 \qquad \text{where} \qquad y_3 = \frac{y_2 - 0.1}{1 + y_2(0.1)}.$$

We proceed in this fashion for n steps, obtaining

(3.9)
$$y_{n+1} = \frac{y_n - 0.1}{1 + (0.1)y_n},$$

until we arrive at a value y_{n+1} which is less than 0.1. The arctan of such a number is easily evaluated by Formula (3.5). Putting together the relations

$$\arctan y_j = \arctan 0.1 + \arctan y_{j+1}, \quad j = 1, 2, \ldots n,$$

we get that

(3.10)
$$\arctan y_1 = n \arctan 0.1 + \arctan y_{n+1}.$$

Observe that since y_{n+1} is less than $y_n - 0.1$, and since $y_1 < 1$, the number of steps needed to complete the process is at most 9. This procedure for calculating arctan is particularly suitable for a computer.

We now turn to calculating sine and cosine. Our first method is based on the differential equations satisfied by sine and cosine,

(3.11)
$$c' = -s, \qquad s' = c,$$

and imitates an approximation method described in Chapter 5 for the exponential function. By definition of derivative,

(3.12)
$$\frac{c(x + h) - c(x)}{h}$$

tends to $c'(x) = -s(x)$ as h tends to 0. Therefore, for small h, (3.12) differs little from $-s(x)$, i.e.,

$$\frac{c(x + h) - c(x)}{h} = -s(x) + \text{small quantity}.$$

Similarly

$$\frac{s(x + h) - s(x)}{h} = c(x) + \text{small quantity}$$

Multiply by h and isolate c and s at $(x + h)$ on the left side:

(3.13)
$$c(x + h) = c(x) - hs(x) + h \times \text{(s.q.)}$$
$$s(x + h) = s(x) + hc(x) + h \times \text{(s.q.)}.$$

For fixed h, we define C and S as functions which satisfy

(3.14)
$$C(x + h) = C(x) - hS(x)$$
$$S(x + h) = S(x) + hC(x).$$

These equations are obtained from (3.13) by omitting the terms containing the small quantities. In addition we require that, at $x = 0$, C and S have the same values as c and s:

(3.15) $$C(0) = 1, \qquad S(0) = 0.$$

Set $x = 0$ in (3.14); using the values (3.15) on the right gives

$$C(h) = 1, \qquad S(h) = h.$$

Setting $x = h$ in (3.14) and using the values above gives

$$C(2h) = 1 - h^2, \qquad S(2h) = 2h.$$

Repeating this process we can clearly determine the values of C and S at all integer multiples of h although, in contrast to the case of the exponential function, we cannot write down a simple formula for $C(nh)$ and $S(nh)$. In the next section we shall show how to use *complex numbers* to express $C(nh)$ and $S(nh)$ by a formula which is not only elegant, but reduces the number of arithmetic operations needed to evaluate $C(nh)$ and $S(nh)$.

We expect $C(nh)$ and $S(nh)$ to be some sort of approximation to $\cos nh$ and $\sin nh$; the smaller h, the better the approximation. To give credence to this claim, we display a table which lists $C(nh)$ and $S(nh)$ with $h = \pi/6n$ and various values of n. This table suggests that, as h tends to 0, $C(nh)$ and $S(nh)$ tend to the exact values of $\cos(\pi/6)$ and $\sin(\pi/6)$, but not particularly rapidly.

The basis of the approximation procedure described above was the closeness of the difference quotients

$$\frac{c(x + h) - c(x)}{h} \quad \text{and} \quad \frac{s(x + h) - s(x)}{h}$$

to the derivatives of c and s at x. According to the discussion at the end of Section 3.8, the above difference quotients are even closer to the derivatives

Table 3.16

n	$C(nh)$	$S(nh)$
10	0.8782	0.5065
50	0.8684	0.5014
100	0.8672	0.500682
1000	0.86614	0.5000685
10000	0.866037	0.50000685
$\vdots$	$\vdots$	$\vdots$
$\to \infty$	$\cos \dfrac{\pi}{6} \cong 0.866025403784$	$\sin \dfrac{\pi}{6} = 0.500000000000$

at the midpoint $x + h/2$ than at the endpoint x. The derivative at the midpoint $x + h/2$ is in turn well approximated by the average of the values at the two endpoints x and $x + h$. Thus, since $c' = -s$, and $s' = c$,

$$\frac{c(x + h) - c(x)}{h} = -\frac{s(x + h) + s(x)}{2} + \text{very small quantity,}$$

$$\frac{s(x + h) - s(x)}{h} = \frac{c(x + h) + c(x)}{2} + \text{very small quantity.}$$

We multiply by h and put all terms involving c and s at $(x + h)$ on the left side:

$$c(x + h) + \frac{h}{2} s(x + h) = c(x) - \frac{h}{2} s(x) + h \times \text{very small quantity,}$$

$$s(x + h) - \frac{h}{2} c(x + h) = s(x) + \frac{h}{2} c(x) + h \times \text{very small quantity.}$$

We can solve for $c(x + h)$ by multiplying the second equation by $h/2$ and subtracting it from the first. After division by $1 + h^2/4$, we get

$$c(x + h) = \frac{1 - h^2/4}{1 + h^2/4} c(x) - \frac{h}{1 + h^2/4} s(x) + h \times (\text{v.s.q.}),$$

where v.s.q. abbreviates very small quantity. Similarly, by eliminating $c(x + h)$, we obtain

$$s(x + h) = \frac{h}{1 + h^2/4} c(x) + \frac{1 - h^2/4}{1 + h^2/4} s(x) + h \times (\text{v.s.q.}),$$

For h fixed, we define C and S as solutions of

$$C(x + h) = \frac{1 - h^2/4}{1 + h^2/4} C(x) - \frac{h}{1 + h^2/4} S(x)$$

(3.17)

$$S(x + h) = \frac{h}{1 + h^2/4} C(x) + \frac{1 - h^2/4}{1 + h^2/4} S(x);$$

we also require that at $x = 0$, C and S agree with c and s:

(3.18) $$C(0) = 1, \quad S(0) = 0.$$

Setting $x = 0$ in (3.17) we can, using the values (3.18), calculate C and S at $x = h$:

(3.19) $$C(h) = \frac{1 - h^2/4}{1 + h^2/4}, \quad S(h) = \frac{h}{1 + h^2/4}.$$

Setting successively $x = h, 2h, 3h$, etc. in (3.17) we can determine the values of C and S at all integer multiples of h. In Table (3.20) below we list $C(nh)$ and $S(nh)$ with $h = \pi/6n$ for various values of n; we expect these values to approximate $\cos(\pi/6)$ and $\sin(\pi/6)$:

Table 3.20

n	$C(nh)$	$S(nh)$
10	0.866085	0.49989
50	0.8660278	0.4999959
100	0.86602600	0.49999896
1000	0.866025409	0.4999999896
10000	0.866025403868	0.499999999908
$\vdots$	$\vdots$	$\vdots$
$\to \infty$	$\cos \dfrac{\pi}{6} \cong 0.866025403784$	$\sin \dfrac{\pi}{6} = 0.500000000000$

Compare this table with Table (3.16); clearly, for the same value of n, (3.20) yields a much closer approximation to cosine and sine than (3.16). Note another advantage of Table (3.20): the relation

$$C^2(nh) + S^2(nh) = 1$$

is approximately satisfied, while for the entries of (3.16), $C^2(nh) + S^2(nh) > 1$. For a further discussion of this point, see Exercise 3.1.

We now turn to another method of computing cosines. By the first addition formula in (1.12), with $\alpha = \beta = x$,

$$\cos 2x = \cos^2 x - \sin^2 x.$$

Using the relation $\cos^2 x + \sin^2 x = 1$, we get (see Exercise 1.1a, b)

$$\text{(3.21)} \qquad \cos 2x = 2\cos^2 x - 1.$$

That is, we can express the cosine of $2x$ as a simple quadratic function of $\cos x$. Repeating this procedure k times, we can express $\cos 2^k x$ in terms of $\cos x$. Clearly this procedure enables us to express the cosine of any angle in terms of the cosine of small angles.

How to compute the cosine of small angles? We saw in (3.19) that for small x, $\cos x$ is approximately

$$\text{(3.22)} \qquad \cos x \cong \frac{1 - x^2/4}{1 + x^2/4}.$$

To compute $\cos a$ for any a, we set $x = 2^{-k}a$, compute $\cos x$ approximately using Formula (3.22), then apply (3.21) k times. We tabulate below the resulting approximations to $\cos a$ for $a = \pi/4$ and $a = \pi/3$, using several values of k.

Table (3.23) shows that, for successive values 1, 5, and 10 for k, our method yields correctly the first 2, 4, and 7 digits, respectively, of $\cos(\pi/4)$ and $\cos(\pi/3)$.

Table 3.23

k	$\cos\dfrac{\pi}{4}$	$\cos\dfrac{\pi}{3}$
1	0.71405	0.5198
5	0.7071347	0.5000809
10	0.707106808	0.5000000721
20	0.7104740175	0.494733013
exact value to 12 decimals	0.7071067811865	0.500000000000

The table shows, however, that when k is increased further to 20, the accuracy of the method drops suddenly and yields no more than 2 digits correctly. A further increase in k would yield still less accurate approximations, and for $k > 23$ the approximate value for $\cos(\pi/4)$ yielded by our method is totally unrelated to the value of $\cos(\pi/4)$. What is the explanation of this anomaly?

To answer this question, take $a \leqslant \pi/4$ and $k = 22$; then

$$x = \frac{a}{2^k} \leqslant \frac{\pi/4}{2^{23}} = 9.36267570730987 \times 10^{-8}$$

so that

$$\frac{x^2}{4} \leqslant 2.19149241000625 \times 10^{-15}$$

and

$$\frac{1 - x^2/4}{1 + x^2/4} \geqslant 0.999999999999999561701517998752.$$

We recall now that in a computer all numbers are represented as finite decimals; in particular in the computer that generated these results, numbers are represented as decimal fractions with 14 digits.* So the above number starting with 14 nines is represented in the machine by $1.-10^{-14}$. This number is *independent* of the value of x, and therefore so is the end result of the doubling algorithm! In other words, no matter which angle a we start with, we can halve it so often (make k so large) that $x/2 = a/2^{k+1}$, when squared, makes the right side of (3.22) uncomputable; the value of k depends upon the number of binary digits representing a number in a given computer.

* This is only approximately true; actually numbers in the computer are represented as binary fractions with 48 binary digits.

We now turn to yet another method for computing cos x and sin x. This method works particularly rapidly for small values of x, so it can be combined advantageously with the method described above.

We recall from Section 3.9 that n times differentiable functions f can be approximated by their Taylor polynomials. Let us recall what the Taylor polynomials of f are: they are the polynomials $t_n(x)$ whose derivatives up to order n at some fixed point a agree with those of f. Taylor's theorem says that for any point x near a

$$(3.24) \qquad |f(x) - t_n(x)| \leq \frac{M(x-a)^n}{n!},$$

where M denotes an upper bound for the absolute value of the n-th derivative of f between a and x. In what follows we shall take the special point a to be 0. Then $t_n(x)$ can be written explicitly as

$$(3.25) \qquad t_n(x) = f(0) + \frac{f'(0)}{1!}x + \frac{f''(0)x^2}{2!} + \cdots + \frac{f^{(n-1)}(0)x^{n-1}}{(n-1)!}.$$

Take $f(x) = \cos x$; the function and its first three derivatives are $\cos x$, $-\sin x$, $-\cos x$, $\sin x$. From the fourth derivative on, the cycle is repeated. The values of f and these successive derivatives at $a = 0$ are 1, 0, -1, 0, repeat. Therefore, the n-th Taylor polynomials of $\cos x$ and $\sin x$ are

$$(3.26) \qquad c_n(x) = 1 - \frac{x^2}{2!} + \frac{x^4}{4!} - \frac{x^6}{6!} + \cdots$$

and

$$(3.27) \qquad s_n(x) = x - \frac{x^3}{3!} + \frac{x^5}{5!} - \frac{x^7}{7!} + \cdots,$$

respectively, where the sums go up to powers of x less than n. The n-th derivatives of cosine and sine are $+$ or $-$ cosine or sine, so *do not exceed 1 in absolute value*. Thus we can take M in inequality (3.24) to be 1, and deduce that $\cos x$ and $\sin x$ differ from the sum (3.26), (3.27) respectively, at most by

$$\frac{x^n}{n!}.$$

It is easy to show that as n gets larger and larger, this quantity tends to zero, so that c_n and s_n give arbitrarily good approximations to $\cos x$ and $\sin x$ if n is taken to be large enough.

For x small, exceedingly good approximations can be obtained even for small n. For instance, for $x < \frac{1}{10}$, and $n = 6$, we have

$$\frac{x^6}{6!} \leq \frac{10^{-6}}{720} < 10^{-8}.$$

346

so that for $x < \frac{1}{10}$

(3.28)
$$c_5 = 1 - \frac{x^2}{2} + \frac{x^4}{24}$$

approximates $\cos x$ with error less than 10^{-8}.

The approximation (3.28) can be used instead of (3.22) in the method described previously for calculating $\cos x$ (see Exercise 3.4; also see the fifth computer program at the end of this volume).

EXERCISES

3.1 (a) Prove that, for $C(x + h)$ and $S(x + h)$ defined by (3.17)

$$C^2(x + h) + S^2(x + h) = C^2(x) + S^2(x).$$

(b) Prove that, for $C(x + h)$ and $S(x + h)$ defined by (3.14),

$$C^2(x + h) + S^2(x + h) = (1 + h^2)[C^2(x) + S^2(x)].$$

3.2 The integral expression for arcsine is

$$\arcsin y = \int_0^y \frac{1}{\sqrt{1 - z^2}}\, dz.$$

Approximate the integral by applying Simpson's rule, after subdividing the interval of integration into n equal parts. Calculate these approximations for $y = 0.2, 0.5$, and 0.9, and for $n = 1, 5, 10, 50$. Compare your answers with tabulated values of arcsin. Explain why the approximations are relatively poor for $y = 0.9$.

3.3 Determine the Taylor polynomials for $1/\sqrt{1 - x^2}$ at $a = 0$. Using them, determine the Taylor polynomials for arcsin y, calculate the value of the Taylor polynomial of degree n for $y = 0.2, 0.5$, and 0.9 and $n = 1, 5, 10, 50$.

3.4 Let a be any number in the range $0 \le a \le \pi/2$. Denote by k the smallest integer such that $x = a2^{-k} < 10^{-1}$.
(a) Show that k does not exceed 5.
(b) Use (3.28) to approximate the value of $\cos x$, and then the doubling formula (3.21) k times to compute $\cos a$. Use this method to find approximations to $\cos \pi/4$, $\cos \pi/3$ and $\cos 2.5$.
(c) Write a computer program which carries out the steps described above for approximating $\cos a$.
(d) Write a computer program in machine language which carries out the algorithm described above to compute $\cos a$. (In machine language dividing a number represented as a floating binary fraction by a power of 2 is simple.)

3.5 Apply Simpson's rule, without any subdivision, to approximate the integral

$$\int_{\pi/6}^{\pi/3} \sin x \, dx.$$

Compare the approximate value with the exact value of this integral, and use this to deduce an approximation to π.

3.6 For a and b close together, $[f(a) - f(b)]/(a - b)$ is well approximated by $[f'(a) + f'(b)]/2$. Use this relation, with $a = \pi/6$, $b = \pi/4$ and $f(x) = \sin x$, to deduce an approximation to π.

3.7 Approximate the solution of

$$A' = B^3, \qquad B' = -A^3$$

$A(0) = 0$, $B(0) = 1$, using one of the methods presented in this section. Find approximately the first positive value of x where $A(x) = 1$.

7.4 Complex numbers

Most people first encounter imaginary and complex numbers as solutions of quadratic equations

$$x^2 + bx + c = 0,$$

that is, zeros z of the function $f(x) = x^2 + bx + c$. The formula for the roots is

$$z = -\frac{b}{2} \pm \frac{\sqrt{b^2 - 4c}}{2}.$$

If the coefficients b, c are such that the quantity under the square root is negative, the roots are *complex*; they can be written as

$$z = -\frac{b}{2} \pm \frac{\sqrt{4c - b^2}}{2} i,$$

where i denotes the square root of -1:

(4.1)
$$i^2 = -1.$$

A complex number z is defined as the sum of a real number and of a real multiple of i,

(4.2)
$$z = x + iy;$$

x is called the *real part* of z, y its *imaginary part*. A complex number whose imaginary part happens to be zero is called (naturally enough) real; a complex number whose real part is zero is called *purely imaginary*.

We now describe a natural way of doing arithmetic with complex numbers: to *add* them, we add their real and imaginary parts separately,

$$(x + iy) + (u + iv) = x + u + i(y + v);$$

similarly for *subtraction*. To *multiply* complex numbers, we make repeated use of the distributive law:

$$(x + iy)(u + iv) = xu + iyu + xiv + iyiv.$$

348

Rewrite xi as ix and yi as iy; this amounts to assuming that multiplication of real numbers and i is commutative. Then, since $i^2 = -1$, we can write the product above as

$$(xu - yv) + i(yu + xv).$$

Note that we have assumed that multiplication by i is distributive.

It is easy to divide a complex number by a real number r:

$$\frac{x + iy}{r} = \frac{x}{r} + i\frac{y}{r}.$$

To express the quotient

$$\frac{x + iv}{u + iv}$$

of two complex numbers as a complex number $s + it$, we apply the trick of multiplying numerator and denominator by the same complex number, so chosen that the new denominator is real. An appropriate choice for the multiplier is $u - iv$; we get

$$\frac{(x + iy)(u - iv)}{(u + iv)(u - iv)} = \frac{xu + yv + i(yu - xv)}{u^2 + v^2}.$$

This is of the desired form $s + it$ with

$$s = \frac{xu + yv}{u^2 + v^2}, \qquad t = \frac{yu - xv}{u^2 + v^2}.$$

Notice that the indicated division by $u^2 + v^2$ can be carried out unless both u and v are zero. In that case the divisor $u + iv$ is zero, so we do not expect to be able to carry out the division.

The usual rules of arithmetic to which we are accustomed from the case of real numbers hold also for complex numbers. To wit, for the complex numbers v, w and z, we have

Associative rule $\qquad v + (w + z) = (v + w) + z$

$$v(wz) = (vw)z,$$

Commutative rule $\qquad v + w = w + v$

$$vw = wv,$$

Distributive rule $\qquad v(w + z) = vw + vz.$

The numbers 0 and 1 are distinguished, inasmuch as adding 0 and multiplying by 1 does not alter a number.

In the case of complex numbers we have an additional operation, called *conjugation*, and denoted by a bar, defined as follows:

The complex conjugate $\bar{z}$ of a complex number $z = x + iy$ is defined by

(4.3)
$$\bar{z} = x - iy.$$

The following rules concerning conjugation are very useful; they can be verified immediately by performing the indicated calculations using the rules of arithmetic for complex numbers.

Rules of conjugation

(a) *Symmetry.* The conjugate $\bar{\bar{z}}$ of $\bar{z}$ is z.
(b) *Additivity.* The conjugate of a sum is the sum of the conjugates of the summands:

$$\overline{z + w} = \bar{z} + \bar{w}.$$

The sum of a complex number and its conjugate is real; in fact,

(4.4)
$$z + \bar{z} = 2(\text{Real part of } z) = 2 \operatorname{Re}\{z\}.$$

(c) *Multiplicativity.* The conjugate of a product is the product of the conjugates of the factors:

$$\overline{zw} = \bar{z}\,\bar{w}.$$

The product of a complex number $z \neq 0$ and its conjugate is real and positive; in symbols,

$$z\bar{z} = x^2 + y^2.$$

Definition. The nonnegative square root of $z\bar{z}$ is called the *absolute value* of z, and is denoted by vertical bars:

(4.5)
$$|z| = \sqrt{z\bar{z}} = \sqrt{x^2 + y^2}.$$

For z real this definition coincides with the usual definition of absolute value. We shall show below that this extension of the notion of absolute value retains all the properties that we are accustomed to.

Properties of absolute value

(a) *Positivity.* The absolute value of a nonzero complex number is positive, and the absolute value of 0 is 0:

$$\text{For } z \neq 0, \quad |z| > 0.$$

(b) *Symmetry.* A complex number and its conjugate have the same absolute value:

$$|\bar{z}| = |z|.$$

350

(c) *Multiplicativity.* The absolute value of the product of two complex numbers is the product of the absolute values of the factors:

(4.6) $$|wz| = |w||z|.$$

(d) *Triangle inequality.* The absolute value of a sum of two complex numbers does not exceed the sum of their absolute values:

(4.7) $$|w + z| \leq |w| + |z|.$$

PROOF. Parts (a) and (b) follow directly from the definition. Part (c) is far from obvious intuitively, so to convince the reader, we first present a few numerical examples.

$$(2 + 5i)(7 + 3i) = -1 + 41i.$$

The absolute values squared of the factors and of the product are

$$4 + 25 = 29, \qquad 49 + 9 = 58, \qquad \text{and} \quad 1 + 1681 = 1682,$$

respectively. Indeed, the last is the product of the first two. Again, for the product

$$(3 + 2i)(4 + 5i) = 2 + 23i,$$

the absolute values squared of the factors and the product are

$$9 + 4 = 13, \qquad 16 + 25 = 41, \qquad \text{and} \quad 4 + 529 = 533.$$

The last number is the product of the first two.

Even if not convinced the reader is probably dazzled by these examples and is ready to move from numerology to a proof. By definition of absolute value of wz,

$$|wz|^2 = wz\,\overline{wz}.$$

Using the multiplicative property of conjugation, the commutative property of multiplication and the definition of $|w|^2$ and $|z|^2$, we get

$$|wz|^2 = wz\,\bar{w}\,\bar{z} = w\bar{w}\,z\bar{z} = |w|^2|z|^2,$$

the multiplicative property of absolute value. In Exercise 4.3 we shall indicate another proof of the multiplicative property.

We preface the proof of part (d) with the observation that *the real part x of any complex number a does not exceed its absolute value.* This is an immediate consequence of writing $|a|$ as $\sqrt{x^2 + y^2}$ and observing that $x \leq \sqrt{x^2 + y^2} = |a|$. We saw before, (4.4), that twice the real part of a can be expressed as $a + \bar{a}$, so we conclude that

(4.8) $$2x = a + \bar{a} \leq 2|a|.$$

Now by definition,

$$|w + z|^2 = (w + z)(\overline{w + z}).$$

351

Using the additive property, $\overline{w + z} = \overline{w} + \overline{z}$, of conjugation and the distributive rule, we can write $|w + z|^2$ as

$$(w + z)(\overline{w} + \overline{z}) = w\,\overline{w} + w\overline{z} + z\overline{w} + z\overline{z} = |w|^2 + w\overline{z} + z\overline{w} + |z|^2.$$

Let us abbreviate the quantity $w\overline{z}$ by a; by the multiplicative and the symmetry properties of conjugation, $z\overline{w} = \overline{a}$, so that the above can be written as

(4.9) $$|w + z|^2 = |w|^2 + a + \overline{a} + |z|^2.$$

By definition, $a = w\overline{z}$, so using properties (c) and (b) of absolute value, we may write

$$|a| = |w\overline{z}| = |w||\overline{z}| = |w||z|.$$

Hence, by (4.8), $a + \overline{a} \le 2|a| = 2|w||z|$. This together with (4.9) gives the estimate

$$|w + z|^2 \le |w|^2 + 2|w||z| + |z|^2$$

for the right side of (4.9). The expression on the right is $(|w| + |z|)^2$, so the above the inequality is equivalent with (4.7), which we set out to prove. $\qquad\square$

We now present a *geometric representation* of complex numbers. This turns out to be a very useful way of thinking about complex numbers, just as it is useful to think of the real numbers as points of the number line. The complex numbers are conveniently represented in a plane, called the *complex number plane*.

First we introduce usual Cartesian coordinates, with an arbitrary point picked as origin and two perpendicular lines through the origin chosen as axes. Each point of the plane is then described by its two Cartesian coordinates x and y. *To each point (x, y) of the Cartesian plane we associate the complex number $x + iy$*. The horizontal axis, $y = 0$, consists of real numbers; it is called the *real axis*. The other, called the *imaginary axis*, consists of purely imaginary numbers.

The notion of complex conjugate has a simple geometric interpretation in the complex plane; see Figure 7.12. The complex conjugate of $x + iy$, $x - iy$, is obtained from $x + iy$ by *reflection across the real axis*.

The geometric interpretation of the absolute value $\sqrt{x^2 + y^2}$ of $x + iy$ is very striking: it is the *distance of $x + iy$ from the origin*.

To visualize geometrically the sum of two complex numbers z and w, move the coordinate system rigidly and parallel to itself, i.e., without rotating it, so that the origin ends up where the point z was originally located; see Figure 7.13. Then the point w will end up where the point $z + w$ was located in the original coordinate system. It follows from this geometric description of addition that the four points 0, w, z, $w + z$ are the vertices of a *parallelogram*; in particular, it follows that the distance of w to $w + z$ equals the distance from 0 to z.

352

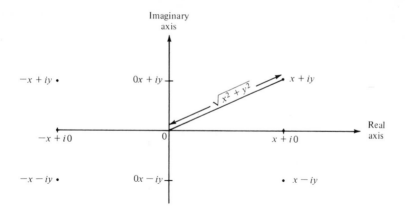

Figure 7.12

Now consider the triangle whose vertices are 0, w and $w + z$. The length of the side 0 to $w + z$ is $|w + z|$; the length of the side 0 to w is $|w|$, and the length of the side w to $w + z$, by the parallelogram interpretation, is $|z|$. According to a famous inequality of geometry, the length of any one side of a triangle does not exceed the sum of the lengths of the other two; therefore

$$|w + z| \leq |w| + |z|.$$

This is precisely our inequality (4.7) and is the reason for its name, the triangle inequality. The present observation can be regarded as a geometric proof of inequality (4.7), while our earlier proof of (4.7) may be regarded as a proof of a theorem about triangles with the aid of complex numbers. In Exercises 4.5 and 4.6 we shall give further examples of how to prove geometrical theorems with the aid of complex numbers.

Suppose p is a complex number whose absolute value is 1; such a p lies on the unit circle. According to the definition given in Section 7.1, the Cartesian coordinates of such a point p are $(\cos \theta, \sin \theta)$, where θ is the

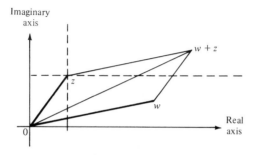

Figure 7.13

353

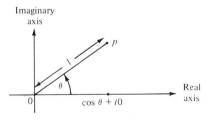

Figure 7.14

radian measure of the angle made by the real axis and the ray from 0 through p; see Figure 7.14. So the complex number p with $|p| = 1$ is

(4.10) $$p = \cos \theta + i \sin \theta.$$

Let z be any complex number; denote its absolute value by r, and define p to be

(4.11) $$p = \frac{z}{r}.$$

Clearly p has absolute value 1, so p can be represented in the form (4.10). Combining (4.10) and (4.11) we get that

(4.12) $$z = r(\cos \theta + i \sin \theta), \quad \text{where } r = |z|.$$

This is called the *polar representation* of complex numbers. The angle θ is called the *argument* of z denoted by arg z; θ *is the angle between the positive real axis and the ray connecting the origin to z.* This geometric characterization of the argument θ of the complex number z is verified by a glance at the location of z and $p = z/|z|$ in the complex plane; see Figure 7.15.

Let z and w be a pair of complex numbers. Represent each in polar form, i.e., z as in (4.12), and

$$w = s(\cos \phi + i \sin \phi), \quad s = |w|.$$

Multiplying these together, we get that

(4.13) $$zw = rs(\cos \theta + i \sin \theta)(\cos \phi + i \sin \phi)$$
$$= rs[(\cos \theta \cos \phi - \sin \theta \sin \phi) + i(\cos \theta \sin \phi + \sin \theta \cos \phi)].$$

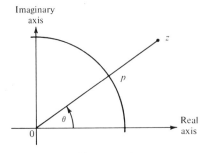

Figure 7.15

Recall from Section 7.1 the addition formulas (1.12) for which we shall give a complete proof in the next section:

$$\cos \theta \cos \phi - \sin \theta \sin \phi = \cos(\theta + \phi)$$
$$\cos \theta \sin \phi + \sin \theta \cos \phi = \sin(\theta + \phi).$$

We use them to rewrite the product formula (4.13) in a particularly simple form. Recalling that r denotes $|z|$ and s denotes $|w|$, we find that

(4.14) $$zw = |z||w|[\cos(\theta + \phi) + i \sin(\theta + \phi)].$$

This formula gives the polar representation of the product zw; it is a symbolic statement of the

Multiplication rule for complex numbers in polar form

(i) The absolute value of the product zw is the product of the absolute values of its factors,

(ii) The argument of the product zw is the sum of the arguments of its factors z and w. In symbols: For

$$z = r(\cos \theta + i \sin \theta), \qquad w = s(\cos \phi + i \sin \phi)$$

with $r = |z|, s = |w|, \theta = \arg z$ and $\phi = \arg w$, we have

(4.15) $$|zw| = |z||w|$$
$$\arg(zw) = \arg z + \arg w.$$

Next, we use the multiplication rule for finding powers of a complex number z. By (4.14), for $z = r(\cos \theta + i \sin \theta)$,

$$z^2 = z \cdot z = r^2(\cos 2\theta + i \sin 2\theta),$$
$$z^3 = z^2 \cdot z = r^3(\cos 3\theta + i \sin 3\theta),$$

and, for any positive integer n, the multiplication rule yields

(4.16) $$z^n = r^n(\cos n\theta + i \sin n\theta).$$

This result, often referred to as de Moivre's theorem, is particularly handy for expressing $\cos n\theta$ as a polynomial in $\cos \theta$, and $\sin n\theta$ as a polynomial in $\sin \theta$ and $\cos \theta$. For, take any complex number p with absolute value 1. Then

(4.17) $$p^n = (\cos \theta + i \sin \theta)^n = \cos n\theta + i \sin n\theta.$$

The binomial theorem yields

$$(\cos \theta + i \sin \theta)^n$$
$$= \cos^n \theta + \binom{n}{1}i \cos^{n-1} \theta \sin \theta - \binom{n}{2}\cos^{n-2} \theta \sin^2 \theta + \cdots + i^n \sin^n \theta$$
$$= \cos^n \theta - \binom{n}{2}\cos^{n-2} \theta \sin^2 \theta + \binom{n}{4}\cos^{n-4} \theta \sin^4 \theta - \cdots$$
$$+ i \sin \theta[\binom{n}{1}\cos^{n-1} \theta - \binom{n}{3}\cos^{n-3} \theta \sin^2 \theta + \binom{n}{5}\cos^{n-5} \theta \sin^4 \theta - \cdots].$$

The binomial coefficients $\binom{n}{k}$ are defined by (9.15) in Chapter 3. When we equate real and imaginary parts in (4.17), we obtain

$$(4.18) \qquad \cos n\theta = \cos^n \theta - \binom{n}{2}\cos^{n-2}\theta \sin^2 \theta$$
$$+ \binom{n}{4}\cos^{n-4}\theta \sin^4 \theta - \cdots,$$

$$(4.19) \qquad \sin n\theta = \binom{n}{1}\cos^{n-1}\theta \sin\theta - \binom{n}{3}\cos^{n-3}\theta \sin^3 \theta$$
$$+ \binom{n}{5}\cos^{n-5}\theta \sin^5 \theta - \cdots.$$

The identity $\sin^2 \theta = 1 - \cos^2 \theta$ enables us to express (4.18) as a polynomial in $\cos\theta$ alone, since $\sin\theta$ appears only to even powers. If n is odd, (4.19) can be written as a polynomial in $\sin\theta$; for some examples, see Exercise 4.10.

Let us find the polar representation for the reciprocal, $1/z$, of a complex number $z = r(\cos\theta + i\sin\theta)$. We know from

$$\left| z \cdot \frac{1}{z} \right| = |z| \left| \frac{1}{z} \right| = 1, \qquad \text{that} \quad \left| \frac{1}{z} \right| = \frac{1}{r}.$$

To find the argument ϕ of $1/z$, compute

$$z \cdot \frac{1}{z} = r(\cos\theta + i\sin\theta) \cdot \frac{1}{r}(\cos\phi + i\sin\phi)$$

$$= \cos(\theta + \phi) + i\sin(\theta + \phi) = 1;$$

the imaginary part is 0, the real part is 1, so

$$\sin(\theta + \phi) = 0, \qquad \cos(\theta + \phi) = 1,$$

and these relations are satisfied when $\theta + \phi = 0$, that is, when $\phi = -\theta$. We conclude: If $z = r(\cos\theta + i\sin\theta)$, then

$$z^{-1} = r^{-1}(\cos(-\theta) + i\sin(-\theta))$$
$$= r^{-1}(\cos\theta - i\sin\theta);$$

and clearly

$$(4.16)_- \qquad z^{-n} = r^{-n}(\cos n\theta - i\sin n\theta).$$

In particular, if p has absolute value 1, then

$$p^{-1} = \bar{p}.$$

After our successful polar representation of positive and negative integer powers of z, we tackle the problem of rational powers, $z^{p/q}$, next. Since $z^{p/q} = (z^{1/q})^p$, we first settle the problem of finding the q-th roots of z.

By analogy to (4.17). we shall tentatively represent a q-th root of z by

$$w_1 = r^{1/q}\left(\cos\frac{\theta}{q} + i\sin\frac{\theta}{q}\right).$$

Indeed, this is a q-th root of z, for its q-th power is

$$(w_1)^q = (r^{1/q})^q \left(\cos q\,\frac{\theta}{q} + i \sin q\,\frac{\theta}{q} \right) = r(\cos \theta + i \sin \theta) = z,$$

but it is not the only q-th root. Another root, different from w_1, is the number

$$w_2 = r^{1/q}\left[\cos\left(\frac{\theta + 2\pi}{q}\right) + i \sin\left(\frac{\theta + 2\pi}{q}\right) \right].$$

It is not difficult to see, from the periodicity of the sine and cosine functions, that the q numbers

(4.20)

$$w_{k+1} = r^{1/q}\left[\cos\left(\frac{\theta}{q} + \frac{2k\pi}{q}\right) + i \sin\left(\frac{\theta}{q} + \frac{2k\pi}{q}\right) \right], \qquad k = 0, 1, \ldots, q - 1$$

yield the q distinct q-th roots of $z \neq 0$. If $z = 0$, they are all equal to zero.

Example. The three cube roots of

$$1 = (\cos 0 + i \sin 0)$$

are

$$\cos\frac{0}{3} + i \sin \frac{0}{3} = 1, \qquad \cos \frac{2\pi}{3} + i \sin \frac{2\pi}{3} = -\frac{1}{2} + i\frac{\sqrt{3}}{2}$$

and

$$\cos \frac{4\pi}{3} + i \sin \frac{4\pi}{3} = -\frac{1}{2} - i\frac{\sqrt{3}}{2}.$$

EXERCISES

4.1 Carry out the following operations with complex numbers:

(a) $(2 + 3i) + (5 - 4i)$ (b) $(3 - 2i) - (8 - 7i)$ (c) $(3 - 2i)(4 + 5i)$

(d) $\dfrac{3 - 2i}{4 + 5i}$ (e) $\dfrac{4 + 5i}{3 - 2i}$ (f) $\dfrac{1 + i}{1 - i}$.

4.2 Find the absolute values of the following complex numbers:

(a) $3 + 4i$ (b) $5 + 6i$ (c) $\dfrac{3 + 4i}{5 + 6i}$ (d) $\dfrac{1 + i}{1 - i}$.

4.3 Verify the identity

(4.21) $$(a^2 + b^2)(c^2 + d^2) = (ac - bd)^2 + (ad + bc)^2.$$

By definition of multiplication of complex numbers,

$$(a + ib)(c + id) = ac - bd + i(ad + bc).$$

Deduce from (4.21) that the absolute value of the product of two complex numbers is the product of their absolute values.

4.4 (a) Show that for any pair of complex numbers z and w,

$$|z + w|^2 + |z - w|^2 = 2|z|^2 + 2|w|^2.$$

(b) Using the parallelogram interpretation of the addition of complex numbers, deduce from (a) that the sum of the squares of the diagonals of a parallelogram equals the sum of the squares of its four sides.

4.5 (a) Show that the argument of $\bar{w}$ is the negative of the argument of w.

(b) Show that the argument of $z\bar{w}$ is the difference of the arguments of z and w.

(c) Show that the argument of z/w is the difference of the arguments of z and w.

(d) Show that the ray from 0 to z is perpendicular to the ray from 0 to w if and only if the number $z\bar{w}$ is purely imaginary.

(e) Let p be a complex number whose absolute value is 1. Show that

$$(p - 1)(\bar{p} + 1)$$

is purely imaginary.

(f) Show that (d) and (e) combined lead to the following geometric theorem, known as Thales' theorem:

Let p be a point on the unit circle around the origin; the ray connecting p to the point 1 on the real axis is perpendicular to the ray connecting p to the point -1 on the real axis. Draw a picture.

4.6 (a) Let p be a complex number whose absolute value is 1. Show that $(p - 1)^2\bar{p}$ is real.

(b) Let p and q be a pair of complex numbers, each with absolute value equal to 1. Show that

$$[(p - 1)(\bar{q} - 1)]^2 \bar{p}\, q$$

is real.

(c) Using part (b) of Exercise 4.5 show that (b) above has the following geometric interpretation; see Figure 7.16:

If p and q lie on the unit circle, then the angle β between the rays from the origin to p and q respectively is twice the angle α between the rays connecting the point $z = 1$ to p and q respectively: $\beta = 2\alpha$.

4.7 Find the real and imaginary parts (3 decimals) of all roots in the following list:

(a) $z^2 - i = 0$ (b) $z^2 - 1 + i = 0$

(c) $z^3 - i = 0$ (d) $z^3 + 1 = 0.$

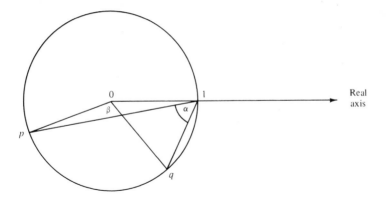

Figure 7.16

4.8 Find the real and imaginary parts (3 decimals) of both roots of the following quadratic equations:
(a) $z^2 + 2iz + 3 = 0$
(b) $z^2 - (1 + i)z + 5 = 0$
(c) $z^2 + z - i = 0$.

4.9 Newton devised a method for finding real solutions of an equation of the form $p(x) = 0$. The method starts with a rough approximation x_0 to a solution x, and then proceeds by producing a sequence $x_1, x_2, \ldots$ of better approximations, according to the formula

(4.22)
$$x_{n+1} = x_n - \frac{p(x_n)}{p'(x_n)}.$$

It was shown in Section 3.10 that, if x_0 is close enough to a solution x which is a simple root in the sense that the derivative $p'(x) \neq 0$, then the sequence of approximations constructed in this fashion converges rapidly to x.

In this exercise we want you to test experimentally Newton's method for finding *complex* roots of an algebraic equation $p(z) = 0$. Take for p the cubic polynomial

$$p(z) = z^3 + z^2 + z - i,$$

so that

$$p'(z) = 3z^2 + 2z + 1.$$

(a) Write a computer program which constructs a sequence $z_1, z_2, \ldots$ according to the recipe (4.22), stopping when both the real and imaginary parts of z_{n+1} and z_n differ by less than 10^{-6}, or when n exceeds 30.
(b) Show that $p(z) = z^3 + z^2 + z - i \neq 0$ if $|z| > 2$ or if $|z| < \frac{1}{2}$. [*Hint*: If $|z| > 2$, the absolute value of the first term is bigger than that of the sum of the last 3; if $|z| < \frac{1}{2}$ the absolute value of the last term is bigger than that of the sum of the first 3.]

(c) Starting with the first approximation

$$z_0 = 0.35 + 0.35i,$$

construct the sequence z_n by Newton's method, and determine whether or not it converges to a solution of $p(z) = 0$.
(d) Try to find all solutions of $p(z) = 0$ by starting with different choices for z. [*Hint*: Remember what you have found in (b)!]
(e) Investigate how Newton's method works for finding zeroes of some other polynomials.

4.10 (a) Write $\cos 4\theta$ as a polynomial in $\cos \theta$.
(b) Write $\sin 5\theta$ as a polynomial in $\sin \theta$.
(c) What are the degrees of the polynomials you obtained in (a) and (b)? What is your guess about the degree of the polynomial $P(\cos \theta) = \cos n\theta$?

4.11 Prove that Formula (4.20) furnishes all q-th roots of z.

4.12 Find all three cube roots of -1 and represent each in the complex plane.

4.13 Show that the two square roots of i are

$$\frac{1}{\sqrt{2}}(1 + i), \qquad -\frac{1}{\sqrt{2}}(1 + i),$$

and represent each in the complex plane.

4.14 Verify that the two square roots of i (see Exercise 4.13) are fourth roots of -1; find the other two fourth roots of -1.

4.15 (a) Find all zeros of the function

$$w(x) = x^4 + x^3 + x^2 + x + 1.$$

[*Hint*: The function $(x - 1)w(x) = x^5 - 1$ vanishes whenever $w(x)$ vanishes.]
(b) Represent all n n-th roots of 1 in the complex plane. Let w be the n-th root of 1 with smallest nonzero argument. Show that the remaining n-th roots are $w^2, w^3, \ldots w^n$. Show also that $w, w^2, \ldots w^{n-1}$ are the $n - 1$ roots of

$$x^{n-1} + x^{n-2} + \cdots + x + 1 = 0.$$

7.5 Isometries of the complex plane

An *isometry* of the complex plane is a mapping $\mathcal{M}$ of the plane onto itself which *preserves distance*. This means that, for every pair of complex numbers z and w,

(5.1) $$|\mathcal{M}(z) - \mathcal{M}(w)| = |z - w|.$$

An isometry is also called a *rigid mapping* or *rigid motion*.

Examples of isometries are *translations*, *rotations*, and *reflections*. Clearly, the *composition* of any number of isometries is again an isometry. Every isometry has an *inverse* mapping which is again an isometry. For example, if $\mathscr{T}$ is translation by the complex number p, then its inverse $\mathscr{T}^{-1}$ is translation by $-p$; if $\mathscr{R}$ is rotation by θ, $\mathscr{R}^{-1}$ is rotation by $-\theta$; and reflection $\mathscr{F}$ across any line is its own inverse, since $\mathscr{F} \circ \mathscr{F}$ is the identity map $\mathscr{I}$.

Theorem. *Every isometry $\mathscr{M}$ can be obtained as the composition of a translation, a rotation and a reflection across the real axis, or as a composition of just a translation and a rotation:*

$$\mathscr{M} = \mathscr{T} \circ \mathscr{R} \circ \mathscr{F} \qquad or \qquad \mathscr{M} = \mathscr{T} \circ \mathscr{R}.$$

Here we include translation by 0 (in which case $\mathscr{T} = \mathscr{I}$, and rotation by zero (in which case $\mathscr{R} = \mathscr{I}$).

PROOF. If the isometry $\mathscr{M}$ maps 0 into $p \neq 0$, we consider the composite

$$\mathscr{N} = \mathscr{T}^{-1} \circ \mathscr{M},$$

where $\mathscr{T}^{-1}$ is the inverse of the translation $\mathscr{T}$ that maps 0 into p. Thus

(5.2) $$\mathscr{N}(0) = \mathscr{T}^{-1}[\mathscr{M}(0)] = \mathscr{T}^{-1}(p) = 0,$$

and

(5.3) $$\mathscr{T}\mathscr{N} = \mathscr{T}\mathscr{T}^{-1} \circ \mathscr{M} = \mathscr{I} \circ \mathscr{M} = \mathscr{M}.$$

Claim. *An isometry $\mathscr{N}$ which maps 0 into 0 is either a rotation, or can be composed of a rotation and reflection across the real axis.*

To prove the claim, suppose that $\mathscr{N}(1) = n$. Then $|n| = 1$, by isometry, and there exists a rotation $\mathscr{R}$ such that $\mathscr{R}(1) = n$. Consider the composite

$$\mathscr{K} = \mathscr{R}^{-1} \circ \mathscr{N}.$$

It leaves the numbers 0 and 1 fixed:

(5.4)
$$\mathscr{K}(0) = \mathscr{R}^{-1} \circ \mathscr{N}(0) = \mathscr{R}^{-1}(0) = 0$$
$$\mathscr{K}(1) = \mathscr{R}^{-1} \circ \mathscr{N}(1) = \mathscr{R}^{-1}(n) = 1.$$

We recall that

(5.5) $$\mathscr{R} \circ \mathscr{K} = \mathscr{N} \qquad and \qquad \mathscr{T} \circ \mathscr{R} \circ \mathscr{K} = \mathscr{M}.$$

To establish our claim we shall show that *an isometry which leaves 0 and 1 fixed is either the identity $\mathscr{I}$ or reflection $\mathscr{F}$ across the real axis.* For any complex number z, we have

$$|\mathscr{K}(z)| = |\mathscr{K}(z) - \mathscr{K}(0)| = |z - 0| = |z|$$

and

$$|\mathscr{K}(z) - 1| = |\mathscr{K}(z) - \mathscr{K}(1)| = |z - 1|,$$

because $\mathscr{K}$ is an isometry satisfying (5.4). Thus z and $\mathscr{K}(z)$ have the same distance from 0, and they have the same distance from 1. It follows (see Figure 7.17a) that $\mathscr{K}(z)$ is either z or $\bar{z}$, the reflection $\mathscr{F}(z)$ of z. We claim that

361

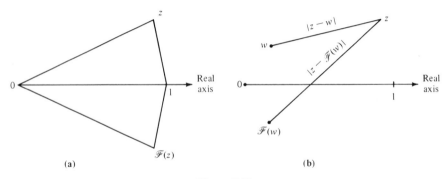

Figure 7.17

either $\mathcal{K}(z) = z$ for all z, or $\mathcal{K}(z) = \mathcal{F}(z) = \bar{z}$ for all z. For, if for one unreal number z (z not on real axis) $\mathcal{K}(z) = z$, and for another unreal number w, $\mathcal{K}(w) = \mathcal{F}(w)$, then

$$|\mathcal{K}(z) - \mathcal{K}(w)| = |z - \mathcal{F}(w)| \neq |z - w|,$$

see Figure 7.17b, and so $\mathcal{K}$ would not be an isometry. We conclude that either

$$\mathcal{K} = \mathcal{I} \qquad \text{or} \qquad \mathcal{K} = \mathcal{F},$$

and hence, by (5.5), that

(5.6) $$\mathcal{N} = \mathcal{R} \qquad \text{or} \qquad \mathcal{N} = \mathcal{R} \circ \mathcal{F}.$$

This proves our claim. It also proves the theorem, since, by (5.3) and (5.6)

$$\mathcal{M} = \mathcal{T} \circ \mathcal{R} \qquad \text{or} \qquad \mathcal{M} = \mathcal{T} \circ \mathcal{R} \circ \mathcal{F}. \qquad \square$$

What we have learned about isometries will enable us to give a new proof of the additive relation (4.14) between the argument of a product and the arguments of its factors. This new proof does *not* use the addition formulas (1.12) for the sine and cosine; on the contrary, it constitutes a new algebraic proof of these addition formulas.

Let n be a complex number of absolute value 1, and denote by $\mathcal{N}$ multiplication by n:

(5.7) $$\mathcal{N}(z) = nz, \qquad |n| = 1.$$

$\mathcal{N}$ is an isometry, because, for any complex z and w,

$$|\mathcal{N}(z) - \mathcal{N}(w)| = |nz - nw| = |n||z - w| = |z - w|.$$

Moreover

$$\mathcal{N}(0) = n \cdot 0 = 0,$$

so $\mathcal{N}$ keeps 0 fixed. Therefore, according to (5.6),

$$\mathcal{N} = \mathcal{R} \qquad \text{or} \qquad \mathcal{N} = \mathcal{R} \circ \mathcal{F},$$

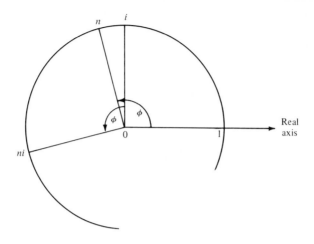

Figure 7.18

where $\mathscr{R}$ is the rotation which carries 1 into the same point as $\mathscr{N}$ does. Now by definition (5.7), $\mathscr{N}$ carries 1 into n. Denote the argument of n by ϕ:

(5.8) $n = \cos\phi + i\sin\phi.$

Therefore $\mathscr{R}$ is rotation by angle ϕ.

We claim now that $\mathscr{N} = \mathscr{R}$; according to (5.6), it suffices to show that $\mathscr{N}(z) = \mathscr{R}(z)$ for a single z different from 0 or 1. A convenient value is $z = i$: by definition (5.7) of $\mathscr{N}$ and the polar decomposition (5.8) of n we get

$$\mathscr{N}(i) = ni = -\sin\phi + i\cos\phi.$$

On the other hand, Figure 7.18 shows that

(5.9) $\mathscr{R}(i) = -\sin\phi + i\cos\phi.$

This proves that

(5.10) $\mathscr{N} = \mathscr{R}.$

Now let z be any complex number; its polar representation is

$$z = r(\cos\theta + i\sin\theta), \qquad r = |z|.$$

By definition of $\mathscr{N}$,

$$\mathscr{N}(z) = nz = (\cos\phi + i\sin\phi)r(\cos\theta + i\sin\theta)$$
$$= r[(\cos\phi\cos\theta - \sin\phi\sin\theta) + i(\cos\phi\sin\theta + \sin\phi\cos\theta)],$$

and by definition of rotation by the angle ϕ,

$$\mathscr{R}(z) = r[\cos(\theta + \phi) + i\sin(\theta + \phi)].$$

363

But by (5.10), $\mathcal{N} = \mathcal{R}$; so the real parts of $\mathcal{N}(z)$ and $\mathcal{R}(z)$ are equal, and their complex parts are equal. This yields

(5.11)
$$\cos\phi\cos\theta - \sin\phi\sin\theta = \cos(\theta + \phi)$$
$$\cos\phi\sin\theta + \sin\phi\cos\theta = \sin(\theta + \phi).$$

Thus complex multiplication and our analysis of isometries yield a proof (promised in Section 7.1) of the trigonometric addition formulas (1.12) used in Section 7.1.

We close this section by showing how to use complex numbers to express the area of a triangle. Let z and w be a pair of complex numbers; consider the triangle formed by 0, z and w, as shown in Figure 7.19a. We claim: *the area of the triangle formed by 0, z, w is given by the formula*

(5.12)
$$A(z, w) = \tfrac{1}{2}\,\mathrm{Im}\{\bar{z}w\}.$$

PROOF. The formula on the right side does not change if we subject z and w to a rotation; for, we have seen that rotation amounts to multiplication by a complex number a of absolute value 1,
$$|a| = 1.$$

So if we set
$$z_1 = az, \qquad w_1 = aw,$$

we get
$$A(z_1, w_1) = \tfrac{1}{2}\,\mathrm{Im}\{\bar{z}_1 w_1\} = \tfrac{1}{2}\,\mathrm{Im}\{\bar{a}\bar{z}aw\}$$
$$= \tfrac{1}{2}\,\mathrm{Im}\{|a|^2\,\bar{z}w\} = \tfrac{1}{2}\,\mathrm{Im}\{\bar{z}w\} = A(z, w).$$

We perform a particular rotation which makes z real and positive. For such z
$$A(z, w) = \tfrac{1}{2}\,\mathrm{Im}\{\bar{z}w\} = \tfrac{1}{2}z\,\mathrm{Im}\{w\}.$$

But this is just the familiar $\tfrac{1}{2}$ base $\times$ altitude formula for the area of a triangle, see Figure 7.19b. □

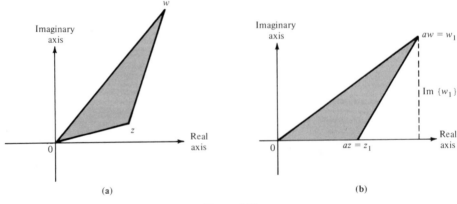

(a)   (b)

Figure 7.19

Remark. Formula (5.12) is a formula for *signed* area, i.e., the value of the area in case the counterclockwise rotation that takes the direction of z into the direction of w is less than π, and $-$area if that rotation is greater than π.

7.6 Complex functions

In this section we shall discuss the relationship of complex numbers to the concept of function. So far in this text, numbers have entered the concept of function in two places: as *input* and *output*. In the first part of this section we shall show how simple it is to replace real numbers by complex ones as output if the input is kept real; such functions are called *complex valued functions with real input*. In the second part we shall discuss some very special but important functions whose input is complex, and whose output is also complex, i.e., complex valued functions with complex input.

The theory of complex valued functions is simple, because it can be reduced at one stroke to the theory of real valued functions. There are two ways of going about it. The first is to observe that everything (well, almost everything) that has been said about real valued functions makes sense when carried over to complex valued functions. "Everything" includes the following notions:

(i) the concept of function itself;
(ii) the operations of adding and multiplying functions, and forming the reciprocal of a function (with the usual proviso that the function should not be zero);
(iii) the concept of a continuous function;
(iv) the concept of a differentiable function and its derivative;
(v) higher derivatives;
(vi) the integral of a function over an interval.

Let us for example review the concept of continuity, discussed in Section 2.7. The intuitive meaning of the continuity of a function f is that, to determine $f(x)$ approximately, approximate knowledge of x is sufficient. The precise version is this:

f is continuous on its interval of definition if, in order for $f(x)$ and $f(y)$ to be so close that

(6.1) $$| f(x) - f(y)| < 10^{-k},$$

it suffices for x and y to be so close that

$$|x - y| < 10^{-m}.$$

The choice of m depends on k.

Both the intuitive and the precise definitions make sense for complex valued functions; note that the absolute value sign occurring in (6.1) has been properly defined for complex numbers. Sums and products of continuous complex valued functions are again continuous, and so is the reciprocal of a

365

nonzero continuous function. If the reader has any twinge of doubt, he should consult Section 2.7 and convince himself that the proofs presented there for the real valued case apply *verbatim* to the complex valued case.

Turn to the concept of differentiation, discussed first in Section 3.1. Starting with the difference quotient

$$f'_h(t) = \frac{f(t+h) - f(t)}{h},$$

the derivative f' is defined as the limit of f'_h as h tends to zero. The precise formulation is as follows:

$$|f'_h(t) - f'(t)| < 10^{-m}$$

holds for all t provided that h is small enough; how small h has to be depends on how large m is.

Again, this definition makes perfectly good sense for complex valued functions, and furthermore, the usual rules for differentiating sums, products and reciprocals of differentiable functions hold, i.e.,

(6.2)
$$(f + g)' = f' + g'$$
$$(fg)' = fg' + f'g$$
$$\left(\frac{1}{f}\right)' = -\frac{f'}{f^2}.$$

The wary reader is urged to consult Section 3.2 to convince himself that the proofs offered there retain their validity for complex valued functions as well.

For complex numbers there is an additional algebraic operation: taking the complex conjugate. Given any complex valued function f, let us denote by $\bar{f}$ the function defined wherever f is, and whose value at any point x is the conjugate of the value of f:

$$\bar{f}(x) = \overline{f(x)}.$$

Here is the rule for differentiating $\bar{f}$: *The conjugate $\bar{f}$ of a differentiable function f is differentiable, and its derivative is the conjugate of the derivative of f:*

(6.3)
$$\bar{f}' = \overline{f'}.$$

To derive this rule we merely observe that, if $\bar{f}$ is the conjugate of f, then the difference quotients $\bar{f}'_h$ of $\bar{f}$ are conjugates of the corresponding difference quotients of f. Since the derivative is the limit of such difference quotients, the same relation must then hold for the derivative. We now give a few illustrations of the rules for differentiation of combinations of complex valued functions.

Example 1. Suppose

$$f(x) = ax + b,$$

a and b arbitrary complex numbers. Then

$$f' = a.$$

Example 2. If

$$f(x) = \frac{1}{x + i},$$

then the rule for differentiating the reciprocal of a function yields

$$f' = -\frac{1}{(x + i)^2}.$$

Example 3. Use the quotient rule for differentiating

$$f(x) = \frac{x}{x^2 + i}$$

and obtain

$$f'(x) = \frac{i - x^2}{(i + x^2)^2}.$$

We remarked at the beginning of this section that there is another way of basing the study of complex valued functions on real valued functions. It consists of looking upon the real and imaginary parts of $f(x)$ separately. That is, write $f(x)$ as

(6.4) $$f(x) = a(x) + ib(x),$$

a being the real part of f, b its imaginary part. We introduce the frequently used symbols $\text{Re}\{z\}$ and $\text{Im}\{z\}$ denoting the real and imaginary parts of a complex number z. Clearly,

$$a(x) = \text{Re}\{f(x)\}, \qquad b(x) = \text{Im}\{f(x)\}$$

is a pair of real valued functions which describes the complex valued function f. All properties of f can be determined from knowing the corresponding properties of a and b. For instance, f is continuous if, and only if, a and b are continuous. Similarly, f is differentiable if, and only if, both a and b are differentiable. Furthermore, *the derivative of f can be expressed in terms of the derivatives of a and b:*

$$f' = a' + ib'.$$

These facts are an immediate consequence of the previously given definitions of the derivative as the limit of difference quotients. For, the real and imaginary parts of f'_h are a'_h and b'_h,

$$f'_h = a'_h + ib'_h.$$

If f'_h tends to a limit, its real and imaginary parts must tend to limits. These limits are a' and b', respectively, and they contribute the real and imaginary parts of f'. Here are some more examples:

Example 4. Let

$$f(x) = \frac{1}{x + i}.$$

To separate f into its real and imaginary parts, we write it as

$$f(x) = \frac{1}{x + i} = \frac{1}{x + i} \cdot \frac{x - i}{x - i} = \frac{x - i}{x^2 + 1}$$

$$= \frac{x}{x^2 + 1} - \frac{i}{x^2 + 1} = a(x) + ib(x).$$

Differentiating $a(x) = x/(x^2 + 1)$ and $b(x) = -1/(x^2 + 1)$, we get

$$a'(x) = \frac{1 - x^2}{(x^2 + 1)^2}, \qquad b'(x) = \frac{2x}{(x^2 + 1)^2}.$$

The expression we found in Example 2 for the derivative of the same function f was

$$f'(x) = -\frac{1}{(x + i)^2}.$$

To see that the real and imaginary parts of f' are $a'(x)$ and $b'(x)$, we write it as

$$f'(x) = -\frac{1}{(x + i)^2} = -\frac{1}{(x + i)^2} \frac{(x - i)^2}{(x - i)^2}$$

$$= \frac{-(x - i)^2}{(x^2 + 1)^2} = \frac{-x^2 + 2xi + 1}{(x^2 + 1)^2}$$

$$= \frac{1 - x^2}{(x^2 + 1)^2} + i \frac{2x}{(x^2 + 1)^2},$$

which is indeed equal to $a' + ib'$ as calculated above.

Example 5. Now consider

$$f(x) = (x + i)^2.$$

If we carry out the indicated squaring, we can split f into its real and imaginary parts:

$$f(x) = x^2 + 2ix - 1 = x^2 - 1 + 2ix = a(x) + ib(x).$$

Differentiating

$$a(x) = x^2 - 1 \qquad \text{and} \qquad b(x) = 2x,$$

we get

$$a'(x) = 2x \qquad \text{and} \qquad b'(x) = 2,$$

so that

$$f'(x) = a'(x) + ib'(x) = 2x + 2i.$$

The function $f(x) = (x + i)^2$, when differentiated directly, yields

$$f'(x) = 2(x + i) = 2x + 2i,$$

the same answer we got before.

If $g(t)$ is a real valued function and f is a complex valued function defined at all values taken on by g, then we can form their composite $f \circ g$, defined as $f(g(t))$. If f and g are both differentiable, so is the composite, and the derivative of the composite is given by the usual chain rule. The proof is the same as in the real valued case.

Example 6. If

$$f(x) = \frac{1}{x + i}, \qquad g(t) = t^2,$$

then

$$f(g(t)) = \frac{1}{t^2 + i}.$$

The derivative of this function can be calculated by the chain rule; using Example 4 for f', we get

$$(f \circ g)' = f'(g(t))g'(t) = -\frac{1}{(t^2 + i)^2} 2t.$$

Example 7. Take f and g to be

$$f(x) = (x + i)^2, \qquad g(t) = t^2.$$

Then

$$f(g(t)) = (t^2 + i)^2 = t^4 + 2it^2 - 1.$$

Differentiating the composite directly yields

$$(f \circ g)' = 4t^3 + 4it.$$

Using the chain rule and the result of Example 5 yields the derivative

$$(f \circ g)' = f'(g(t))g'(t) = 2(t^2 + i)2t,$$

the same as before.

369

Splitting a complex valued function into its real and imaginary parts is suitable for defining the integral of a complex valued function. For $f = a + ib$, we simply set

$$(6.5) \qquad \int_p^q f(t)dt = \int_p^q a(t)dt + i \int_p^q b(t)dt;$$

i.e., *the integral of f is the sum of the integrals of the real and imaginary parts of f.* This follows from Section 4.2 according to which the integral of the real valued functions a and b are defined as the limits of approximating sums

$$I_{approx}(a, S) = \sum a(t_j)|S_j| \qquad \text{and} \qquad I_{approx}(b, S) = \sum b(t_j)|S_j|,$$

where $\sum S_j = S$ is a partition of the interval of integration, and t_j is some point of S_j. The corresponding sum for f,

$$I_{approx}(f, S) = \sum f(t_j)|S_j|,$$

is the sum of the above approximating sum for a plus i times the approximating sum for b:

$$I_{approx}(f, S) = I_{approx}(a, S) + iI_{approx}(b, S).$$

Comparing this with (6.5) we conclude that $I_{approx}(f,S)$ tends to $\int f(t)dt$; that is, the integral of a complex valued function can be defined in terms of approximating sums. This bears out our original contention that most of the theory we have developed for real valued functions applies verbatim to complex valued functions.

Since the integral of complex valued functions can be defined in the same way as the integral of real valued functions, they share these properties:

Additivity

$$\int_p^q f(t)dt + \int_q^r f(t)dt = \int_p^r f(t)dt.$$

Linearity

$$\int_p^q kf(t)dt = k \int_p^q f(t)dt, \qquad \text{for constant } k,$$

$$\int_p^q [f(t) + g(t)]dt = \int_p^q f(t)dt + \int_p^q g(t)dt.$$

Fundamental theorem

$$\frac{d}{dx} \int_p^x f(t)dt = f(x).$$

Another property of the integral of real valued functions is the

Upper bound property

If $|f(t)| \leq M$ *for every* x *in the interval of integration* S, *then*

(6.6)
$$\left| \int_p^q f(t)dt \right| \leq M|S|.$$

This inequality, too, remains true for complex valued functions, and for the same reason: the analogous estimate holds for the approximating sums, with | | as defined for complex numbers.

It is not possible to draw the graph of a complex valued function in the usual sense; yet it is possible to represent complex valued functions graphically in another way that sheds light on the behavior of such functions. *The graphical representation of a complex valued function is obtained by plotting in the complex plane all the values taken on by the function.*

Example 8. Graph the function

$$f(t) = (i + 1)t + 2$$

on the interval [0, 1]. A partial list of the values of f is given below.

t	0	0.2	0.4	0.6	0.8	1.0
$f(t)$	2	$2.2 + 0.2i$	$2.4 + 0.4i$	$2.6 + 0.6i$	$2.8 + 0.8i$	$3 + i$

Let us plot these numbers in the complex plane; see Figure 7.20. The plot suggests that the values of f lie on a straight line whose endpoints are 2 and $3 + i$. This is indeed so; we leave it to the reader to supply a proof.

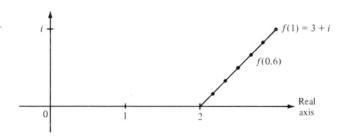

Figure 7.20

Example 9. Graph $f(t) = 1/(t + i)$ on the interval [0, 1]. A partial list of the values of f is given below:

t	0	0.2	0.4	0.6	0.8
$f(t)$	$-i$	$0.192 - 0.962i$	$0.345 - 0.862i$	$0.441 - 0.735i$	$0.488 - 0.610i$

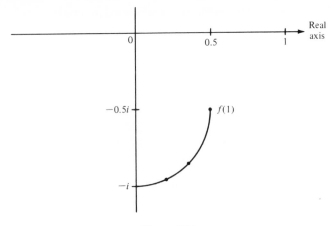

Figure 7.21

These values are plotted in Figure 7.21. The plot suggests that the values of f lie along a smooth curve connecting the points $-i$ and $0.5 + 0.5i$. The nature of this smooth curve is further investigated in Exercise 6.6.

We now turn to the function

(6.7)
$$C(t) = \cos t + i \sin t.$$

Its graph in the complex plane lies on the unit circle. Clearly C is differentiable, and its derivative can be obtained by differentiating the real and imaginary parts of C;

$$C'(t) = -\sin t + i \cos t$$

A moment's observation discloses that

(6.8)
$$C' = iC.$$

We now recall, from Chapter 5 on the exponential function, that $e^{at} = P(t)$ satisfies the differential equation

$$P' = aP.$$

Comparing this with (6.8) we are drawn to the conclusion that $C(t)$ is e^{it}. Unable to resist *we define the exponential of a pure imaginary number, it,* to be

(6.9)
$$e^{it} = \cos t + i \sin t.$$

Note that the input into the exponential function here is no longer real, but pure imaginary. It is now a short step to the exponential of any complex number $z = x + iy$. The functional equation of the exponential function suggests that

$$e^{x+iy} = e^x e^{iy};$$

combining this with (6.9), we arrive at this *definition of the exponential of a complex number: for x, y real*

(6.10) $e^{x+iy} = e^x[\cos y + i \sin y].$

The exponential function as defined above has complex input, complex output, and it has all the usual properties of the exponential function, as we shall now show.

(i) Functional equation

(6.11) $e^{z+w} = e^z e^w.$

Setting

$$z = x + iy, \qquad w = u + iv,$$

and using the definition (6.10) of the exponential function, we get

$$e^z e^w = e^x[\cos y + i \sin y]e^u[\cos v + i \sin v].$$

According to the functional equation of the exponential function for real inputs, $e^x e^u = e^{x+u}$. Similarly, using the addition formulas (5.11) for sine and cosine, we conclude that

$$[\cos y + i \sin y][\cos v + i \sin v] = \cos(y+v) + i \sin(y+v).$$

Combining these two facts we conclude that $e^z e^w = e^{z+w}$, as asserted.

The addition formulas for sine and cosine, used in the above derivation, are *equivalent* to the relation

(6.12) $e^{iy} e^{iv} = e^{i(y+v)}$

if we define e^{it} by (6.9). Relation (6.12) is concise and crystal clear, whereas the addition formulas for sine and cosine are less transparent.

(ii) The differential equation

For any complex number c

$$P(t) = e^{ct}$$

satisfies

(6.13) $P'(t) = cP(t).$

PROOF. Set $c = a + ib$, a and b real. By definition of complex exponentials,

$$e^{ct} = e^{at} e^{ibt}.$$

Abbreviate e^{at} by $A(t)$, e^{ibt} by $B(t)$; so

$$P = AB.$$

A is a real exponential function and so satisfies

$$A' = aA$$

373

while B, a purely imaginary exponential, was defined so that it would satisfy

$$B' = ibB.$$

So

$$P' = (AB)' = A'B + AB' = aAB + AibB = (a + ib)AB = cP,$$

as asserted. ▢

(iii) Series representation for complex exponentials

We saw at the end of Chapter 5, Equation (3.24), that the exponential function can be represented by its Taylor series:

(6.14) $$e^z = 1 + z + \frac{z^2}{2} + \cdots + \frac{z^n}{n!} + \cdots.$$

We shall now show that this series representation holds also for complex inputs z. For the sake of simplicity, we prove this only for pure imaginary z, i.e., for z of the form ib, b real. Substituting $z = ib$ into the series on the right in (6.14) we get

(6.15) $$e^{ib} = 1 + ib - \frac{b^2}{2} + \cdots + \frac{(ib)^n}{n!} + \cdots.$$

The powers i^n have a period of 4, i.e. we have

$$i^0 = 1, \qquad i^1 = i, \qquad i^2 = -1, \qquad i^3 = -i,$$

and then the whole pattern is repeated over and over again. This shows that the terms of odd order in series (6.15) (i.e., the 1-st, 3-rd, 5-th etc.) are real, and the even ones (the 2-nd, 4-th, 6-th, etc.) are pure imaginary. Thus the real and imaginary parts of the series (6.15) are

(6.15)$_{re}$ $$1 - \frac{b^2}{2} + \frac{b^4}{24} - \cdots + \frac{(-1)^m b^{2m}}{(2m)!} + \cdots$$

and

(6.15)$_{im}$ $$b - \frac{b^3}{6} + \frac{b^5}{120} + \cdots + \frac{(-1)^m b^{2m+1}}{(2m+1)!} + \cdots.$$

Comparing these series with series (3.26) and (3.27) given at the end of Section 7.3 for $\cos b$ and $\sin b$, we recognize (6.15)$_{re}$ as $\cos b$ and (6.15)$_{im}$ as $\sin b$. So the series (6.15) is just $\cos b + i \sin b$, which is what e^{ib} is supposed to be.

Set $t = 2\pi$ in the definition (6.9) of the exponential function, and observe that

$$e^{i2\pi} = \cos 2\pi + i \sin 2\pi = 1.$$

374

More generally, since $\cos 2\pi n = 1$, $\sin 2\pi n = 0$,

(6.16)
$$e^{i2\pi n} = \cos 2\pi n + i \sin 2\pi n = 1$$

for any integer n.

Now set $t = \pi$ in the definition (6.9). Since $\cos \pi = -1$, $\sin \pi = 0$, we get that

$$e^{i\pi} = \cos \pi + i \sin \pi = -1.$$

This can be rewritten in the form

(6.17)
$$e^{i\pi} + 1 = 0.$$

It is worth pointing out to those interested in number mysticism that this relation contains the most important numbers and symbols of mathematics:

$$0, \quad 1, \quad i, \quad \pi, \quad e, \quad + \quad \text{and} \quad =.$$

The polar form

$$z = r[\cos \theta + i \sin \theta], \qquad r = |z|,$$

discussed in Section 7.5, can be expressed in a particularly simple way with the aid of the exponential notation:

(6.18)
$$z = re^{i\theta}, \qquad r = |z|, \qquad \theta = \arg z.$$

Having defined the exponential function for complex exponents, we can try to define the logarithmic function on the set of complex numbers as the inverse of the exponential function. Using the form (6.18) and introducing the typographically more convenient notation

$$\exp\{w\} \quad \text{for } e^w,$$

we write

$$z = r \exp\{i\theta\} = \exp\{\log r\}\exp\{i\theta\} = \exp\{\log r + i\theta\}.$$

Thus the exponential of $\log r + i\theta$ is z, and so it can be labelled by z:

(6.19)
$$\log z = \log r + i\theta, \qquad r = |z|, \qquad \theta = \arg z;$$

for complex input z, this function has complex output.

Is this the only possible definition of $\log z$, or is there a number other than $\log r + i\theta$ which, when e is raised to it, produces z? We claim that all numbers of the form

(6.20)
$$\log r + i\theta + i2\pi n, \quad n \text{ integer},$$

have that property; for, according to the functional equation for the exponential function and relation (6.16),

$$\exp\{\log r + i\theta + i2\pi n\} = \exp\{\log r + i\theta\}\exp\{i2\pi n\} = z.$$

In Exercise 6.4 we ask the reader to verify that (6.20) are all the numbers whose exponential is z, i.e., which deserve the name log z. Thus we see that log z is a *multivalued* "function," with infinitely many values for each z. This is not surprising; we had already encountered multivalued "functions" at the end of Section 6.5 in the shape of the m-th roots $z^{1/m}$ of z.

Just as in the real case, we can deduce the functional equation of the logarithm from the functional equation (6.11):

$$e^{z+w} = e^z e^w.$$

Taking logs of both sides, we get

$$z + w = \log(e^z e^w).$$

We set $e^z = u$, $e^w = v$; then

$$z = \log u, \qquad w = \log v,$$

and the above can be rewritten as

(6.21) $$\log u + \log v = \log(uv).$$

Write u, v and uv in their polar forms:

$$u = r(\cos \theta + i \sin \theta)$$
$$v = s(\cos \phi + i \sin \phi)$$
$$uv = t(\cos \chi + i \sin \chi).$$

Using the definition (6.19) of the logarithm, we can rewrite (6.21) as

$$\log r + i\theta + \log s + i\phi = \log t + i\chi.$$

Equating the real and imaginary parts, we get

$$\log r + \log s = \log t$$
$$\theta + \phi \quad = \quad \chi.$$

These are nothing but the rules (4.15) for multiplying complex numbers. Thus the functional equation for complex logarithms is merely another way of stating the multiplication rule for complex numbers in polar form.

Suppose z is a complex valued function of the real variable t. Then $|z|$ and arg z also are functions of t. So, if we write z in polar form

(6.22) $$z = r(\cos \theta + i \sin \theta),$$

r and θ are functions of t. We can express dz/dt by differentiating (6.22) with respect to t. Using the rule for differentiating a product and the chain rule, we get

$$\frac{dz}{dt} = \frac{dr}{dt}(\cos \theta + i \sin \theta) + r(-\sin \theta + i \cos \theta)\frac{d\theta}{dt}.$$

Using the polar form (6.22), we can rewrite this as

(6.23)
$$\frac{dz}{dt} = \frac{dr}{dt}\frac{z}{r} + iz\frac{d\theta}{dt}.$$

The definition (6.19) of log is

$$\log z = \log r + i\theta;$$

differentiating this with respect to t we get

(6.24)
$$\frac{d}{dt}\log z = \frac{1}{r}\frac{dr}{dt} + i\frac{d\theta}{dt}.$$

Dividing (6.23) by z and, comparing the resulting expression with the above formula, we obtain

(6.25)
$$\frac{d}{dt}\log z = \frac{1}{z}\frac{dz}{dt}.$$

This looks suspiciously like an instance of the chain rule, with

$$\frac{d\log z}{dz} = \frac{1}{z}.$$

But this is of course nonsense, for we cannot differentiate with respect to a complex variable z. Or can we? For an answer, turn to any book on complex variables.

In conclusion we list the handful of functions which we have defined for complex inputs:

$$e^z, \qquad \log z, \qquad z^{1/m} \quad \text{for } m = 1, 2, \ldots .$$

Of course, we can build out of these, by the processes of addition, multiplication and composition, a bewildering array of new functions. In Exercise 6.7 we shall pursue this interesting subject.

EXERCISES

6.1 Differentiate the following functions of t:

(a) $\sqrt{i + t}$

(b) $\dfrac{1}{t - i} + \dfrac{1}{t + i}$

(c) $\exp\{it^2\}$

(d) $\log(t + i)$.

6.2 Define the functions C and S by

$$C(t) = \cos it, \qquad S(t) = -i\sin it.$$

(a) Show that

$$C' = S, \qquad S' = C,$$

(b) Express C and S in terms of the exponential function.

377

6.3 Using the identity

$$\frac{1}{t^2 + 1} = \frac{1}{2i}\left[\frac{1}{t - i} - \frac{1}{t + i}\right],$$

deduce that

$$\arctan t = \frac{1}{2i}\log\left(\frac{t - i}{t + i}\right).$$

6.4 (a) Prove that all complex numbers c which satisfy $e^c = 1$ are of the form $c = 2\pi ni$, n an integer.

(b) Let z be any complex number; prove that all complex numbers w which satisfy $e^w \doteq z$ are of the form

$$w = \log r + i\theta + 2\pi ni, \quad \text{where } r = |z|, \quad \theta = \arg z.$$

6.5 Prove that the series expansion (6.14) is valid for any complex number z. [*Hint*: Use $z = x + iy$, $e^z = e^x \cdot e^{iy}$ and (6.15).]

6.6 Show that the graph of the function $f(t) = 1/(t + i)$ lies on the circle of diameter 1 centered at $-i/2$.

6.7 In this rather long exercise we shall present, in the form of a series of problems, the rudiments of the notion of differentiability for functions $f(z)$ defined for complex values of z.

Definition. A complex valued function $f(z)$, defined for $z = a$ and for all complex values of z near a is called *differentiable at a* if the limit

(6.26)
$$\lim_{h \to 0} \frac{f(a + h) - f(a)}{h}$$

exists. The limit is called the *complex derivative* of f, and is denoted by $f'(a)$.

Our definition of derivative of functions with complex input *and* output is the same as in the real case, except that real numbers are replaced by complex numbers. Thus the limit relation (6.26) means that for any given m

(6.27)
$$\left|\frac{f(a + h) - f(a)}{h} - f'(a)\right| < 10^{-m}$$

for all nonzero *complex h* for which $|h|$ is small enough; how small depends on m.

(a) Show that the function $f(z) = z$ is differentiable.

(b) Show that the sum and product of two differentiable functions are differentiable, and that the usual rules for differentiating sums and products are valid.

(c) Show that every polynomial

$$p(z) = a_n z^n + a_{n-1}z^{n-1} + \cdots + a_1 z + a_0$$

is differentiable, and that p' is given by the usual formula.

378

(d) Show that the reciprocal of a differentiable, nonzero function is differentiable, and that the usual rule for differentiating the reciprocal of a function applies. Note that a consequence of the product rule and the reciprocal rule is that a rational function, i.e., a function r of the form

$$r(z) = \frac{p(z)}{q(z)},$$

p and q polynomials, is differentiable at every point where q is nonzero.

We now pose two slightly different versions of the chain rule for differentiating composite functions.

(e) Suppose f and g are complex valued functions of complex input, and that g is differentiable at a, and f differentiable at $b = g(a)$. Show that the composite $f \circ g$, defined as $f(g(z))$, is differentiable at a, and its derivative is given by the usual chain rule.

(f) Suppose $g(t)$ is a complex valued function of the *real* variable t, and that f is a complex valued function of *complex* inputs. Suppose g is differentiable at $t = a$ and f is differentiable at $b = g(a)$. Show that the composite $f(g(t))$ is a complex valued function of the real variable t, differentiable at $t = a$.

(g) Show that, if $g(t)$ is a differentiable complex valued function of a real variable t, then $[g(t)]^n$ is likewise differentiable, and

(6.28) $$(g^n)' = ng^{n-1}g'.$$

(h) Show that the complex valued function

$$f(z) = e^z$$

of the complex variable z is differentiable at $z = 0$, and that its derivative at $z = 0$ equals 1.

(i) Show, using the functional equation for the complex exponential function, that e^z is everywhere differentiable, and that

(6.29) $$(e^z)' = e^z.$$

(j) Let $g(t)$ be a differentiable complex valued function of the real variable t. Show that

(6.30) $$(e^g)' = e^g g'.$$

We now turn to the logarithmic function; in our discussion we have defined $\log z$ as a multivalued "function," see (6.20), which is unsuitable for forming the derivative. We must make a choice among the infinitely many possibilities. We restrict z to the right half of the complex plane, i.e., we take $z = x + iy$ with $x > 0$. For such z we set

(6.31)$_1$ $$r = |z|, \qquad \eta = \arctan\frac{y}{x}, \qquad \text{so} \quad -\frac{\pi}{2} \le \eta \le \frac{\pi}{2},$$

and define the *single valued* function

(6.31)$_2$ $$\log z = \log r + i\eta.$$

(k) Show that log z, as defined above, is differentiable at $z = 1$.

(l) Show that log z, as defined above, satisfies the functional equation for the log function. Using this functional equation, show that log z is differentiable at every point z in the right half plane, and that

(6.32)
$$(\log z)' = \frac{1}{z}.$$

(m) Let $g(t)$ be a complex valued function of the real variable t whose values have positive real part. Show that

(6.33)
$$(\log g(t))' = \frac{g'}{g}.$$

(n) Show that, if the real part of z is positive,

$$\exp\{\tfrac{1}{2}\log z\}$$

is a square root of z. Denoting this square root by $\sqrt{z}$, show that

(6.34)
$$(\sqrt{z})' = \frac{1}{2}\frac{1}{\sqrt{z}}.$$

(o) For any complex number c and complex z with positive real part, define z^c by

(6.35)
$$z^c = e^{c \log z}.$$

Show that

(6.36)
$$(z^c)' = cz^{c-1}.$$

(p) Show that no real valued function $f(z)$ with complex input z is differentiable in the sense of Definition (6.26).

6.8 Compute the complex integrals:

(a) $\displaystyle\int_0^1 e^{is}\, ds$

(b) $\displaystyle\int_0^{\pi/2} (\cos s + i \sin s)ds$

(c) $\displaystyle\int_1^2 \log(s - i)ds$

(d) $\displaystyle\int_0^1 (a + ib)ds, \quad a, b,$ constants.

7.7 Polar coordinates

We saw in Section 7.4 that there are two ways of describing a complex number. The first is in terms of its real and imaginary parts,

$$z = x + iy;$$

we might call this the *Cartesian form* of z. The second is in terms of its absolute value r and argument θ,

$$z = r(\cos \theta + i \sin \theta);$$

this is called the *polar form*. In this section we shall show briefly that the second kind of description, the polar description, is as useful in analytic geometry as it is for complex numbers.

Let P be a point in the plane, x and y its Cartesian coordinates in some coordinate system; i.e., x and y are the signed distances of P from the two coordinate axes, respectively.

The *polar coordinates* of P, r and θ, are defined as follows:
r, called the *radial distance*, is the distance of P to the origin.
θ, called the *polar angle*, is the radian measure of the angle made by the ray from the origin through P and the positive x-axis.

As Figure 7.22 indicates, θ measures the angle through which the positive x-axis has to be rotated in the counterclockwise direction until it coincides with the ray from the origin through P.

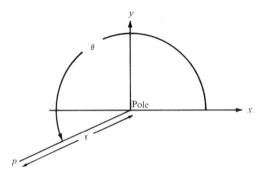

Figure 7.22   Polar coordinates; r, θ. The pole is at the origin, the polar axis coincides with the positive x-axis.

The formulas expressing rectangular coordinates in terms of polar coordinates are simple; they follow from the definition of cosine and sine:

(7.1) $$x = r \cos \theta, \qquad y = r \sin \theta.$$

Polar coordinates can be expressed just as simply in terms of rectangular ones:

(7.2) $$r = \sqrt{x^2 + y^2}$$

is the Pythagorean formula for distance. To express θ in terms of x and y we note that

(7.3) $$\tan \theta = \frac{y}{x}.$$

Polar coordinates are useful for describing pointsets in the plane. This is, of course, what rectangular coordinates are good for, but in many cases the polar description turns out to be simpler. The most striking example is the

381

circle of radius c around the origin. The equation describing it in rectangular coordinates is

$$x^2 + y^2 = c^2,$$

whereas the equation describing it in polar coordinates is

$$r = c,$$

clearly much simpler. Here are some further examples.

Example 1. The equation of a straight line in rectangular coordinates is

(7.4) $$ax + by + c = 0.$$

Using the relation (7.1) between rectangular and polar coordinates we can write (7.4) as

(7.5) $$r(a \cos \theta + b \sin \theta) + c = 0.$$

We see from Figure 7.23 that for any point (r, θ) on line l, the relation

(7.5)′ $$\frac{p}{r} = \cos(\theta - \alpha)$$

holds; α is the angle between the x-axis and the segment $0Q$ perpendicular to l, p is the length $0Q$. Equation (7.5)′ is almost as simple as (7.4). There is a relation between the angle α and the coefficients a, b in (7.4). To see this, use the addition formula for the cosine to write the above relation as

$$p = r \cos \alpha \cos \theta + r \sin \alpha \sin \theta.$$

Comparing this with (7.5), we can make the identification

$$a = \cos \alpha, \qquad b = \sin \alpha \qquad \text{and} \qquad p = -c,$$

provided $a^2 + b^2 = 1$ in (7.4). This can always be achieved by dividing (7.4) by $\sqrt{a^2 + b^2}$ and renaming the resulting coefficients. In this normalized equation, the constant term gives the distance from the origin to the line.

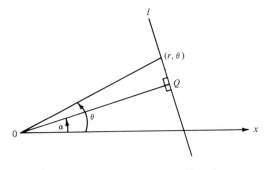

Figure 7.23 $\quad 0Q = p = r \cos(\theta - \alpha)$.

Example 2. The equation of the circle of radius c whose center is located at (a, b), in rectangular coordinates, is

$$(x - a)^2 + (y - b)^2 = c^2;$$

squaring we get, after a slight rearrangement,

$$x^2 + y^2 - 2xa - 2yb = c^2 - a^2 - b^2.$$

Substituting (7.1) for x and y we get

$$r^2 - 2ar \cos \theta - 2br \sin \theta = c^2 - a^2 - b^2.$$

Take the special case that the circle passes through the origin, which is the case when $a^2 + b^2 = c^2$. Then we can divide the above equation by r and write it in the simple form

$$r - 2a \cos \theta - 2b \sin \theta = 0.$$

Example 3. The polar equation of a conic can be derived from this definition of a conic in terms of its eccentricity e: The set of all points P such that the ratio of the distance from P to the focus to the distance from P to a fixed line has the constant value e. In Figure 7.24, the focus is at 0, the fixed vertical line,

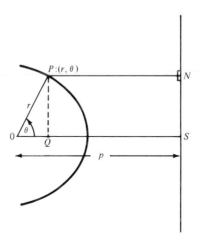

Figure 7.24

called directrix, has polar equation $p = R \cos \phi$ and intersects the horizontal axis at point S. The eccentricity is the constant ratio

$$e = \frac{P0}{PN},$$

where N is the point where the horizontal line through P intersects the directrix. The foot of the perpendicular from P to $0S$ is denoted by Q. Now $P0 = r$ and $PN = QS = 0S - 0Q = p - r \cos \theta$; so for all $P:(r, \theta)$ on the conic,

$$e = \frac{r}{p - r \cos \theta}.$$

Solving this for r, we obtain

(7.6)
$$r = \frac{ep}{1 + e \cos \theta}.$$

If we had taken the focus at 0, but the corresponding directrix to the left of the focus, the same derivation would have led to the equation

(7.6)'
$$r = \frac{ep}{1 - e \cos \theta};$$

(7.6)' follows from (7.6) also by the substitution $\pi - \theta$ for θ.

Example 4. To transform the cartesian equation of the parabola

$$y^2 = 1 - 2x$$

to polar coordinates, we add x^2 to both sides and use (7.1) obtaining

$$r^2 = x^2 + y^2 = 1 - 2x + x^2 = (1 - x)^2 = (1 - r \cos \theta)^2.$$

Solving for r, we obtain the form (7.6) with $e = p = 1$.

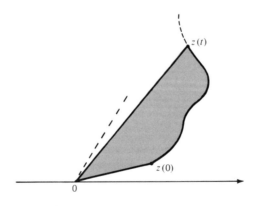

Figure 7.25a

We shall use Formula (5.12) to study the area between a curve $z(t)$ in the complex plane and the segments connecting its endpoints to the origin. Suppose as t increases, the point $z(t)$ moves in a counterclockwise direction

384

about the origin; see Figure 7.25a. We denote this area by $A(t)$ and claim that, if $z(t)$ is differentiable,

(7.7)
$$\frac{d}{dt} A(t) = \frac{1}{2} \text{Im}\left\{\bar{z}(t) \frac{d}{dt} z(t)\right\}.$$

PROOF. The difference $A(t + h) - A(t)$ is the area of the sliver shown in Figure 7.25b. This region is almost a triangle, except that its short side is curved, not straight. If we replace the segment of the curve by a straight line we obtain a triangle whose area differs from $A(t + h) - A(t)$ by the area A_h of the region contained between the curve and its secant. As Figure 7.25c shows, the curve stays in a rectangle whose length is k,

$$k = |z(t + h) - z(t)|$$

and whose width is d, the *maximum* oscillation of the curve in the direction perpendicular to the secant during the interval $[t, t + h]$. It follows that

$$A_h \leq kd.$$

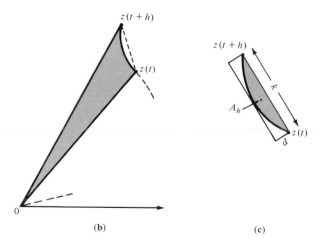

(b)   (c)

Figures 7.25b and 7.25c

Since z is continuous, d tends to zero as h tends to zero. Since z is differentiable,

$$\frac{k}{h} = \frac{|z(t + h) - z(t)|}{h}$$

is bounded by, say, M:

$$k \leq Mh.$$

Therefore $A_h \leq Mhd$, so we can write

$$A_h = hs(h),$$

where s denotes a quantity that tends to 0 as h tends to 0. So

$$A(t + h) - A(t) = A(z(t), z(t + h)) + hs(h).$$

Using Formula (5.12) we get

(7.8) $$A(t + h) - A(t) = \tfrac{1}{2} \operatorname{Im}\{\bar{z}(t)z(t + h)\} + hs(h).$$

Since the absolute value of a complex number is real, we subtract $\tfrac{1}{2} \operatorname{Im}\{|z(t)|^2\}$ $= \tfrac{1}{2} \operatorname{Im}\{\bar{z}(t)z(t)\} = 0$ from the right side of Equation (7.8) and obtain

$$A(t + h) - A(t) = \tfrac{1}{2} \operatorname{Im}\{\bar{z}(t)[z(t + h) - z(t)]\} + hs(h).$$

Dividing by h, we obtain

$$\frac{A(t + h) - A(t)}{h} = \frac{1}{2} \operatorname{Im}\left\{\bar{z}(t)\frac{z(t + h) - z(t)}{h}\right\} + s(h).$$

As h tends to zero, the above equation tends to Equation (7.7). □

Differentiating (7.7) leads to an interesting formula for the second derivative of A. Denoting differentiation with respect to t by prime, we get

(7.9)
$$\begin{aligned}
A'' &= \tfrac{1}{2} \operatorname{Im}\{\bar{z}z'\}' = \tfrac{1}{2} \operatorname{Im}\{\bar{z}'z' + \bar{z}z''\} \\
&= \tfrac{1}{2} \operatorname{Im}\{|z'|^2\} + \tfrac{1}{2} \operatorname{Im}\{\bar{z}z''\} = \tfrac{1}{2} \operatorname{Im}\{\bar{z}z''\}.
\end{aligned}$$

We shall have occasion to use the results (5.12), (7.7), and (7.9) in the discussion of mechanics in Section 7.8.

EXERCISES

7.1 Rewrite the following equations in polar coordinates:
(a) $x^2 - y = 0$ (b) $xy = 1$
(c) $x + 2y - 3 = 0$ (d) $x^2 - y^2 = 1.$

7.2 Rewrite the following equations in rectangular coordinates:

(a) $r^2 \cos 2\theta = 1$ (b) $r^2 = r \cos \theta$ (c) $e^r = \cos \theta.$

7.3 (a) Write the equation of the parabola in Exercise 7.1(a) in polar coordinates, choosing the focus as origin of the coordinate system.
(b) Write the equation of the hyperbola in Exercise 7.1(b) in polar coordinates, choosing one of the foci as origin of the coordinate system.

7.4 Write the equation of a straight line through the origin in polar coordinates.

7.5 Suppose the eccentricity e is less than 1, so that the conic (7.6) is an ellipse. Use the relation $e = c/a$, where a is the semimajor axis and c is half the distance between the foci of the ellipse, to transform (7.6) to its Cartesian form

$$\frac{(x + c)^2}{a^2} + \frac{y^2}{b^2} = 1.$$

After a coordinate translation, the standard form

$$\frac{x^2}{a^2} + \frac{y^2}{b^2} = 1$$

can be achieved.

7.6 Consider the graph, in polar coordinates, of the function

$$r = f(\theta), \qquad 0 < \theta < 2\pi.$$

Denote by $A(\theta_1, \theta_2)$ the area between the curve and the rays from the origin making angles θ_1 and θ_2 with the horizontal axis, see Figure 7.26a.

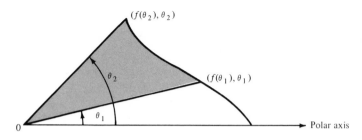

Figure 7.26a

(a) Show that

$$\frac{(\theta_2 - \theta_1)}{2} m^2 \le A(\theta_1, \theta_2) \le \frac{(\theta_2 - \theta_1)}{2} M^2,$$

where m and M denote the minimum and maximum of $f(\theta)$ for $\theta_1 \le \theta \le \theta_2$.
[*Hint*: Look at the circular sectors contained in, and containing $A(\theta_1, \theta_2)$.]

(b) Show that

$$A(\theta_1, \theta_2) + A(\theta_2, \theta_3) = A(\theta_1, \theta_3).$$

(c) Deduce from (a) and (b) that

(7.10) $$A(\theta_1, \theta_2) = \frac{1}{2} \int_{\theta_1}^{\theta_2} f^2(\theta)d\theta.$$

[*Hint*: Recall the two basic properties of integrals from Chapter 4.]

7.7 (a) Verify that a straight line at distance p from the origin and whose normal makes an angle α with the horizontal has cartesian equation

$$(\cos \alpha)x + (\sin \alpha)y = p$$

and polar equation

(7.11) $$r = f(\theta) = \frac{p}{\cos(\theta - \alpha)}.$$

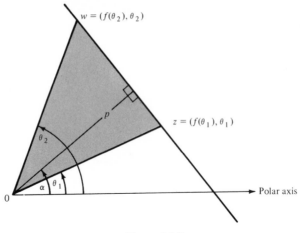

Figure 7.26b

(b) Let z and w be the points $(f(\theta_1), \theta_1)$ and $(f(\theta_2), \theta_2)$ on the line (7.11), see Figure 7.26b. Calculate the area $A(\theta_1, \theta_2)$ of triangle $0zw$ using Formula (7.10), and obtain

(7.12) $\qquad A(\theta_1, \theta_2) = \tfrac{1}{2}p^2[\tan(\theta_2 - \alpha) - \tan(\theta_1 - \alpha)],$

and interpret this result as $\tfrac{1}{2}$ base $\times$ altitude, with p as altitude.

7.8 Write the complex numbers z and w in their polar forms

$$z = r_1(\cos \theta_1 + i \sin \theta_1), \quad w = r_2(\cos \theta_2 + i \sin \theta_2).$$

(a) Verify that Formula (5.12) yields the area

(7.13) $\qquad A(z, w) = \tfrac{1}{2}r_1 r_2 \sin(\theta_2 - \theta_1).$

(b) For points $z = (f(\theta_1), \theta_1)$, $w = (f(\theta_2), \theta_2)$ on the line whose polar equation is

$$r = f(\theta) = \frac{p}{\cos(\theta - \alpha)},$$

see Equation (7.11) in Exercise 7.7, we found that the area of triangle $0zw$ is given by Equation (7.12). Substitute the values $f(\theta_1)$ and $f(\theta_2)$ for r_1 and r_2, respectively, and transform (7.13) to Equation (7.12). [*Hint*: Write $\sin(\theta_2 - \theta_1) = \sin[(\theta_2 - \alpha) + (\alpha - \theta_2)]$ and use the addition formulas.]

7.8 Two-dimensional mechanics

In Section 3.6 we have discussed the basic concepts of one-dimensional mechanics; particle, mass, position, velocity, acceleration, and force. A particle moving along a line under the influence of a force satisfies Newton's law,

$$\text{mass} \times \text{acceleration} = \text{force}.$$

In mathematical notation

(8.1)
$$mx'' = f,$$

where $x(t)$ is the position of a particle on the line at time t, and prime denotes differentiation with respect to t. Force is a real valued quantity; when it is positive, it is directed forward on the x-axis, when negative, backward.

In Volume II we shall present a thorough-going discussion of mechanics in two and three dimensions, employing the concept of vector. In this section we give a brief indication how mechanics of particles in two dimensions can be treated in terms of complex valued functions.

Our indication is brief indeed: *everything is the same as in one dimension except that position, velocity, acceleration and force are complex valued functions.* That is, the *position* of a particle moving in the plane is a complex valued function of time $z(t)$. Velocity v, also complex valued, is the derivative of position with respect to t:

(8.2)
$$v = \frac{dz}{dt}.$$

Acceleration a, also complex valued, is the derivative of velocity:

(8.3)
$$a = \frac{dv}{dt}.$$

Newton's law of motion is

(8.4)
$$mz'' = f,$$

where the mass m is a positive real quantity, and f is *force*, a complex valued function. The absolute value of f is the *magnitude* of the force, and the direction from the origin 0 to f is its *direction*. We give some examples:

Example 1. *A particle moving under zero force.* Equation (8.4) is

$$mz'' = 0.$$

All solutions of this equation are of the form

$$z(t) = z_0 + v_0 t.$$

This shows that a particle in the absence of a force travels in a straight line at constant speed $|v_0|$.

Example 2. *A constant force,* $f = f_0$. Equation (8.4) is

$$mz'' = f_0.$$

All solutions are of the form

$$z(t) = z_0 + v_0 t + \frac{f_0}{2m} t^2.$$

Example 3. *A linear force $f = kz$, k some complex constant.* Equation (8.4) is

$$(8.5) \qquad\qquad mz'' = kz.$$

The reader can verify that all functions of the form

$$(8.6) \qquad z(t) = a\, \exp\left\{\sqrt{\frac{k}{m}}\,t\right\} + b\, \exp\left\{-\sqrt{\frac{k}{m}}\,t\right\}$$

are solutions of this equation.

Now consider any force function $f = f(z)$ which has the property that $f(z)$ is always parallel to z;

$$(8.7) \qquad\qquad f(z) = a(z)z, \quad a \text{ real valued.}$$

We claim that *the path of a particle moving under such a force sweeps out equal areas during equal time intervals*; that is, the *area $A(t)$* (see Figure 7.27) satisfies

$$(8.8) \qquad\qquad A(t) = A_0 + kt.$$

PROOF. According to Formula (7.9)

$$A'' = \tfrac{1}{2}\,\text{Im}\{\bar{z}z''\}.$$

Since z satisfies Newton's law (8.4), $z'' = f/m$, and so

$$A'' = \frac{1}{2m}\,\text{Im}\{\bar{z}f\}.$$

By assumption, f is of form (8.7); therefore

$$A'' = \frac{1}{2m}\,\text{Im}\{\bar{z}az\} = \frac{1}{2m}\,\text{Im}\{a|z|^2\} = 0.$$

This shows that the second derivative of A is zero; we have shown in Examples 6 and 7 of Section 3.3 that such a function is linear; this completes the proof of (8.8). $\qquad\square$

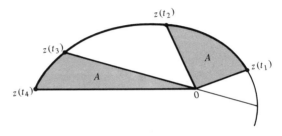

Figure 7.27 $A = k(t_2 - t_1) = k(t_4 - t_3)$.

We now introduce a particular force of the form (8.7),

$$(8.9) \qquad f(z) = -\frac{mgz}{|z|^3}, \qquad g > 0.$$

The force law (8.9) is universally called the inverse square law, because the magnitude of this force,

$$(8.10) \qquad |f(z)| = \frac{mg|z|}{|z|^3} = \frac{mg}{|z|^2}$$

is proportional to the inverse of the square of the distance $|z|$ of the point z to the origin. The constant of proportionality is the product of the mass m on which the force acts, and another positive constant g. According to Newton's law of universal gravitation, (8.9) is the force exerted by a mass placed at the origin on a particle of mass m located at z. The constant g itself is proportional to the mass M placed at the origin:

$$g = MG.$$

Note that this force is *attractive*, i.e., the force draws the particle at z towards the origin.

Newton's law of motion (8.4) for a particle in the gravitational force field (8.9) of a mass placed at the origin is,

$$(8.11) \qquad z'' = -\frac{gz}{|z|^3}.$$

The force exerted on a particle of mass m located at z by a mass M located at a is

$$(8.12) \qquad \frac{-mMG(z - a)}{|z - a|^3}.$$

Suppose n masses $M_1, \ldots, M_n$ are located at fixed positions $a_1, \ldots, a_n$. The force exerted by each on a particle of mass m at z is given by (8.12). According to Newton, when a number of forces are acting on a particle, *the effective force is their sum*. So the effective force of all masses $M_1, \ldots, M_n$ on the particle at z is m times the acceleration

$$(8.13) \qquad z'' = -G \sum_{j=1}^{n} \frac{M_j(z - a_j)}{|z - a_j|^3}.$$

We return now to Equation (8.11) governing a particle moving under the gravitational force of a single mass placed at the origin. Multiply Equation (8.11) by $\bar{z}'$; we get

$$\bar{z}'z'' = -g\frac{z\bar{z}'}{|z|^3}.$$

The real parts of both sides are equal:

(8.14)
$$\frac{\bar{z}'z'' + z'\bar{z}''}{2} = \frac{-g}{2|z|^3}(z\bar{z}' + \bar{z}z').$$

The left side is the derivative of

(8.15)
$$\tfrac{1}{2}\bar{z}'z' = \tfrac{1}{2}|z'|^2.$$

We claim that the right side is the derivative of

(8.16)
$$\frac{g}{|z|}.$$

For

(8.17)
$$\frac{d}{dt}\frac{1}{|z|} = -\frac{1}{|z|^2}\frac{d|z|}{dt},$$

and

(8.18)
$$\frac{d|z|}{dt} = \frac{d(z\bar{z})^{1/2}}{dt} = \frac{1}{2}\frac{1}{|z|}(z'\bar{z} + z\bar{z}').$$

Substituting (8.18) into (8.17), we get

$$\frac{d}{dt}\frac{1}{|z|} = -\frac{1}{2}\frac{1}{|z|^3}(z'\bar{z} + z\bar{z}').$$

Since (8.15) and (8.16) are the derivatives of the left and right sides of (8.14), we conclude that their difference is a constant

(8.19)
$$\frac{1}{2}|z'|^2 - \frac{g}{|z|} = E.$$

The quantity $\tfrac{1}{2}|z'|^2$ is the *kinetic energy* of the particle of unit mass at z; $-g/|z|$ is its *potential energy* in the gravitational field of the mass placed at the origin. Their sum is the *total mechanical energy* of the particle. Relation (8.19) expresses the constancy of this total energy throughout the motion, an instance of a *law of conservation of energy*.

In Exercise 8.1 the reader is asked to deduce from Equation (8.13) the following conservation of energy law for a particle moving in the gravitational field of n masses:

(8.20)
$$\frac{1}{2}|z'|^2 - G\sum_{j=1}^{n}\frac{M_j}{|z - a_j|} = E, \quad E \text{ constant.}$$

In Volume II we shall present a very general form of the law of conservation of energy for a particle moving in a force field.

392

A very interesting and useful consequence of the energy conservation law (8.20) is this observation: suppose that *the total energy of a particle in the gravitational field of n masses is negative; then the particle stays in a bounded portion of space.* For, suppose that the particle got further and further away, i.e., $|z|$ were not bounded. When $|z|$ is large, the potential energy $-G \sum M_j/|z - a_j|$ is small; since the kinetic energy is non-negative, for z large enough we would violate (8.20), the law of conservation of energy.

We return to Equation (8.11) governing the motion of a particle in the gravitational field of a single mass at the origin. It turns out that these motions can be determined explicitly, they are conics, with one focus located at the origin. In Exercise 8.2 the reader will be led to derive this result. Suppose the total energy of the particle is negative; then according to the argument presented above, the path of the particle is confined to a bounded portion of the plane. Since the only conic sections which are confined to a bounded portion of the plane are ellipses,* it follows that the path of a particle with negative total energy is an *ellipse*. This is the celebrated *second law of Kepler on the orbits of planets around the sun.* Kepler's first law is that equal areas are swept out during equal times; we saw earlier that this law is true for any force of the form (8.7), not just the inverse square law. In Exercise 8.3 we discuss Kepler's third law.

EXERCISES

8.1 From (8.13) we deduce that the force F of gravitational attraction for a particle of mass m located at z surrounded by n masses M_j located at $a_j, j = 1, 2, \ldots, n$, satisfies

(8.13)′
$$\frac{F}{m} = -G \sum_{j=1}^{n} \frac{M_j(z - a_j)}{|z - a_j|^3}.$$

Show that the total energy E of a particle moving in a gravitational field is (8.20):

$$E = \frac{1}{2}|z'|^2 - G \sum_{j=1}^{n} \frac{M_j}{|z - a_j|}.$$

[*Hint*: Compare (8.13) with (8.13)′ and then multiply (8.13)′ by $\bar{z}'$.]

8.2 In this exercise, you will deduce Kepler's second law from Newton's gravitational law.
(a) Use the polar form $z = fe^{i\theta}$ for the complex position $z(t)$ of a particle and compute $z''(t)/z(t)$, considering f, θ to be functions of time t.
(b) Divide Newton's gravitational law (8.11),

$$z'' = -\frac{gz}{|z|^3}, \quad g \text{ constant},$$

* Circles are included, since they are special cases of ellipses whose major and minor axes are equal.

by z and, noting that the ratio is real, equate it to the real part found in (a), obtaining

(8.21)
$$\frac{f''}{f} - \theta'^2 = -\frac{g}{|z|^3} = \frac{g}{f^3}.$$

Equate the imaginary parts of the ratio z''/z found in (a) and (b) and obtain

(8.22)
$$\text{Im}\left\{\frac{z''}{z}\right\} = 0 = \frac{2f'}{f}\theta' + \theta''.$$

(c) Show that this is equivalent to

(8.23)
$$(\tfrac{1}{2}f^2\theta')' = 0,$$

and deduce from the above relation Kepler's first law of equal areas in equal time:

(8.24)
$$\tfrac{1}{2}f^2\theta' = k.$$

(d) Now consider the composite $f = f(\theta(t))$; use the chain rule, $f' = (df/d\theta)\theta'$, together with (8.24) to obtain $f' = (2k/f^2)df/d\theta$. Again apply the chain rule to verify

(8.25)
$$f'' = -\frac{4k^2}{f^2}\frac{d^2 1/f}{d\theta^2}.$$

(e) Eliminate the second time derivative, f'', between (8.21) and (8.25) and use (8.24) to eliminate θ' from the resulting expression to obtain the differential equation

(8.26)
$$\frac{d^2(1/f - g/4k^2)}{d\theta^2} = -(1/f - g/4k^2).$$

(f) Verify that

(8.27)
$$\frac{1}{f} - \frac{g}{4k^2} = A\cos\theta + B\sin\theta, \qquad A \text{ and } B \text{ any constants,}$$

is a solution of (8.26).

(g) Set $B = 0$ in (8.27) and solve for f; show that the solution can be written in the form (7.6):

(8.28)
$$f(\theta) = \frac{ep}{1 + e\cos\theta}$$

where

(8.29)
$$ep = \frac{4k^2}{g}, \qquad e = \frac{4k^2 A}{g}$$

(h) Show that the semi-major axis d of an ellipse of form (8.28) is

$$d = \frac{ep}{1 - e^2}.$$

[*Hint*: Since $f(\theta)$ is the distance from the focus, the major axis is the sum of the maximum and minimum of $f(\theta)$.]

For ep given by (8.29) this becomes

(8.30)
$$d = \frac{4k^2/g}{1 - e^2}$$

8.3 In this problem we discuss

Kepler's third law. *The square of the periods T, of revolution of any two planets are proportional to the cubes of the semimajor axes d, of their respective orbits. In symbols,*

$$\frac{d^3}{T^2} = \text{constant}$$

for all planets.

(a) Using the cartesian form for the ellipse with semimajor axis d and semiminor axis b,

$$\frac{x^2}{d^2} + \frac{y^2}{b^2} = 1,$$

find that the area is πdb. [*Hint*: See Example 3 of Section 2 and Exercise 2.10.]

(b) Show that the constant k in Kepler's first law is

$$k = \frac{\pi db}{T} = \frac{\pi d}{T} d\sqrt{1 - e^2}.$$

(c) Using the above result and (8.30), show that the constant in Kepler's third law is given in terms of the constant g in Newton's inverse square law by

(8.31)
$$\frac{d^3}{T^2} = \frac{g}{4\pi^2}.$$

8.4 (a) To convince yourself that the constant g in (8.31) does not depend upon extraneous factors such as the relative volumes of orbiting bodies, complete the following table by computing d^3/T^2 for each planet.

	Period of revolution T (days)	Semimajor axes d(km)
Mercury	88.027	$5.790 \cdot 10^7$
Venus	224.7	$1.08241 \cdot 10^8$
Earth	365.256	$1.496 \cdot 10^8$
Moon	27.322*	$3.84403 \cdot 10^5$
Mars	687.047	$2.279 \cdot 10^8$
Jupiter	11.86 years	$7.783 \cdot 10^8$
Saturn	29.46 years	$1.427 \cdot 10^9$
Uranus	84.01 years	$2.8696 \cdot 10^9$
Neptune	164.8 years	$4.496 \cdot 10^9$
Pluto	247.7 years	$5.900 \cdot 10^9$

* Sideral month, see part (b) of this exercise.

(b) Compute the earth–moon constant d^3/T^2 considering the earth fixed with the moon in orbit with a period equal to the sideral month (the synodic month, 29.5 days from full moon to full moon, includes the extra time required for the moon to "catch up" to the same relative earth–moon position because of the motion of the earth in orbit around the sun). If you are now convinced that the acceleration (force per unit mass) experienced by two bodies under mutual gravitational pull is independent, say, of the volume of the bodies, you are probably led to the following natural question: Is the acceleration experienced by a person jumping on the surface of the earth governed by the same law of attraction (8.11) which governs the motion of the planets? In part (c) we answer this question.

(c) The average value of the gravitational constant, as measured on the earth's surface, is roughly 32.17 ft/sec^2 = 9.8054 m/sec^2. Compute the magnitude of the acceleration experienced by the *moon* due to earth's pull. Use the radius of the earth $r = 6.3784 \cdot 10^6$ meters and the distance to the moon $R = 384.4 \cdot 10^6$ meters. Compare your answer (9.89086 m/sec^2) with the accepted value of the gravitational constant for earth.

The conclusion one might reasonably draw from this and the previous problem is that Newton's law is a *universal law* describing how all bodies, human or otherwise, pull on each other.

8.5 In this problem we wish to compute the curve $z(t)$ in the complex plane which describes the motion of a particle trapped in a gravitational field.

Introduce the complex valued velocity $w = u + iv$ in the universal gravitational law (8.13):

(8.32) $$w'(t) = z''(t) = -G \sum_1^n \frac{M_j(z - a_j)}{|z - a_j|^3}.$$

We have seen in Section 3.8 that the derivative (8.32) may be well approximated for h small enough by a central difference quotient, i.e., by

$$\frac{w(t + h/2) - w(t - h/2)}{h} = w'(t) + \text{small quantity.}$$

Rewrite this, after multiplying by h and neglecting the term $h \times$ (small quantity), as

(8.33) $$w\left(t + \frac{h}{2}\right) = w\left(t - \frac{h}{2}\right) + hw'(t).$$

In order to solve for the complex velocity at the advanced time $t + h/2$, $w'(t)$ must be evaluated from (8.32). This requires a knowledge of the complex position $z(t)$; approximate $z'(t) = w(t)$ by the central difference quotient

$$\frac{z(t + h) - z(t)}{h} = w\left(t + \frac{h}{2}\right) + \text{small quantity,}$$

then

(8.34)
$$z(t + h) = z(t) + hw\left(t + \frac{h}{2}\right)$$

if we neglect the term $h \times$ small quantity.

The quantities $z(h)$, $w'(h)$, $w(3/2h)$, $z(2h)$, etc., may be computed from (8.34), (8.32), (8.33), (8.34), etc.

In order to get the computation started use, for one time only, the forward difference approximation to the acceleration at $t = 0$:

(8.33)′
$$\frac{w(h/2) - w(0)}{h/2} = w'(0).$$

(a) Using the method outlined above write a computer program using complex arithmetic and compute the orbit of a particle of mass $m = 1$, with initial conditions

$$z(0) = 0.5 + i0, \qquad w(0) = 0 + i1.75,$$

moving about a body with mass $M = 1 = G$, located at $a = 0 + i0$. Use a step size $h = 0.01$, printing the position $z(nh)$, $n = 10, 20, \ldots, 1000$. Check the results to see if the orbit satisfies Kepler's first law. The orbit shown in Figure 7.28 was computed by this method.

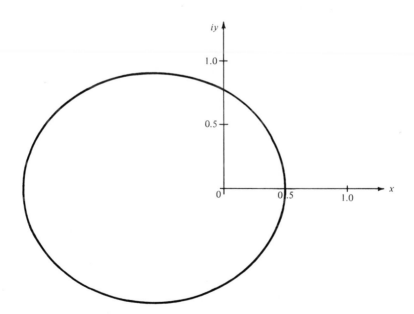

Figure 7.28   Orbital motion about sun at $0 + i0$. Initial conditions $z(0) = 0.5 + i0$, $w(0) = 0 + i1.75$.

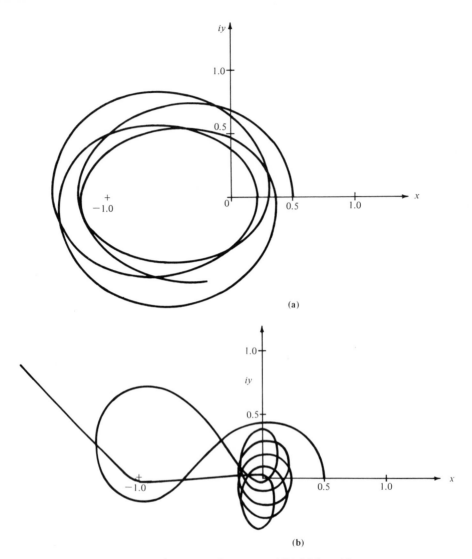

(a)

(b)

Figure 7.29 Planetary motions around two suns. All initial positions are at $z(0) = 0.5 + i0$. Paths vary with initial velocity, $w(0) = 0 + iv$, where v lies in the interval $1.75/\sqrt{2} < v \leq 1.75$.

(b) Use the above method to compute the orbit of a particle attracted by the pair of stars located at

$$a_1 = 0 + i0, \qquad a_2 = -1 + i0$$

in the complex plane. The particle has initial conditions

$$z(0) = 0.5 + i0, \qquad w(0) = 0 + i1.291884090.$$

What is the shape of this orbit? Does it satisfy Kepler's first law? Compare your results with the orbits computed by this method and shown in Figure 7.29.

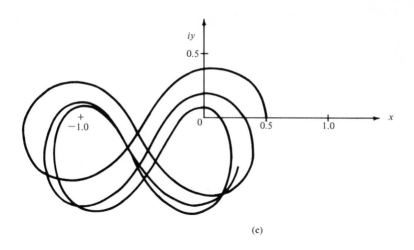

(c)

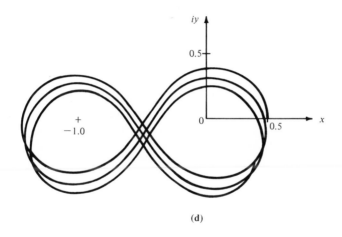

(d)

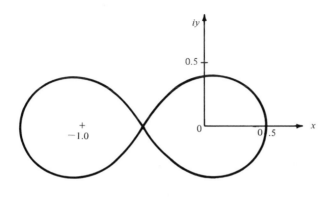

(e)

8 Vibrations

Most people realize that *sound*—its generation, transmission and perception—is vibration. For this reason alone it stands as a very important subject. But vibrations are more general and pervasive than mere sound, and constitute one of the fundamental phenomena of physics. The reason is the *mechanical stability* of everyday objects from bells and horns to the basic constituents of matter. Mechanical stability means that when an object is distorted by an outside force, it springs back into its original shape when released. This is accomplished by *restoring forces* inherent in any object worthy of the name. Restoring forces work in a peculiar way: they not only bring the object back to its original shape, but tend to overcorrect and distort it in the opposite direction. This is again overcorrected, and so ad infinitum, leading to a vibration around equilibrium.

In this chapter we shall explain this process in detail in exceedingly simple situations. The first six sections are devoted to mechanical systems, the seventh to so-called self-induced oscillations. Section 8.8 deals briefly with electric circuits.

8.1 The differential equation governing vibrations of a simple mechanical system

In Section 3.6 we stated the fundamental concepts of one-dimensional mechanics, i.e., the mechanics of particles moving along a straight line under the influence of forces. These concepts are: *particle, mass, position, velocity, acceleration, and force.*

Position of the particle along a line is specified by a single real number x. The position of the particle changes in time, so it is a *function* of the time t. The derivative of position with respect to t is the velocity of the particle,

400

denoted usually as $v(t)$. The derivative of velocity is called *acceleration*, and is denoted by $a(t)$:

$$x' = v, \qquad v' = x'' = a.$$

The mass of the particle, denoted by m, does not change throughout the motion.

Newton's law of motion says that

(1.1) $$f = ma$$

where f is the total force acting on the particle, m is the mass and a the acceleration. To put teeth into Newton's law, we have to be able to calculate the total force acting on the particle. According to Newton, the total force acting on a particle is the sum of all the various forces acting on it. In this section we shall deal with two kinds of forces: *restoring force* and *friction force*. We shall describe them in the following specific context.

Imagine a piece of elastic string (rubber band, elastic wire) placed in a vertical position with its endpoints fastened and a mass attached to its middle; see Figure 8.1a. In this position the mass is at rest. Now displace the

Figure 8.1 (a) Position of static equilibrium; (b) position of displaced rubber band.

mass to one side, as shown in Figure 8.1b. In this position the elastic string exerts a force on the mass. It is clear, to anyone who ever shot paper clips with a rubber band, that

(i) *this force acts in the direction opposite to the displacement, tending to restore the mass to its predisplaced position.*
(ii) *the greater the magnitude of the displacement, the greater the magnitude of this force.*

A force with these two properties is called a *restoring force*.

401

We turn next to describing the force of friction. Friction can be caused by various mechanisms, one of which is air resistance. As anyone who has ever bicycled at high speed knows,

(i) *the force of air resistance acts in the direction opposite to the direction of motion;*
(ii) *the greater the velocity, the greater the force of resistance.*

Any force with these two properties is called a *friction force.*

In order to turn the verbal descriptions of these two kinds of forces into mathematical descriptions, we denote a restoring force by f_{re}, which is to be regarded as a function of the position x. Properties (i) and (ii) can be expressed as follows in the language of functions:

(1.2)$_i$
$$f_{re}(x) \begin{cases} <0 & \text{for } x > 0 \\ =0 & \text{for } x = 0 \\ >0 & \text{for } x < 0, \end{cases}$$

and

(1.2)$_{ii}$ $\qquad\qquad\qquad f_{re}(x)$ is a decreasing function of x.

Figure 8.2 illustrates a function with these properties.

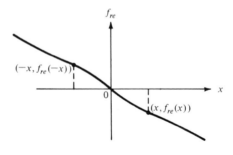

Figure 8.2

Many restoring forces, such as the one exerted by a rubber band, have yet a third property, symmetry:

(iii) *Displacements by the same magnitude but in opposite directions generate restoring forces which are equal in magnitude but opposite in direction.*

This property of $f_{re}(x)$ can be expressed in the following way:
(iii) f_{re} *is an odd function of x,* i.e.,

(1.2)$_{iii}$ $\qquad\qquad\qquad f_{re}(-x) = -f_{re}(x).$

402

We turn now to friction force, which we shall denote as f_{fr} and which we shall regard as a function of velocity. The properties of a friction force can be expressed as follows:

(1.3)$_i$
$$f_{fr}(v) = \begin{cases} <0 \text{ for } v > 0 \\ =0 \text{ for } v = 0 \\ >0 \text{ for } v < 0. \end{cases}$$

(1.3)$_{ii}$ $\qquad\qquad\qquad f_{fr}(v)$ is a decreasing function of v.

The graph of a typical friction force is shown in Figure 8.3. This graph displays yet another property common to most friction forces, their symmetry:

(iii) *The magnitude of the force f_{fr} depends only on the magnitude of the velocity.*

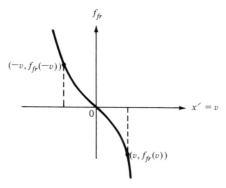

Figure 8.3

In terms of the function f_{fr} this, together with (1.3)$_{ii}$, means f_{fr} is an odd function, i.e.,

(1.3)$_{iii}$ $\qquad\qquad\qquad f_{fr}(-v) = -f_{fr}(v).$

The total force f is the sum of the individual forces,

$$f = f_{re} + f_{fr};$$

the additivity of forces follows from experimental fact 1 of Section 3.6 and Newton's law of motion (1.1). With this decomposition of force, (1.1) may be written as

(1.4) $\qquad\qquad\qquad ma - f_{re}(x) - f_{fr}(v) = 0.$

Since v and a are the first and second derivatives of x respectively, (1.4) is a *differential equation*

(1.4)' $\qquad\qquad\qquad mx'' - f_{fr}(x') - f_{re}(x) = 0$

for x as function of t. Solutions of this differential equation describe all possible motions of a particle driven by a restoring force and restrained by a friction force.

8.2 Dissipation and conservation of energy

This section will be devoted to the mathematics of extracting information about solutions of (1.4). There are two sets of information that can be extracted: qualitative and quantitative.

We start with a trick. Multiply Equation (1.4) by v, obtaining

$$(2.1) \qquad mva - vf_{re}(x) - vf_{fr}(v) = 0.$$

We can say something pertinent about each of the three terms in (2.1). Since, according to (1.3)$_i$, the sign of $f_{fr}(v)$ is opposite to that of v, it follows that $-vf_{fr}(v)$, the third term in (2.1), is positive, except when $v = 0$. By dropping this positive term from (2.1) we convert the equality into the inequality

$$(2.2) \qquad mva - vf_{re}(x) \le 0.$$

Recalling that a, the acceleration, is the derivative v' of v, we can rewrite the first term in (2.2) as mvv'. We recognize this as the derivative of $mv^2/2$,

$$(2.3) \qquad mva = \frac{d}{dt}(\tfrac{1}{2}mv^2).$$

Recalling that v is x', the derivative of x, we can rewrite the second term of (2.2) as $-x'f_{re}(x)$.

Let us introduce the function $p(b)$ as the integral of $-f_{re}$,

$$(2.4) \qquad p(b) = -\int_0^b f_{re}(x)\,dx.$$

By the fundamental theorem of calculus, the derivative of p is $-f_{re}$,

$$(2.5) \qquad \frac{d}{dx}p(x) = -f_{re}(x),$$

and obviously, from (2.4),

$$(2.6) \qquad p(0) = 0.$$

Using the chain rule, we can now express the second term in (2.2), $-x'f_{re}(x)$, as the derivative of $p(x(t))$ with respect to t:

$$(2.7) \qquad \frac{d}{dt}p(x(t)) = -vf_{re}(x).$$

Substituting expressions (2.3) and (2.7) for the first and second terms in (2.2), we obtain

$$(2.8) \qquad \frac{d}{dt}[\tfrac{1}{2}mv^2 + p(x)] \le 0.$$

According to the ·monotonicity criterion, a function whose derivative is ≤0 is decreasing. We use "decreasing" to mean "not increasing." So we conclude that the function

$$\tfrac{1}{2}mv^2 + p(x)$$

decreases with time. This function, and both terms appearing in it, have physical meaning: The quantity $mv^2/2$ is called *kinetic energy*, and the quantity $p(x)$ is called *potential energy*. The sum of kinetic and potential energies is called the *total energy*. In this terminology the result we have derived is the

Law of decrease of energy. *The total energy of a particle moving under the influence of a restoring force and a friction force decreases with time.*

Suppose there is no friction force, i.e., f_{fr} is zero. Then inequality (2.8), which was derived from Equation (2.1) by neglecting the term involving f_{fr}, becomes an equality:

(2.8)′ $$\frac{d}{dt}\left[\tfrac{1}{2}mv^2 + p(x)\right] = 0.$$

A function whose derivative is zero for all t is constant; so we have proved the

Law of conservation of energy. *In the absence of friction, the total energy of a particle moving under the influence of a restoring force does not change with time.*

Observe that the only property of the restoring force we have used in deriving the laws of energy flow is that the restoring force is a function of position alone.

We called the function p defined by (2.4) the potential energy; what is the justification for this name? It is based on the concept of *work*, of which an easily visualized example is hoisting a load with the aid of a rope over a pulley, see Figure 8.4. How much work is required depends on the weight of the load and on the difference between its initial height and the height to which it has

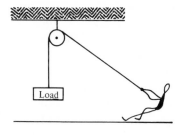

Figure 8.4

405

to be hoisted. The following facts are suggested in our example by the intuitive notion of work:

(i) The amount of work done is proportional to the *vertical* distance h by which the load is elevated.
(ii) The amount of work done is proportional to the weight f of the load.

Accordingly we *define* the work W done in elevating a load of weight f by the vertical distance h to be

(2.9)
$$W = fh.$$

In lifting a load we work against the force of gravity. We claim that the work performed does not depend on the nature of the force against which we are working, since in Newton's theory we can't tell (without looking) what kind of force we are working against. So *we take* (2.9) *to be the definition of work done in moving against any kind of a force f through a distance h in the direction opposing the force.*

So far in our intuitive analysis, we have tacitly assumed that f is positive. For f negative we still define work to be given by Formula (2.9), but we owe the reader an explanation. It is based on a yet unmentioned law of Newton:

Law of action and reaction. Whenever one particle exerts a force f_1 on a second particle, the second particle exerts a force f_2 on the first one which is equal in magnitude but opposite in direction to the first force:

$$f_2 = -f_1.$$

When two particles are pushing against each other, according to Definition (2.9), the work W_1 done by the first is $f_2 h$, since the first particle is pushing against the force f_2, and the work W_2 done by the second is $f_1 h$. The sum of these two is

$$W_1 + W_2 = f_1 h + f_2 h = (f_1 + f_2)h$$

which, according to the law of action and reaction, is zero. So thanks to using Definition (2.9) for both positive and negative forces, we deduce another instance of a law of conservation of energy. When two particles are pushing against each other, the sum of the work done by *both* equals zero.

How much work is done in moving an object through an interval S against a variable force f, i.e., a force f which is of different magnitude at different points of S and may even reverse its direction? In this case f is a function defined on S; let us denote the work done by $W(f, S)$.

What kind of function is W of S? Suppose S is divided into two disjoint intervals $S = S_1 + S_2$. Since moving the object across S means moving it first across S_1, then across S_2, it follows that the total work is the sum of the work done in accomplishing the separate tasks:

(2.10)
$$W(f, S) = W(f, S_1) + W(f, S_2).$$

How does W depend on f? Clearly, if at every point of S the force f stays below some value M, then the work done in pushing against f is less than the work done in pushing against a constant force of magnitude M. Likewise, if the force f is greater than m at every point of S, then pushing against f requires more work than pushing against a constant force of magnitude m. The work done by pushing against a constant force is given by Formula (2.9). Thus, if the force f lies between the bounds

(2.11) $$m \leq f(x) \leq M \quad \text{for } x \text{ in } S,$$

then

(2.12) $$mh \leq W(f, S) \leq Mh \quad \text{where } h = |S|.$$

Looking back at Section 4.2 on integration, we recognize (2.10) as the *additive property* and (2.11), (2.12) as the *upper and lower bound property*. As we have shown there, these two properties characterize W as the integral of f over S:

(2.13) $$W(f, S) = \int_a^b f(x)dx, \qquad S = [a, b].$$

Let us return to the particle acting under a restoring force f_{re}. To move a particle against this force, we have to exert a force on the particle which, according to Newton's law of action and reaction, is equal in strength but opposite in direction, i.e., the force $f = -f_{re}$. A force of this magnitude will exactly counterbalance the restoring force and will, according to Newton's law of motion, keep the acceleration of the particle zero. Using Formula (2.13) we conclude that the work done in moving against the restoring force across the interval $[0, a]$ is

$$\int_0^a -f_{re}(x)dx.$$

This is precisely how the function p was defined and shows that $p(a)$ is the work done in moving the particle from 0 to a; one can think of this work as being stored as energy, hence the name potential energy.

Let us now look at our law of conservation of energy. It asserts that the sum

(2.14) $$\tfrac{1}{2}mv^2 + p(x) = E$$

remains constant throughout the motion; see (2.8)'. We have identified the second term $p(x)$ as potential energy; the first term is called kinetic energy. Throughout the motion, energy always ebbs and flows from one form to the other, but their sum E remains constant. We shall now use the law of conservation of energy to study details of the motion, at first qualitatively, then quantitatively. For this purpose we first study what kind of function of position the potential energy p is. According to (2.5), the derivative of p with respect to x is $-f_{re}$ and according to (1.2)$_i$, $-f_{re}(x)$ is positive for x positive, negative for x negative. According to the monotonicity criterion, this means that p is increasing for $x > 0$, decreasing for $x < 0$. Since by (2.6), $p(0) = 0$, it follows that $p(x)$ is positive for all $x \neq 0$.

Property (1.2)$_{ii}$ says that $-f_{re}$ is increasing; a function whose derivative is an increasing function is called *convex*, see Section 3.8. So we conclude that p is convex.

Property (1.2)$_{iii}$ says that f_{re} is an odd function. A function whose derivative is odd is itself even (see Exercise 3.5c of Chapter 3); so p is even.

We summarize these facts: *The potential p, corresponding to a restoring force which has all three properties (1.2), is an even, convex function of x, zero at $x = 0$ and positive everywhere else.* Figure 8.5 pictures the graph of such a potential.

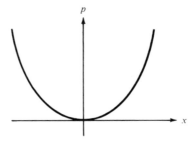

Figure 8.5

8.3 Vibration without friction

We shall study the time history of a particle moving subject to a restoring force in the absence of friction. Suppose we start the motion by displacing the particle to the position $x = -b$, $b > 0$, so that initially its velocity is zero (until we let go). The total energy imparted thereby to the system is, by (2.14), $p(-b) = E$. Upon being released, the particle starts moving toward the position $x = 0$. For negative x, $p(x)$ decreases with x; therefore, during this initial phase of the motion, the potential energy decreases. It follows then, from the law of conservation of energy, that the kinetic energy, $\frac{1}{2}mv^2$, increases, so that the particle gains speed during this phase of the motion. The potential energy reaches its minimum at $x = 0$; as soon as the particle swings past $x = 0$, its potential energy starts increasing, and its kinetic energy decreases accordingly. This state of affairs persists until the particle reaches the position $x = b$. At this point its potential energy equals $p(b)$; and since p is an even function, $p(b) = p(-b) = E$ is the total energy. Therefore at this point the kinetic energy is zero, and b is the right endpoint of the interval through which the particle moves. After reaching $x = b$ the particle turns around and describes a similar motion from right to left until it returns to its original position $x = -b$. Its velocity at this point is zero, so everything is just as it was at the beginning of the motion. Therefore the *same* pattern is repeated all over again; such motion is called *periodic*, and the time T taken

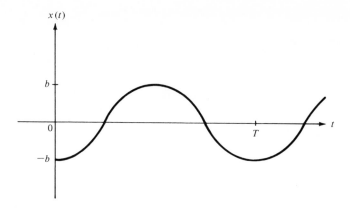

Figure 8.6

by the particle to return to its original position is called the *period* of the motion. The mathematical expression of periodicity is

$$x(t + T) = x(t),$$

and the graph of $x(t)$ is shown in Figure 8.6. We now turn from this qualitative description of motion to a quantitative description, which we also shall deduce from the law of conservation of energy (2.14),

$$\tfrac{1}{2}mv^2 + p(x) = E.$$

The motion was initiated by displacing the particle so that $x(0) = -b$, $v(-b) = 0$, $-b = $ initial displacement. By (2.14) these conditions imply that

(3.1) $$E = p(-b) = p(b) = \text{constant}.$$

Solving (2.14) for v, we get

(3.2) $$v = \sqrt{\frac{2}{m}(E - p(x))}.$$

In the first phase of the motion, $0 \le t \le T/2$, x is an increasing function of time, therefore $v = x'$ is positive, so that the positive square root is to be taken in (3.2). Since $x(t)$ is monotonic during this interval, we can express t as a function of x. According to the rule for differentiating the inverse of a function, the derivative of t with respect to x is

$$\frac{dt}{dx} = \frac{1}{dx/dt};$$

dx/dt is velocity, and it is related to x by (3.2). Therefore we deduce that

(3.3) $$\frac{dt}{dx} = \sqrt{\frac{m}{2(E - p(x))}}.$$

409

According to the fundamental theorem of calculus

(3.4) $$t(y_2) - t(y_1) = \int_{y_1}^{y_2} \sqrt{\frac{m}{2(E - p(x))}} \, dx.$$

The integral on the right expresses the time it takes for the particle to move from position y_1 to position y_2 during the first phase of the motion. Take, in particular, $y_1 = -b$ and $y_2 = b$; these positions are reached at $t = 0$ and $t = T/2$, respectively, therefore we have from (3.4) that

(3.5) $$T = \int_{-b}^{b} \sqrt{\frac{2m}{E - p(x)}} \, dx.$$

According to (3.1) the value of the total energy E is $p(b)$. This shows that as x approaches $-b$ or b, the difference $E - p(x)$ tends to zero. This makes the integrand in (3.5) tend to ∞. as x approaches the endpoints; in other words, this integral is improper in the sense explained in Section 4.7. Such an integral is defined by evaluating the integral over a subinterval, and taking the limit as the subinterval approaches the original interval. For an efficient method for evaluating (3.5) refer to Exercise 6.2. In the following section, we describe a particular $p(x)$ which allows explicit evaluation of (3.5).

8.4 Linear vibrations without friction

Suppose that the restoring force is a *differentiable* function of x. According to the basic tenet of differential calculus, *a differentiable function can be well approximated over a short interval by a linear function*. We have seen earlier that the motion is confined to the interval $-b \leq x \leq b$, $-b$ the initial displacement. For small $b, f_{re}(x)$ can be well approximated over $[-b, b]$ by a linear function; see Figure 8.7. It is reasonable to expect that, if we replace the true restoring force by its linear approximation over the small interval $[-b, b]$, the characteristic properties of motions with small displacements will not change drastically.

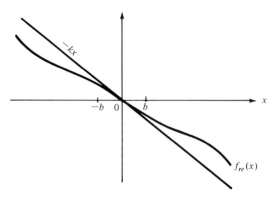

Figure 8.7

So, consider the *linear* restoring force

(4.1)
$$f_{re}(x) = -kx.$$

The positive constant k measures the *stiffness* of the elastic medium exerting the force, i.e., the larger k is, the greater the resistance to the displacement. For this reason, k is called the *stiffness constant*. Clearly the linear f_{re} described in (4.1) has all properties (1.2). The corresponding potential

$$p(x) = -\int_0^x f_{re}(s)ds = -\int_0^x -ks\ ds$$

is the quadratic function

(4.2)
$$p(x) = \frac{k}{2}x^2.$$

Let us substitute it into the Formula (3.5) for the period of the motion. Using the fact that $E = p(b)$, we get

$$T = \int_{-b}^{b} \sqrt{\frac{4m}{kb^2 - kx^2}}\ dx.$$

Performing the change of variable $x = by$, we get

$$T = \int_{-1}^{1} \sqrt{\frac{4m}{kb^2 - kb^2y^2}}\ b\ dy = 2\sqrt{\frac{m}{k}} \int_{-1}^{1} \frac{dy}{\sqrt{1 - y^2}}.$$

We recall from Section 7.2 that the function $1/\sqrt{1 - y^2}$ is the derivative of arc sine y; so, according to the fundamental theorem of calculus,

$$\int_{-1}^{1} \frac{1}{\sqrt{1 - y^2}}\ dy = \arcsin 1 - \arcsin(-1) = \frac{\pi}{2} - \left(-\frac{\pi}{2}\right) = \pi.$$

Substituting this into the above formula for T, we obtain

(4.3)
$$\boxed{T = 2\pi \sqrt{\frac{m}{k}}.}$$

This remarkable formula shows how the period of the motion depends on the data:

1. The period is independent of the size of the initial displacement, provided the initial displacement is small enough to warrant approximating f_{re} by a linear function.

2. The period is proportional to $\sqrt{m/k}$.

411

Statement 1 is a quantitative version of what our physical intuition tells us: Increasing the mass m slows down the motion, tightening the elastic string (which is the same as increasing the stiffness constant k) speeds up the motion.

If the restoring force is not a linear function of displacement, it follows from Formula (3.5) that the period definitely depends on the initial displacement; a more precise statement is contained in Exercise 6.1.

A periodic motion is often called a *vibration*; any portion of such motion lasting a full period is called a *cycle*. The number of cycles per unit time is called *frequency*, i.e.,

$$(4.3)' \qquad \text{frequency} = \frac{1}{\text{period}} = \frac{1}{2\pi}\sqrt{\frac{k}{m}}.$$

The most striking manifestation of vibration is caused by the *pressure waves* transmitted through the air to the ears of a nearby observer who perceives them as *sound*. The pitch of the sound is determined by the number of pressure pulses per unit time reaching the eardrum, and this number is the frequency of the vibrating source of the sound. A piece of metal, when struck by a hammer, vibrates. We know from every day observation (alas) that the pitch of the sound generated does *not* depend on how hard the metal has been struck, although the loudness of the sound does. On the other hand the sound generated by a plucked rubber band has a twangy quality, indicating that the pitch changes as the displacement changes. We conclude that the elastic force which acts in metal when slightly displaced from equilibrium is a (nearly) linear function of displacement, while the force exerted by a rubber band is a nonlinear function of displacement.

In order to study vibrations without friction under a linear restoring force in some detail, we solve the differential equation of motion and study the properties of its solution. Substituting $f_{fr}(v) = 0$ and $f_{re}(x) = -kx$ into Newton's law of motion in the form (1.4), we get

$$ma + kx = 0.$$

Since $a = x''(t)$, we can rewrite this, after dividing by m, as

$$(4.4) \qquad x''(t) + \frac{k}{m} x(t) = 0.$$

The initial displacement is $-b$, and the velocity at the moment of release is zero:

$$(4.5) \qquad x(0) = -b, \qquad x'(0) = v(0) = 0.$$

Equation (4.4) is a differential equation for $x(t)$. Solutions of this equation have the following property, known as the

Principle of linearity

(i) *If x is a solution of (4.4) and c is a constant then cx(t) also is a solution.*
(ii) *If x_1 and x_2 are two solutions, then $x_1 + x_2$ also is a solution.*

The truth of these statements rests on two of the basic properties of differentiation:

$$(cx)' = cx'$$

and

$$(x_1 + x_2)' = x_1' + x_2'.$$

We leave the proof of the principle of linearity to the reader; see Exercise 4.1.

We can combine the two properties listed under linearity: *Suppose that x_1 and x_2 are two solutions of (4.4) and c_1 and c_2 are two constants. Then $c_1 x_1 + c_2 x_2$ also is a solution of (4.4).*

We shall use the principle of linearity to prove the

Uniqueness theorem. *A solution x(t) of (4.4) is completely determined if, at $t = 0$, its value x(0) and the value of its derivative x'(0) are specified.*

The physial content of this statement is that Newton's law plus the initial position and velocity of the motion completely determine the subsequent course of the motion.

PROOF. Suppose that x_1 and x_2 both are solutions, and $x_1(0) = x_2(0)$, $x_1'(0) = x_2'(0)$. According to the principle of linearity their difference $x = x_1 - x_2$ also is a solution, and

$$x(0) = 0, \qquad x'(0) = 0.$$

It follows that the total initial energy of the motion $x(t)$,

$$\frac{m}{2} [x'(0)]^2 + \frac{k}{2} [x(0)]^2,$$

is 0. According to the law of conservation of energy, the total energy of the motion remains zero; and since both terms are non-negative, this can be true only if $x(t)$ and $x'(t)$ are both 0 for all t. Since $x = x_1 - x_2$, this proves that x_1 and x_2 are the same for all t. ☐

Next we derive a formula for the position $x(t)$ and velocity $v(t)$ of a particle moving under the influence of a linear restoring force in the absence of friction. Denote by $x_1(t)$ the solution of

(4.4)$_1$ $$x_1'' + x_1 = 0,$$

413

which at $t = 0$ satisfies

(4.5)$_1$ $\qquad\qquad\qquad x_1(0) = -1, \qquad x_1'(0) = 0.$

The function

(4.6) $\qquad\qquad\qquad x(t) = bx_1\left(\sqrt{\dfrac{k}{m}}\,t\right),$

where the argument of x_1 now is $\sqrt{k/m}\,t$, satisfies

$$x'' = b\,\frac{k}{m}\,x_1''.$$

From (4.4), $x_1'' = -x_1$, and from (4.6),

$$x_1\left(\sqrt{\frac{k}{m}}\,t\right) = \frac{x(t)}{b}.$$

It follows that $x'' = -(k/m)x$, i.e., x satisfies (4.4). Likewise we can verify that x satisfies (4.5).

Since the function x_1 is independent of the physical constants k, m of the motion, Formula (4.6) tells us how x depends on b, m and k. In particular, let T_1 denote the period of x_1:

$$x_1(t + T_1) = x_1(t).$$

It follows from the definition (4.6) of x that

$$x\left(t + \sqrt{\frac{m}{k}}\,T_1\right) = bx_1\left(\sqrt{\frac{k}{m}}\,t + T_1\right)$$

$$= bx_1\left(\sqrt{\frac{k}{m}}\,t\right) = x(t),$$

i.e., that the period T of x is related to the period T_1 of x_1 by

(4.7) $\qquad\qquad\qquad T = T_1\sqrt{\dfrac{m}{k}}\,.$

This is exactly Formula (4.3) if T_1 is identified as 2π. Another way of finding T_1 is to determine the function x_1; the differential equation (4.4)$_1$ asserts that the second derivative of x_1 is $-x_1$. This property is the hallmark of the functions sine and cosine. The second condition imposed on x_1 is (4.5)$_1$, and this is satisfied by $-\cos t$. So we conclude that $x_1(t) = -\cos t$; the period of $-\cos t$ is indeed 2π. With the determination of x_1 and (4.6) we have

(4.8) $\qquad\qquad\qquad x(t) = -b\cos\sqrt{\dfrac{k}{m}}\,t$

as the solution of (4.4) with initial values (4.5).

414

The velocity can be obtained by differentiation:

$$(4.9) \qquad v(t) = x' = b\sqrt{\frac{k}{m}} \sin\sqrt{\frac{k}{m}}\, t.$$

So the position and velocity of a particle moving under the influence of a restoring force and no friction is, for small initial displacements, described by the cosine and sine functions, respectively. It turns out that the same is true for all vibrating systems, no matter how complicated. And this is one reason why sine and cosine are such important functions.

The function x_1 describes motion where the constants m, k, and b all have the value 1. The total energy in this case is

$$\tfrac{1}{2}v_1{}^2(t) + \tfrac{1}{2}x_1{}^2(t);$$

and, in view of $(4.5)_1$, its value at $t = 0$ is $\tfrac{1}{2}$. Since

$$x_1(t) = -\cos t, \qquad v_1(t) = x_1'(t) = \sin t,$$

the total energy is

$$\tfrac{1}{2}\sin^2 t + \tfrac{1}{2}\cos^2 t.$$

It follows from the law of conservation of energy that this quantity remains at its initial value $1/2$ for all t. In Section 7.1, we reached the same conclusion using the *Pythagorean theorem*. So it turns out, rather surprisingly, that *the basic theorems of geometry and mechanics are related.*

EXERCISES

4.1 Let $x_1(t)$ be a solution of the n-th order differential equation

$$(4.10) \qquad A_n x^{(n)}(t) + \cdots + A_2 x''(t) + A_1 x'(t) + A_0 x(t) = 0,$$

where the A_i are constants and $x^{(k)}$ denotes the k-th derivative of $x(t)$.
(a) Show that $cx_1(t)$, c any constant, is a solution of (4.10).
(b) If $x_2(t)$ is another solution of (4.10) show that $y(t) = x_1(t) + x_2(t)$ is a solution of (4.10).
(c) Deduce from (a) and (b) that the principle of linearity holds for Equation (4.4).

4.2 Suppose that the coefficients A_0, A_1, ... A_n in (4.10) are functions of t; are the assertions made in parts (a) and (b) of Exercise 4.1 still valid? Justify your answer.

8.5 Linear vibrations with friction

We now turn to the study of motion with friction. We shall assume that the restoring force is a differentiable function of x, and that the friction force is a differentiable function of v. Furthermore, we shall restrict our study to motions where displacement and velocity are relatively small, i.e., so small that both, f_{re} and f_{fr} are so well approximated by linear functions that we might as well take them to be linear:

(5.1)
$$f_{re}(x) = -kx$$

and

(5.2)
$$f_{fr}(v) = -hv.$$

The positive constant k was called the *stiffness constant*; the positive constant h is called the *friction constant*. Substituting these forces into Newton's equation of motion (1.4), we get

$$ma + hv + kx = 0.$$

Since $v = x'$ and $a = v' = x''$, this equation can be rewritten as

(5.3)
$$mx'' + hx' + kx = 0.$$

This is a differential equation that we shall try to solve by assuming that the solution function is of exponential form

(5.4)
$$x(t) = e^{rt}, \quad r \text{ constant.}$$

This function has first and second derivatives

$$x'(t) = re^{rt}, \qquad x''(t) = r^2 e^{rt}.$$

Substituting these into (5.3), we get

$$mr^2 e^{rt} + hre^{rt} + ke^{rt} = 0,$$

and factoring out the exponential,

$$(mr^2 + hr + k)e^{rt} = 0.$$

Since the exponential factor is never zero, the sum in the parentheses must be zero:

(5.5)
$$mr^2 + hr + k = 0.$$

This is a quadratic equation for r, whose solutions are

(5.6)
$$r_\pm = -\frac{h}{2m} \pm \frac{\sqrt{h^2 - 4mk}}{2m}.$$

There are two cases, depending on the sign of the quantity under the square root.

(5.7)$_\text{I}$
$$h^2 - 4mk < 0,$$

or

(5.7)$_\text{II}$
$$h^2 - 4mk \geq 0.$$

In case I the roots are complex, in case II they are real. We first consider case I. Denoting by w the quantity

(5.8)
$$\frac{1}{2m}\sqrt{4mk - h^2} = w,$$

the roots (5.6) can be written as

$$r_{\pm} = -\frac{h}{2m} \pm iw,$$

and the exponential (5.4) takes the form

$$x_{\pm}(t) = \exp\{r_{\pm}t\} = \exp\left\{-\frac{h}{2m}t \pm iwt\right\}.$$

In Chapter 7 we gave an interpretation of complex exponentials; it is

(5.9)
$$x_{\pm}(t) = \exp\left\{-\frac{h}{2m}t\right\}(\cos wt \pm i \sin wt).$$

We have shown that complex exponentials satisfy the customary rules for differentiation, so that the functions x_+ and x_- are solutions of (5.3). From these complex valued functions we shall construct real valued solutions of (5.3) for the position, as a function of time, of a particle moving under the influence of linear restoring and friction forces, with the aid of the

Principle of linearity

(i) *If $x(t)$ is a complex valued solution of (5.3) and c is a complex constant, then cx also is a solution.*
(ii) *If x_1 and x_2 are solutions, so is $x_1 + x_2$.*

This principle, just as in the previous case of Equation (4.4), follows from the basic properties of differentiation which are valid for complex valued functions as well. We can combine the two statements of linearity into one: *If x_1 and x_2 are two solutions of (5.3), and c_1 and c_2 are two numbers, then $c_1x_1 + c_2x_2$ also is a solution.*
In particular, use the solutions (5.9),

$$x_1 = x_+, \qquad x_2 = x_-,$$

taking first the two numbers $c_1 = c_2 = 1/2$, then the two numbers $c_1 = 1/2i$, $c_2 = -1/2i$. By the linearity principle, the resulting functions

$$\frac{x_+ + x_-}{2} = \exp\left\{-\frac{h}{2m}t\right\}\cos wt \quad \text{and} \quad \frac{x_+ - x_-}{2i} = \exp\left\{-\frac{h}{2m}t\right\}\sin wt$$

are also solutions. By applying the linearity principle again, we see that any combination of the form

(5.10) $x(t) = \exp\left\{-\dfrac{h}{2m}t\right\}(A \cos wt + B \sin wt)$, A and B constants,

is again a solution of (5.3). We ask the reader to show, in Exercise 5.3, that the constants A and B can be so chosen that $x(t)$ takes the initial values (4.5).

The function x described in (5.10) is the product of a trigonometric and an exponential function. The trigonometric function is periodic, with period $2\pi/w$, and the exponential function tends to 0 as t tends to infinity. The exponential function diminishes by the factor

(5.11) $$\exp\left\{-\dfrac{h}{2m}\right\}$$

per unit time. This is called the *decay rate* of $x(t)$. Such motion is called a *damped vibration*. Its graph is sketched in Figure 8.8.

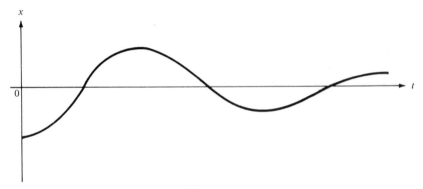

Figure 8.8

We now turn to case II defined by condition (5.7)$_{\text{II}}$. The special case of equal roots, $h = 2\sqrt{mk}$, will be discussed in Exercise 5.2. For $h > 2\sqrt{mk}$, the roots r_+ and r_-, see Equation (5.6), are both real and furnish two distinct real exponential solutions, $\exp\{r_+ t\}$ and $\exp\{r_- t\}$. According to the principle of linearity, any combination of them,

(5.12) $x(t) = A_+ \exp\{r_+ t\} + A_- \exp\{r_- t\}$,

is also a solution. We shall choose the constants A_+ and A_- so that the initial displacement is $x(0) = -b$ and the initial velocity $x'(0) = v(0) = 0$:

(5.13) $\begin{aligned} x(0) &= A_+ + A_- = -b, \\ v(0) &= r_+ A_+ + r_- A_- = 0. \end{aligned}$

A_+ and A_- are easily determined from these relations.

418

Both roots r_+ and r_- given by (5.6) are negative; consequently both exponentials which make up $x(t)$ as given by (5.12) tend to zero as t tends to ∞. Of the two negative roots, r_- has the greater magnitude:

$$|r_-| > |r_+|.$$

From this and the second relation in (5.13) we conclude

$$|A_+| > |A_-|,$$

and also that for $t > 0$

$$\exp\{r_+ t\} > \exp\{r_- t\}.$$

Hence the term $A_+ \exp\{r_+ t\}$ in (5.12) is always greater than $A_- \exp\{r_- t\}$; as t tends to ∞, the first term becomes very much greater than the second. This shows that the decay of $x(t)$ is governed by the decay rate of the first term. That decay rate is

(5.14) $$\exp\{r_+\}.$$

The difference between case I and case II is that in case I the force of friction is not strong enough to keep the particle from swinging back and forth, although it does diminish the magnitude of successive swings. In case II friction is so strong compared to the restoring force that it slows down the particle so much that it never swings over to the other side. This motion is called *overdamped*: it is graphed in Figure 8.9.

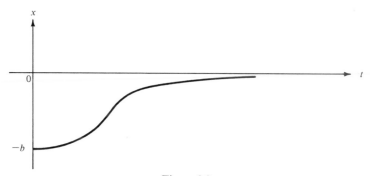

Figure 8.9

In both case I and case II, motion decays to zero as t tends to ∞. We now investigate the rates of this decay, given by Formulas (5.11) and (5.14), respectively as $\exp\{-h/2m\}$ and $\exp\{r_+\}$. The logarithms of these decay rates are called *coefficients of decay* and are denoted by the symbol l. Using Formula (5.6) for r_+ and criterion (5.7) for distinguishing between cases I and II, we have the following formula for the coefficient of decay l:

(5.15) $$l = \begin{cases} -\dfrac{h}{2m} & \text{for } h < 2\sqrt{mk} \quad \text{Case I} \\[2ex] \dfrac{-h + \sqrt{h^2 - 4mk}}{2m} & \text{for } 2\sqrt{mk} < h \quad \text{Case II} \end{cases}$$

419

We shall study how l varies as the friction constant h changes while m and k remain fixed.

Properties of l as a function of h

 (i) $l(h)$ is a continuous function for $0 \leq h$.
 (ii) $l(h)$ is a decreasing function of h for $0 \leq h < 2\sqrt{mk}$.
 (iii) $l(h)$ is an increasing function of h for $2\sqrt{mk} < h$.
 (iv) $l(h)$ reaches its minimum value at

(5.16)
$$h = 2\sqrt{mk}.$$

PROOF

 (i) is true because, at the point $h = 2\sqrt{mk}$ where case I joins case II, the two formulas for l furnish the same value.
 (ii) is true because, for $0 \leq h < 2\sqrt{mk}$, the derivative l' of l is $-1/2m$, a negative quantity.
 (iii) is true because, for $h > 2\sqrt{mk}$, the derivative l' of l is positive. To see this, form the derivative of l:

$$\frac{dl}{dh} = \frac{1}{2m}\left[-1 + \frac{h}{\sqrt{h^2 - 4mk}}\right].$$

 Since $h > \sqrt{h^2 - 4mk}$, the fraction in the square bracket is > 1 so that dl/dh is positive.
(iv) is a consequence of the first three propositions. $\square$

Note that the function $l(h)$ is continuous and achieves a minimum at $h = 2\sqrt{mk}$; it is not differentiable at $h = 2\sqrt{mk}$, as can be seen from Figure 8.10. As the graph indicates, $l(h)$ tends to zero as h tends to ∞; see Exercise 5.5 concerning this point.

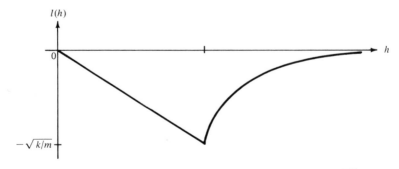

Figure 8.10 Coefficient of decay l is minimum at $h = 2\sqrt{mk}$.

Knowing the value (5.16) of h that minimizes l is important. For example, in a bouncing automobile, the springs provide a restoring force and the shock absorbers provide frictional damping. Actually the shock is absorbed by the springs; the role of the shock absorbers is to dissipate the energy resulting from a sudden displacement. In a properly designed car the shock absorbers should have the property that they provide the critical amount of friction (5.16).

EXERCISES

5.1 The function

$$x(t) = \exp\{-(h/2m)t\}\cos wt$$

represents motion under a linear restoring force and linear friction.

(a) Show that the interval between two successive times when $x(t) = 0$ has length π/w; and

(b) show that the time interval between two successive local maxima is $2\pi/w$.

5.2 Consider the equation of motion

$$mx'' + hx' + kx = 0,$$

and suppose that h has the critical value $2\sqrt{mk}$.

(a) Show that $\exp\{rt\}$ and $t\exp\{rt\}$ are both solutions of the equations of motion, where r is the root of the quadratic equation

$$mr^2 + hr + k = 0.$$

(b) Show that any function x of the form

$$x(t) = (A + Bt)e^{rt}$$

is a solution of the equation of motion. Choose the constants A and B so that x and x' have the initial values

$$x(0) = 1, \qquad x'(0) = 0.$$

5.3 Find a solution $x(t)$ of the equation of motion

$$x'' + x' + x = 0$$

which, at $t = 0$, satisfies

$$x(0) = 1, \qquad x'(0) = 0.$$

5.4 Newton's equation of motion for a particle at the end of a vertical spring (see Figure 8.11) under the influence of the restoring force of the spring, friction and gravity is

$$mx'' + hx' + kx = mg.$$

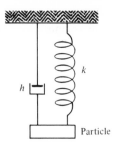

Figure 8.11

Here the displacement x is measured as positive downward; m, h and k and g are positive constants.

Show that as t tends to ∞, $x(t)$ tends to the equilibrium position

$$x(\infty) = \frac{gm}{k}.$$

Note that the equilibrium position is proportional to the mass m. [*Hint*: Look at Statement 5 in the summary at the end of Section 8.6.]

5.5 Show that as friction h tends to ∞, the logarithm of the rate of decay,

$$-\frac{h}{2m} + \frac{\sqrt{h^2 - 4mk}}{2m},$$

tends to zero. [*Hint*: Write the square root as $h\sqrt{1 - 4mk/h^2}$ and use an approximation to $\sqrt{1 - K}$ for small K.]

8.6 Linear systems driven by an external force

Next we study the motion of particles under the influence of a restoring force, friction and a *driving force* f_d presented as a known function of time. This is a frequently occurring situation; examples of it in nature are

(i) the motion of the eardrum driven by pressure pulses in the air,
(ii) the motion of a magnetic diaphragm under an electromagnetic force,
(iii) the motion of air in the resonating cavity of a violin under the force exerted by a vibrating violin string,
(iv) the motion of a building under the force exerted by the wind or tremors in the earth.

Of course these examples are much more complicated than the case of a single particle which we shall investigate.

Newton's law of motion governing a single particle says that

$$(6.1) \qquad mx'' = f_{re}(x) + f_{fr}(v) + f_d(t).$$

We assume that the motion is so slow and the displacement so small that both f_{re} and f_{fr} might as well be taken to be linear functions, as in (5.1), (5.2):

$$f_{re}(x) = -kx, \qquad f_{fr}(v) = -hv.$$

We also assume that the driving force is exerted by some other vibrating system, and that this force is proportional to the displacement which occurs in that other system. Denote the frequency of the driving force by $q/2\pi$; then form

(6.2) $$f_d(t) = F \cos qt,$$

where F is a constant. Substituting these forces into (6.1) we get the equation

(6.3) $$mx'' + hx' + kx = F \cos qt.$$

We start by establishing a simple relation between any two solutions of this equation. Let x_0 be a solution, then

(6.3)$_0$ $$mx_0'' + hx_0' + kx_0 = F \cos qt.$$

Subtracting this equation from (6.3), we get

(6.4) $$m(x - x_0)'' + h(x - x_0)' + k(x - x_0) = 0;$$

i.e., the difference of any two solutions of (6.3) is a solution of Equation (6.4). Now Equation (6.4), for $x - x_0$, is just Equation (5.3) governing the motion of particles subject only to a restoring force and a friction force. We know quite a bit about solutions of that equation, and in particular, we know that all solutions tend to zero as t tends to ∞. This shows that for large t, $x - x_0$ is insignificantly small, i.e., that *for large t any two solutions of* (6.3) *differ by very little.*

We shall find a solution of Equation (6.3) by the following trick. We shall look for complex valued solutions z of the complex equation

(6.5) $$mz'' + hz' + kz = Fe^{iqt}.$$

Denote by $x = x(t)$ the real part of z; since the real parts of the two sides of (6.5) are just the two sides of (6.3), x, the real part of z, is a solution of (6.3). Equation (6.5) has a complex valued solution z of the form

(6.6) $$z(t) = Ae^{iqt}$$

with derivatives

$$z' = Aiqe^{iqt} \quad \text{and} \quad z'' = -Aq^2 e^{iqt}.$$

Substituting these into (6.5) we get, after division by e^{iqt},

$$A[-mq^2 + ihq + k] = F.$$

We find A from this equation and substitute its value into (6.6), obtaining

(6.7) $$z(t) = \frac{F}{-mq^2 + ihq + k} e^{iqt}.$$

The real part x of z, denoted by $\mathrm{Re}\{z\}$, is the solution we are after:

(6.8) $$x(t) = \mathrm{Re}\left\{ \frac{F}{-mq^2 + ihq + k} e^{iqt} \right\}.$$

423

How is the motion $x(t)$ related to the driving force? We assert that

(i) *The frequency of $x(t)$ is the same as the frequency of the driving force.*

(ii)

(6.9)
$$\max|x(t)| = \frac{\max|f_d(t)|}{|-mq^2 + ihq + k|}.$$

Fact (i) stares us in the face if we look at Formula (6.8). To prove part (ii) we note first that $\max|f_d(t)| = F$; so to prove (6.9) we have to show that, for all t,

(6.10)
$$|x(t)| \leq \frac{F}{|-mq^2 + ihq + k|},$$

and that the sign of equality holds for some t. The inequality follows from the following sequence of observations: To determine the absolute value of z, we use the property of complex numbers that the absolute value of a product (and quotient) of complex numbers is the product (and quotient) of their absolute values. So using the definition (6.7) of z and the fact that $\exp\{iqt\}$ has absolute value 1, we get

(6.11)
$$|z(t)| = \frac{F}{|-mq^2 + ihq + k|}.$$

The absolute value of the real part of z does not exceed the absolute value of z:

$$|x(t)| = |\text{Re}\{z(t)\}| \leq |z(t)|.$$

Combining this with (6.11), we deduce inequality (6.10). To deduce that equality holds for some values of t, observe that, since the argument of $\exp\{iqt\}$ takes on all values, $z(t)$ is real for some values of t.

$\text{Max}|f_d(t)|$ is called the *amplitude of the driving force*, $\max|x(t)|$ is called the *amplitude of the vibration* caused by the driving force. The importance of (6.9) is that it relates the amplitude of the driving force to the amplitude of the resulting vibration. The two amplitudes are proportional, the ratio $R(q) = \max|x|/\max|f_d|$ being

(6.12)
$$R(q) = \frac{1}{|-mq^2 + ihq + k|}.$$

In many ways the most interesting question is: for what value of q is $R(q)$ the largest? Clearly, $R(q)$ tends to zero as q tends to ∞, so $R(q)$ has a maximum. We shall calculate the position and value of the maximum; it occurs at the same place where

(6.13)
$$\frac{1}{[R(q)]^2} = |-mq^2 + ihq + k|^2 = (k - mq^2)^2 + h^2q^2$$

is minimum. At a minimum the derivative of (6.13) is zero; that derivative is

$$4mq(mq^2 - k) + 2h^2q,$$

and clearly it is zero at $q = 0$. To find other possible zeros we divide by q and set the remaining factor equal to zero:

$$4m(mq^2 - k) + 2h^2 = 0.$$

After rearrangement we get

$$q^2 = \frac{k}{m} - \frac{h^2}{2m^2}.$$

If the quantity on the right is negative, the equation cannot be satisfied and we conclude that the maximum of R is reached at $q = 0$. If, however, the quantity on the right is positive, then

$$(6.14) \qquad q_r = \sqrt{\frac{k}{m} - \frac{h^2}{2m^2}}$$

is a possible candidate for the value where $R(q)$ achieves its maximum. A direct calculation shows that

$$(6.15) \qquad R(q_r) = \frac{1}{h} \frac{1}{\sqrt{\dfrac{k}{m} - \dfrac{h^2}{4m^2}}}.$$

It is not hard to check (see Exercise 6.3) that $R(q_r)$ is greater than $R(0) = 1/k$. So for $h < \sqrt{2mk}$, the graph of $R(q)$ looks qualitatively like the sketch in Figure 8.12. The graph of R is called the *response curve* of the vibrating system.

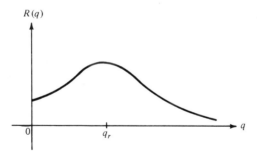

Figure 8.12

The significance of the maximum at q_r is that, *among all driving forces of the form $F \cos qt$, the one with $q = q_r$ causes the motion with the largest amplitude.* This phenomenon is called *resonance*, and $q_r/2\pi$ is called the *resonant frequency*. Resonance is particularly striking if friction is small, i.e., if h is small, for then $R(q_r)$ is so large that even a low amplitude driving force will, at the resonant frequency, cause a motion of large amplitude. A known

425

dramatic example of this kind of resonance is the shattering of a wine glass by a musical note pitched at the resonant frequency of the glass. We conclude this section with a

Summary

For motion under a restoring force without friction

1. Total energy is conserved.
2. All motions are periodic.
3. All motions with relatively small amplitude have approximately the same period.

For motion under a restoring force with friction

4. Total energy is decreasing.
5. All motion decays to zero at an exponential rate.
6. There is a critical value of the friction coefficient which maximizes the rate at which solutions decay to zero.

For motions under a linear restoring force, linear friction and a sinusoidal driving force

7. All motions tend toward a sinusoidal motion with the same frequency as the driving force.
8. If friction is not too large, there is a resonant frequency.

Actually we have proved (5) and (6) only for a linear restoring force and linear friction. In Section 8.7 we study motions of a particle subject to a linear restoring force but nonlinear friction. Extensions of the theory to more general situations will be treated in Volume II.

EXERCISES

6.1 This problem is about vibrations governed by nonlinear restoring forces, with no friction. Linearity of a function $g(x)$ means that $g(x + d) - g(x)$ is independent of x; if g is interpreted as the force pushing against a spring, this property means that to compress that spring by an additional distance d takes the same amount of force no matter how far the spring is already compressed. We define a spring, and its restoring force, to be *hard*, if the additional force needed to compress the spring already displaced by the amount x by an additional amount d *increases* with x. A restoring force is called *soft* if this amount *decreases* with x. In mathematical language: a function $g(x)$ is hard or soft depending on whether

$$g(x + d) - g(x), \qquad d > 0,$$

is an increasing or decreasing function of x for $x > 0$.
(a) Prove that g is hard if and only if g' is an increasing function, and soft if and only if g' is a decreasing function.

(b) Show that if g is hard and $g(0) = 0$, $g(x)/x$ is an increasing function of x. Show that if g is soft and $g(0) = 0$, $g(x)/x$ is a decreasing function of x.

If $g(x) = -f_{re}(x)$, the potential energy function $p(x)$ is defined by

$$p'(x) = g(x), \qquad p(0) = 0.$$

According to Formula (3.5), the period of motion after an initial displacement by the amount b is

(6.16) $$T(b) = 4\sqrt{\frac{m}{2}} \int_0^b \frac{1}{\sqrt{p(b) - p(x)}}\, dx.$$

Theorem. *If $g = -f_{re}$ is hard, $T(b)$ is a decreasing function of b, and if $g = -f_{re}$ is soft, $T(b)$ is an increasing function of b.*

We break up the proof of this theorem into 4 steps, each of which is to be carried out by the reader:

(c) Introduce $y = x/b$ as a new variable of integration in the expression (6.16) for $T(b)$ and derive the formula

(6.17) $$T(b) = 4\sqrt{\frac{m}{2}} \int_0^1 \frac{b}{\sqrt{p(b) - p(yb)}}\, dy.$$

(d) Using the fact that $p' = g$, show that

(6.18) $$\frac{p(b) - p(yb)}{b^2} = \int_{yb}^b \frac{g(x)}{b^2}\, dx = \int_y^1 \frac{g(zb)}{b}\, dz.$$

(e) Using the result of (b) and Formula (6.18), show that, for y fixed,

$$\frac{p(b) - p(yb)}{b^2}$$

is an increasing function of b if g is hard, and a decreasing function if g is soft.

(f) Applying the result of (e) to the integral representation (6.17) of $T(b)$ prove the theorem.

In less formal language the theorem says that a hard spring vibrates faster if the initial displacement is increased whereas a soft spring vibrates more slowly if the initial displacement is increased. Does this result agree with the reader's physical intuition?

6.2 This problem develops a method for evaluating efficiently the singular integrals occurring in (6.16). These integrals are of the form

(6.19) $$\int_0^b \frac{1}{\sqrt{K(x)}}\, dx,$$

where $K(x)$ is positive for $0 \le x < b$, $K(b) = 0$, but $K'(b) < 0$. Such a function K can be factored as follows:

(6.20) $$K(x) = (b - x)H(x),$$

where $H(x)$ is positive for $0 \le x \le b$. By means of this factored form of K, integral (6.19) can be written as

(6.21)
$$\int_0^b (b - x)^{-1/2} H(x)^{-1/2} \, dx.$$

(a) Show, using integration by parts, that the integral (6.19) is equal to

(6.22)
$$2\sqrt{\frac{b}{H(0)}} - \int_0^b \frac{\sqrt{b - x}}{H^{3/2}} H' \, dx.$$

The integral occurring in (6.22) is no longer singular.

(b) Show, using integration by parts, that (6.22) is equal to

(6.23)
$$2\sqrt{\frac{b}{H(0)}} - \frac{2}{3}\sqrt{\left(\frac{b}{H(0)}\right)^3} H'(0) + \frac{2}{3}\int_0^b \sqrt{\left(\frac{b - x}{H}\right)^3}\left[H'' - \frac{3}{2}\frac{H'^2}{H}\right] dx.$$

The integrand in (6.23) is differentiable in the whole interval $[0, b]$.
   Consider the restoring force

(6.24)
$$f_{re}(x) = -x - 2x^3.$$

The corresponding potential energy is

$$p(x) = \frac{x^2}{2} + \frac{x^4}{2}.$$

The period of motion of a unit mass $(m = 1)$ under this force after an initial displacement b is, according to (6.16), equal to

(6.25)
$$T(b) = 4 \int_0^b \frac{1}{\sqrt{b^2 + b^4 - x^2 - x^4}} \, dx.$$

This is an integral of the form (6.19), with

(6.26)
$$K(x) = b^2 + b^4 - x^2 - x^4.$$

(c) Show that the function K in (6.26) has the factorization

$$K(x) = (b - x)[b + x + b^3 + b^2 x + bx^2 + x^3].$$

(d) Evaluate $T(b)$ by applying Simpson's rule to the integral in (6.23) into which (6.24) can be transformed according to (b). Choose for b the values 0.2, 0.5, and 1.0, and choose the number of subdivisions to be $n = 5, 10, 20$.

(e) Using linear theory, calculate the period of small amplitude vibrations under the restoring force (6.24).

(f) Show that $g = -f_{re}$ defined by (6.24) is hard. Verify that the periods $T(0)$, $T(0.2)$, $T(0.5)$, and $T(1)$ form an increasing sequence, in accordance with the theorem in Exercise 6.1.

6.3 Prove that $R(q_r)$, as defined by (6.15), is greater than $R(0) = 1/k$; i.e., that

$$\frac{1}{h\sqrt{\dfrac{k}{m} - \dfrac{h^2}{4m^2}}} > \frac{1}{k}.$$

[*Hint*: Show that the inequality to be proved is equivalent to

$$k^2 - h^2\left(\frac{k}{m} - \frac{h^2}{4m^2}\right) \ge 0,$$

and then verify directly this last inequality.]

6.4 Write the solution $x(t)$ defined by (6.8) in the form

(6.27) $$x(t) = R(q)F \cos(qt - p),$$

where R is defined by (6.12).

(a) Show that p satisfies

(6.28) $$\sin p = hqR.$$

The quantity p is called the *phase* of the motion relative to the driving force $F \cos qt$.

(b) How are the maxima of the driving force and of $x'(t)$ related to each other?

(c) Show that as q goes from 0 to ∞, p goes from 0 to π.

(d) Determine the value of p for $q = \sqrt{k/m}$.

(e) Show that the work done by any driving force $F(t)$ on a moving particle $x(t)$ during the time interval $(0, T)$ is

(6.29) $$\int_0^T F(t)\frac{dx}{dt}\,dt.$$

(f) Show that the work done by any driving force $F \cos qt$ on the moving particle (6.27) during a single period, i.e., during $(0, 2\pi/q)$, is equal to

$$\pi RF^2 \sin p.$$

(g) Show, using relation (6.28) that $\overline{W}$, the *average work during a period*, i.e., the work divided by the length of a period $2\pi/q$, is

(6.30) $$\overline{W} = \frac{F^2}{2h}\sin^2 p.$$

(h) Show that for m, k, h and F fixed, $\overline{W}$ is largest for the frequency

$$q = \sqrt{\frac{k}{m}}.$$

Show that the value of $\overline{W}$ for this value of q is

(6.31) $$\overline{W} = \frac{F^2}{2h}.$$

(i) Show that for m, k, F and q fixed, and q *not* equal to $\sqrt{k/m}$, $\overline{W}$ tends to zero as h tends to zero.

Things to note. It follows from Formula (6.30) that $\overline{W}$ is positive. That is, on the average, the driving force pumps energy into the vibrating system; this energy is dissipated by friction.

According to part (i), for $q \ne \sqrt{k/m}$, W, the rate at which energy is pumped into the vibrating system, tends to zero as h, the coefficient of friction, tends to zero. This is another expression of energy conservation.

For $q = \sqrt{k/m}$, on the other hand, $\overline{W}$ tends to ∞ as h tends to zero. This is because q is very near the *resonant frequency* given by Formula (6.14).

(j) Show, using $q = \sqrt{k/m}$, that the driving force is *in phase* with the velocity x', i.e., that velocity and driving force are at all times in the same direction.

6.5 Find a solution of

$$x'' + x' + x = \cos t$$

which, at $t = 0$, satisfies

$$x(0) = 0, \qquad x'(0) = 0.$$

8.7 An example of nonlinear vibration

In this section we shall investigate the motion of a particle acting under the influence of two forces. One is a restoring force that depends linearly on displacement; the other is a velocity-dependent force which is *not* friction, inasmuch as the direction of the force is *not* always in the direction opposite to the direction of motion. More precisely, we shall consider velocity-dependent forces $f(v)$ with the following properties:

(7.1)
    (i) f is an odd function of v, $f(-v) = -f(v)$;
    (ii) $f(v)$ is positive for $0 < v < v_0$;
    (iii) $f(v)$ is negative and a decreasing function of v for $v > v_0$.

Figure 8.13 shows the graph of this kind of function f. A force of this kind represents *negative friction* in the range $-v_0 < v < v_0$; such negative friction can be realized in nature as *aerodynamical lift*. For electrical circuits, a term of this kind corresponds to negative resistance experimentally realizable as *feedback*.

A particle moving under the influence of such forces satisfies Newton's law

(7.2)
$$mx'' + kx - f(v) = 0.$$

We have shown in Section 8.2 how to derive an energy equation from the law of motion: multiply (7.2) by $v = x'$ and write the resulting equation

$$mvv' + kxx' = vf(v)$$

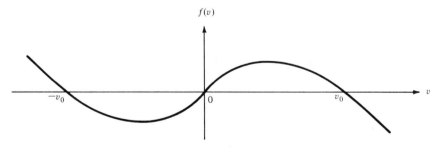

Figure 8.13   A friction force exhibiting negative friction in the interval $-v_0 < v < v_0$.

in the form

(7.3) $$\frac{d}{dt}\left(\frac{m}{2}v^2 + \frac{k}{2}x^2\right) = vf(v).$$

The quantity $mv^2/2$ is kinetic energy, $kx^2/2$ potential energy, their sum the total energy. As we saw in Section 8.2, where f was a friction force, v and $f(v)$ had opposite signs, so that $vf(v)$ was negative, except for $v = 0$. So in that case it followed from (7.3) that total energy decreases; the explicit and detailed analysis presented in Section 8.5 shows that total energy tends to zero at an exponential rate. For the kind of force contemplated in this section the story is quite different.

According to (7.1), $vf(v)$ is positive in the range $(-v_0, v_0)$, except for $v = 0$, and negative when $|v| > v_0$. So if the particle has initial speed less than v_0, its total energy will start to increase. As the total energy gets larger and larger, one of two things must happen: either the kinetic energy $mv^2/2$ gets large, or the potential energy $kx^2/2$ gets large (or both). In the first case, as soon as $mv^2/2$ exceeds $mv_0^2/2$, the function $vf(v)$ becomes negative and by (7.3) the total energy starts decreasing. In the second case, when $kx^2/2$ is large, the restoring force $-kx$ also is large, and this starts to drive the particle towards a position of low potential energy. In this process potential energy is converted to kinetic energy which brings the particle to the previous situation of high kinetic energy, resulting in a decrease of total energy. This crude analysis indicates that during motion energy increases and decreases alternately, and that the particle displays *oscillatory motion*.

We now test this tentative conclusion by numerical experiments; for purposes of the experiment we choose for f the cubic polynomial

(7.4) $$f(v) = h(1 - v^2)v,$$

h some parameter which is positive. Clearly f given by (7.4) has properties (7.1), with $v_0 = 1$. For this choice of f, Equation (7.2) takes the form

(7.5) $$mx'' - h(1 - v^2)v + kx = 0, \qquad v = x'.$$

This equation governs current in a diode; the mathematical theory of solutions of (7.5) has been investigated extensively by van der Pol, a mathematician at the Eindhoven Laboratory of Phillips. In recognition of this, (7.5) is called van der Pol's equation.

We now present some numerical solutions of Equation (7.5). The parameters m and k are chosen equal to 1, and h is taken as 0.1. Figures 8.14a, b are graphs of solutions $x(t)$ as functions of t; Figures 8.15a, b present $x(t)$, $v(t)$ as curves in the x, v-plane.

We have shown in Section 8.4 that solutions of Equation (4.4) are uniquely determined once the initial values, $x(0)$, $v(0)$ are specified. This result also holds true for Equation (7.5); the proof presented in Section 8.4 can be modified to cover this case and the details are sketched in Exercise 7.3. A geometrical formulation of this result is that, through any point of the x, v-plane,

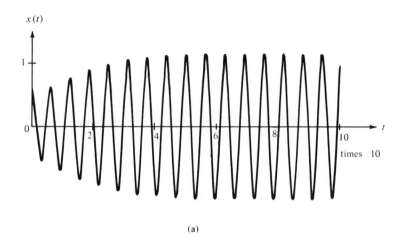

(a)

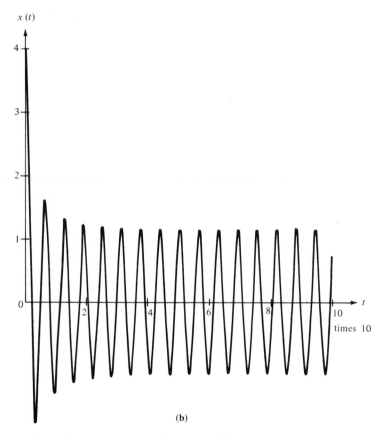

(b)

Figure 8.14   Solutions $x(t)$ of Equation (7.5), as curves in the t, x-plane, with initial values (a) $x(0) = 0.5$, $v(0) = 0$ and (b) $x(0) = 4$, $v(0) = 0$, and parameters $m = k = 1$ and $h = 0.1$.

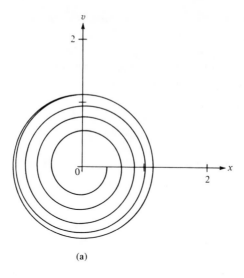

(a)

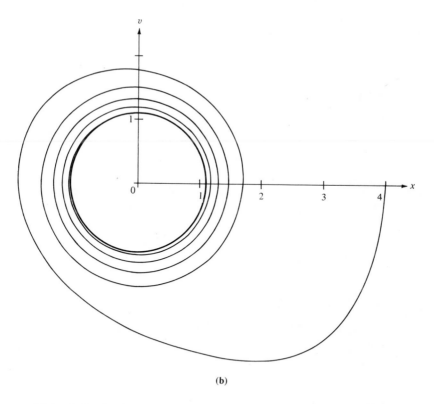

(b)

Figure 8.15 Solutions $x(t)$, $v(t)$ of Equation (7.5), as curves in the x, v-plane, with initial values (a) $x(0) = 0.5$, $v(0) = 0$ and (b) $x(0) = 4$, $v(0) = 0$, and parameters $m = k = 1$ and $h = 0.1$.

there passes exactly one solution curve $x(t)$, $v(t)$ of (7.5). The graphical evidence of Figures 8.15a, b suggests that all solution curves tend toward a single closed curve in the x, v-plane, no matter where they started. This curve, indicated in dark in the figures, is called the *limit cycle* for Equation (7.5). Thus numerical evidence indicates that *all solutions of* (7.5) *tend to the limit cycle.*

A solution which starts at a point on the limit cycle stays on the limit cycle. After a certain time T this solution will traverse the whole limit cycle and come back to its starting point. Since solutions are uniquely determined by their initial values, if we continue the solution beyond this time T it will merely retrace its steps; that is

$$x(t + T) = x(t), \qquad v(t + T) = v(t).$$

In words: *the limit cycle represents a periodic solution of Equation* (7.5). In view of this we can rephrase our previous result so: *Every solution of* (7.5) *tends to the periodic solution represented by the limit cycle.*

Thus our numerical evidence, combined with a bit of analysis, indicates not only that the van der Pol equation has a periodic solution, but also that this solution is extremely stable. In those applications where the periodic solution represents the operating mode of a device, this strong stability is an exceedingly desirable feature.

We now shall present a theoretical analysis of the limit cycle for h small. As h tends to 0, Equation (7.5) tends to

(7.6) $$mx'' + kx = 0.$$

It is reasonable to expect that, as indicated by the numerical evidence, the periodic solution of (7.5) tends to a periodic solution of (7.6). As we have seen in Section 4, *all* solutions of (7.6) are periodic, with period $2\pi\sqrt{k/m}$. This leaves us with the problem of determining which solution of (7.6) is the limit of the periodic solutions of (7.5) as h tends to 0. The remainder of the section is devoted to answering this question.

For simplicity we take $m = 1$ and $k = 1$ in (7.5) and (7.6). According to the discussion in Section 8.4, solutions of (7.6) which, at $t = 0$, satisfy

$$v(0) = 0$$

are of the form

(7.7) $$x(t) = a \cos t, \qquad v(t) = a \sin t, \qquad a = \text{constant.}$$

Let $x(t)$, $v(t)$ be the periodic solution of (7.5), adjusted so that $v(0) = 0$. We write x, v in polar form:

(7.8) $$x = b \cos w, \qquad v = b \sin w,$$

where b and w are defined by

$$b = \sqrt{x^2 + v^2}, \qquad w = \arctan \frac{v}{x}.$$

Since x and v are functions of t, so are b and w. We now derive differential equations for b as function of t. It is convenient to use the energy equation (7.3); setting $m = 1$, $k = 1$ and taking $f(v) = h(1 - v^2)v$, we get, after substituting the polar form (7.8) for x and v:

$$\frac{d}{dt}\left(\frac{b^2}{2}\right) = hb^2 \sin^2 w(1 - b^2 \sin^2 w).$$

Carrying out the differentiation and dividing by b, we get

(7.9)
$$b' = hb \sin^2 w(1 - b^2 \sin^2 w).$$

Since (7.8) is the periodic solution of (7.5), with period $T = T(h)$, b is periodic. So

$$\int_0^T b' \, dt = b(T) - b(0) = 0.$$

Using relation (7.9) for b' we get, after dividing by h, that

(7.10)
$$\int_0^T b \sin^2 w(1 - b^2 \sin^2 w)dt = 0.$$

Suppose now that, as h tends to zero, the periodic solution of (7.5) tends to one of the periodic solutions of (7.6), say to

$$x_0(t) = a_0 \cos t, \qquad v_0(t) = a_0 \sin t.$$

This means that, as h tends to 0, the function $b(t)$ tends to a_0, and the function $w(t)$ tends to t. Since relation (7.10) holds for every h, we deduce from the convergence theorem for integrals in Section 4.2 that it also holds if we replace b and w by their limits a_0 and t, and if we replace $T(h)$ by its limit 2π. Thus

(7.11)
$$a_0 \int_0^{2\pi} [\sin^2 t - a_0^2 \sin^4 t]dt = 0.$$

From what we have shown in Exercise 2.3 of Chapter 7,

$$\int_0^{2\pi} \sin^2 t \, dt = \pi, \qquad \int_0^{2\pi} \sin^4 t \, dt = \frac{3\pi}{4}.$$

Substituting this into (7.11) we get, after dividing by π that

$$a_0 - \tfrac{3}{4}a_0^3 = 0.$$

The nonzero solution of this is

$$a_0 = \frac{2}{\sqrt{3}}.$$

The numerical evidence of Figure 8.5a, b bears out the fact that, for h small, the limit cycle lies close to the circle of radius $2/\sqrt{3}$.

EXERCISES

7.1 Consider the generalized van der Pol equation

(7.12) $$x'' - hg(v) + x = 0, \qquad v = x',$$

where $g(v)$ has properties (7.1), and h is a positive parameter.

(a) Show that as h tends to 0, the only possible limit of periodic solutions of (7.12) is

$$x(t) = a_0 \cos t, \qquad v(t) = a_0 \sin t,$$

where $a_0 > 0$ satisfies the equation

(7.13) $$\int_0^{2\pi} g(a_0 \sin t)\sin t \, dt = 0.$$

(b) Deduce from (7.1) that, for a_0 small, the integral in (7.13) is positive, and that, for a_0 large, it is negative. Deduce from this that there is a positive value of a_0 which satisfies (7.13).

7.2 In this exercise the reader is asked to perform a numerical experiment. Define

$$g(v) = \begin{cases} v(1 - v) & \text{for } 0 < v < 1 \\ (1 - v) & \text{for } 1 \le v \end{cases}$$

and

$$g(v) = -g(-v) \qquad \text{for } v < 0.$$

Construct solutions of (7.12) and ascertain that they tend to a periodic solution for $h = 0.1$.

    [*Hint*: Introduce $v = x'$ in (7.12) and solve the resulting set of equations

(7.14) $$\begin{aligned} v' &= hg(v) - x \\ x' &= v. \end{aligned}$$

Use the *one*-sided difference quotients

$$\frac{v(t + k) - v(t)}{k} = v'(t) + \textit{small quantity},$$

$$\frac{x(t + k) - x(t)}{k} = x'(t) + \textit{small quantity}$$

as approximations to the derivatives of v and x; with these formulas, for k small enough, we may neglect $k \times$ *small quantity* and obtain

(7.15) $$\begin{aligned} v(t + k) &= v(t) + kv'(t), \\ x(t + k) &= x(t) + kx'(t). \end{aligned}$$

The obvious next step would be to reapply the above equations and advance the solution to $t + 2k$. However, without much additional work, as we saw in Chapter 5, when we were determining the behavior of the exponential function, a more rapidly converging method can be constructed by using *centered* difference quotients as approximations to derivatives. The idea is to use the above solutions $v(t + k)$, $x(t + k)$ to evaluate temporary estimates of v' and x' at $t + k$. For k small enough, the average value of the derivatives, evaluated at the endpoints of the interval $[t, t + k]$, is a good approximation to the value of the derivative at the midpoint of the interval. Therefore we write

$$\frac{v(t + k) - v(t)}{k} = \frac{1}{2}[v'(t) + v'(t + k)] + very \ small \ quantity,$$

$$\frac{x(t + k) - x(t)}{k} = \frac{1}{2}[x'(t) + x'(t + k)] + very \ small \ quantity.$$

By neglecting the terms $k \times very \ small \ quantity$, we find that

(7.16)
$$v(t + k) = v(t) + \frac{k}{2}[v'(t) + v'(t + k)]$$

$$x(t + k) = x(t) + \frac{k}{2}[x'(t) + x'(t + k)]$$

are better approximation formulas than (7.15) for the solution of (7.12).]
  Verify this result using both (7.15) and (7.16) with the same stepsize k.

7.3 Let x and y be the solutions of Equation (7.2), with $m = 1, k = 1$:

$$x'' + x = f(v), \qquad v = x'$$
$$y'' + y = f(w), \qquad w = y'.$$

(a) Show that the difference

$$d = x - y$$

satisfies the equation

$$d'' + d = f(v) - f(w).$$

(b) Show that d satisfies

$$d'' + d = gd',$$

where g is the derivative of f with respect to v at some point.

(c) Show that

$$\frac{d}{dt}\left(\frac{1}{2}d'^2 + \frac{1}{2}d^2\right) = gd'^2.$$

(d) Define $E = \frac{1}{2}d'^2 + \frac{1}{2}d^2$, and denote by G the maximum of g. Show that

$$\frac{d}{dt}E \leq 2GE.$$

(e) Show that

$$E(t)e^{-2Gt}$$

is a decreasing function of t.

(f) Show that if $E(0) = 0$, then $E(t) = 0$ for all $t > 0$. Deduce from this that if the two solutions x and y have the same initial values

$$x(0) = y(0), \qquad v(0) = w(0),$$

then

$$x(t) = y(t) \quad \text{for all } t > 0.$$

437

8.8 Electrical systems

For the sake of comparison, and without too much explanation, we now state the equation governing the electric current I as a function of time in a simple electric circuit consisting of an inductance, capacitance and resistance, and possibly some source of electromotive force such as a battery or dynamo; see Figure 8.16. *Current*, denoted by I, is the rate at which electric charge is conducted in the circuit in one direction (in the direction of the arrow in the figure). Since charge does not accumulate anywhere (except on the plates of the capacitance), the current is the same at all points of the circuit. The current does, however, change with time; we shall derive a differential equation which governs the way the current I varies in time.

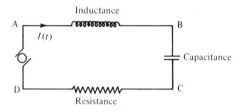

Figure 8.16

At any given time, there is between any two points, say A and B of the circuit a *potential difference,** defined as *the work needed to bring a unit charge from one point* B *to the other point* A.

According to one of the laws formulated by *Kirchhoff* concerning circuits, work is conserved. That is, if C is a third point in the circuit, *the potential difference between* A *and* C *is the sum of the potential differences between* A *and* B, *and between* B *and* C.

The potential difference depends on the pair of points, and on time as well. We state now and explain the nature of the potential difference across the terminals of each of the three types of circuit elements that we are considering:

(i) *Self-inductance* resists change in the current. To bring about a change, a potential difference is needed across the terminals A and B. For linear self-inductances the potential difference is

(8.1) $$L\frac{dI}{dt}.$$

Self-inductance is analogous to inertia in mechanics, and L, the inductance, is analogous to mass.

---

* The negative of the potential difference is called *voltage drop*.

(ii) A *capacitor* accumulates *electric charge*; such an accumulation of charge can be maintained only by a potential difference between the points B and C whose magnitude increases with the total accumulated charge. For a linear capacitor, the potential difference is

(8.2)
$$C \int_0^t I(s)ds.$$

C is a positive constant called *capacitance*; the effect of the capacitor is analogous to a restoring force in mechanics, and C is analogous to the stiffness constant k.

(iii) Electric *resistance* opposes the passage of current. Current of magnitude I can be maintained through a resistance only if there is a potential difference between the points C and D which increases with the current I. For linear resistors the potential difference is

(8.3)
$$RI.$$

R is called the resistance; the effect of the resistor is analogous to friction in mechanics, and R is analogous to the friction coefficient h. Electric resistance dissipates energy by converting it into heat—hence electric blankets. Mechanical friction also dissipates energy by converting it into heat—hence boy scouts rubbing sticks of wood together.

If there are no further sources or sinks of potential in the circuit, the potential difference between A and D is zero. According to Kirchhoff's law, the potential difference between A and D is the sum of the potential differences between A and B, B and C, C and D. Since these latter amounts are given by (8.1), (8.2), and (8.3), we conclude that

(8.4)
$$L\frac{dI}{dt} + C \int_0^t I(s)ds + RI = 0.$$

Differentiate this relation with respect to t, using the fundamental theorem of calculus to differentiate the second term, and obtain

(8.5)
$$LI'' + CI + RI' = 0.$$

If we identify L with m, C with k and R with h, we unmask Equation (8.5) as Equation (5.3) governing the motion of a particle under a linear restoring force and linear friction.

Suppose there is a *dynamo* between points D and C of the circuit, creating a potential difference between D and A equal to $M \sin qt$. Then Kirchhoff's law, which before led to (8.4), leads now to

$$L\frac{dI}{dt} + C \int_0^t I(s)ds + RI = M \sin qt.$$

Differentiating with respect to t, we get

(8.6)
$$LI'' + CI + RI' = qM \cos qt.$$

439

Identifying L, C, R as before with m, k, h, and M with F/q, we recognize that (8.6) is identical with (6.3), the equation governing the motion of a vibrating particle under friction and driven by a sinusoidal driving force.

So, *the mathematical theory of oscillating electric circuits is identical with the mathematical theory of vibrating mechanical systems.*

Note. Just as a linear restoring force and linear friction are merely approximations to nonlinear forces, so (8.1), (8.2), and (8.3) are merely approximations to nonlinear relations. These nonlinear laws are a source of new phenomena; see Section 8.7 and Volume II.

Population dynamics and chemical reactions 9

In this chapter calculus is used to study the evolution of populations—animal, vegetable, or mineral. About half the material is devoted to formulating the laws governing population changes in the form of differential equations, and the other half to studying their solutions. Only in the simplest cases can this be accomplished by obtaining explicit formulas for solutions. In cases where explicit solutions are not to be had, relevant qualitative and quantitative properties of solutions can nevertheless be deduced directly from the equations, as our examples will show. By using numerical methods, one can generate extremely accurate approximations to any specific solution of a differential equation. These may lead to the answers we seek or suggest trends that, once perceived, can often be deduced logically from the differential equations.

Theoretical population models have become more and more useful in such diverse fields as the study of epidemics and of the distribution of inherited traits. Yet the most important application, the one about which the public needs to be informed in order to make intelligent decisions, is to demography, the study of human populations. Indeed, the proper study for man is mankind.

In Section 9.1, we discuss the behavior of solutions of first order differential equations arising in the simplest population models. The important concepts of stable and unstable steady states are explained.

In Section 9.2, we treat three population models. The first, Verhulst's modification of Malthusian theory, includes an inhibiting effect of overpopulation and predicts a levelling off to a stable steady state in contrast to Malthusian theory which predicts exponential blowup. The second model takes into account the depressing effect of underpopulation on birthrate and shows that, if the population is reduced below a certain threshold it

441

inevitably tends to extinction. The third is Volterra's celebrated model of two species of fish, one a predator, the other its prey, in the same environment. Volterra's theory predicts a periodic variation around a semistable steady state of these populations. The theory indicates that if fishing is intensified moderately, the edible fish population will actually increase.

Section 9.3 is an elementary introduction to the theory of chemical reactions. This subject is of enormous interest to chemical engineers and to theoretical chemists. It also plays a central role in two topics which have recently been in the center of public controversy: emission by automobile engines and the deleterious effect on ozone of the accumulation of fluorcarbon compounds in the stratosphere.

9.1 The differential equation

$$(1.1) \qquad \frac{d}{dt} N(t) = R(N)$$

A great part of the elementary theory of population growth and chemical kinetics involves solutions $N(t)$ of various differential equations which are special cases of equations of form (1.1). To prepare the way for discussing these subjects in Sections 9.2 and 9.3, we present in this section the main facts about solutions of Equation (1.1).

Assume that $R(N)$ is a continuous function, different from zero; then we can divide both sides of (1.1) by $R(N)$, obtaining

$$(1.2) \qquad \frac{1}{R(N)} \frac{dN}{dt} = 1.$$

By the fundamental theorem of calculus, there is a function $Q(N)$ whose derivative is $1/R(N)$, i.e.,

$$(1.3) \qquad \frac{d}{dN} Q(N) = \frac{1}{R(N)}.$$

If $N(t)$ satisfies (1.1), then by the chain rule,

$$\frac{dQ}{dt} = \frac{dQ}{dN} \frac{dN}{dt} = \frac{1}{R(N)} \frac{dN}{dt} = 1.$$

According to Section 3.3 (cf. Example 6), a function with constant derivative is linear; so

$$(1.4) \qquad Q(N(t)) = t + c, \qquad c \text{ a constant.}$$

It follows from (1.3) that dQ/dN is not zero, and from the continuity of R that dQ/dN does not change sign. Therefore $Q(N)$ is monotonic and hence invertible. This means that (1.4) can be solved for N:

$$(1.5) \qquad N(t) = Q^{-1}(t + c).$$

The constant c can be related to the initial value $N(0) = N_0$ by setting $t = 0$ in (1.4); this yields

$(1.4)_0$ $$Q(N_0) = c.$$

With this determination of c, the function $N(t)$ in (1.5) is a solution of (1.1) with initial value

$(1.5)_0$ $$N(0) = N_0 = Q^{-1}(c).$$

We now turn to the more interesting case where $R(N)$ vanishes at some points. We require that the function $R(N)$

(i) be differentiable

(ii) have simple zeros; M is called a *simple zero* of $R(N)$ if

(1.6) $$R(M) = 0 \quad \text{and} \quad \frac{dR}{dN}(M) \neq 0.$$

We claim that $R(N)$ *changes sign as N passes a simple zero M of R.* To prove this claim, we use the assumption (1.6) that dR/dN is not zero at M. By continuity, $dR/dN \neq 0$ at all points sufficiently close to M, so $R(N)$ is monotonic in an interval containing M as an interior point, and hence changes sign as N crosses the point M where R vanishes.

We shall show next that, if the initial value N_0 of $N(t)$ is not a zero of $R(N)$, then $N(t)$ is never a zero of R. We shall accomplish this by showing that, if $R(N_0) < 0$ and if M is the largest zero of R less than N_0, then $N(t)$ decreases monotonically toward M *but never reaches M*. Similarly, if $R(N_0) > 0$ and K is the smallest zero of R greater than N_0, then $N(t)$ increases monotonically toward K *but never reaches K*.

If $R(N_0) \neq 0$, then Formulas (1.5), $(1.5)_0$ furnish a uniquely determined solution of Equation (1.1) with initial value N_0 as long as $R(N(t)) \neq 0$. By assumption $R(N_0) \neq 0$; suppose, for the sake of definiteness, that $R(N_0) < 0$. Then, by (1.1), $dN/dt < 0$, and $N(t)$ is and remains a decreasing function of t until $N(t)$ crosses a zero of R which, we claim, will never happen. Since N decreases from N_0, it can approach only those zeros of R which are less than N_0. Among those, let M be the closest to N_0, that is,

$$R(M) = 0, \quad M < N_0, \quad \text{and } R(N) < 0 \quad \text{in } (M, N_0).$$

Denote the deviation of N from M by F (see Figure 9.1a):

$$F(t) = N(t) - M.$$

Setting $N = M + F$ into (1.1), we get

(1.7) $$\frac{dF}{dt} = \frac{dN}{dt} = R(N) = R(M + F).$$

Next we apply the mean value theorem to $R(N)$:

$$R(M + F) = R(M) + F \frac{dR}{dN}(M_1), \quad M_1 \text{ between } M \text{ and } M + F.$$

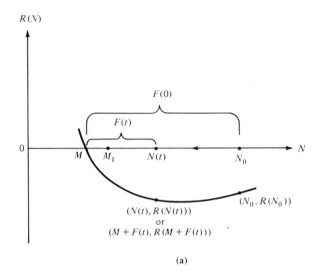

(a)

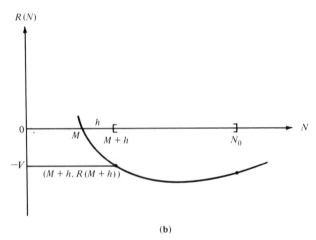

(b)

Figure 9.1

Since $R(M) = 0$, we have

(1.8)
$$R(M + F) = R'(M_1)F,$$

where

(1.8)'
$$R'(M_1) = \frac{dR}{dN}(M_1).$$

Substituting (1.8) into (1.7) gives

(1.9)
$$\frac{dF}{dt} = R'(M_1)F, \quad M_1 \text{ between } M \text{ and } M + F.$$

Denote the minimum of dR/dN on the interval $[M, N_0]$ by $-H_0, H_0 > 0$; then $R'(M_1) \geq -H_0$. Dividing both sides of (1.9) by F and using the lower bound $-H_0$ for $R'(M_1)$, we obtain the inequality

$$\text{(1.10)} \qquad \frac{1}{F}\frac{dF}{dt} \geq -H_0.$$

We recognize the left side of (1.10) as the derivative of $\log F$ with respect to t and rewrite (1.10) as

$$\text{(1.11)} \qquad \frac{d}{dt}[\log F + H_0 t] \geq 0.$$

By the monotonicity criterion, we conclude that the function $\log F + H_0 t$ increases with t. So, for any positive input t, its value is larger than it was at $t = 0$:

$$\log F(t) + H_0 t \geq \log F(0)$$

or, equivalently,

$$\log F(t) \geq \log F(0) - H_0 t.$$

Since the exponential function is an increasing function, it follows that

$$e^{\log F(t)} \geq e^{\log F(0) - H_0 t},$$

whence

$$\text{(1.12)} \qquad F(t) \geq F(0)e^{-H_0 t}, \qquad t > 0.$$

This shows that the positive decreasing function $F(t)$ remains positive forever, and therefore that $F(t) + M = N(t)$ never reaches the zero M of R.

Next we show that, although $N(t)$ never reaches M in a finite time, it reaches M in infinite time in the sense that

$$\lim_{t \to \infty} N(t) = M.$$

The intuitive argument for this is as follows: Let h be any positive time interval, and let T be a time interval during which $N(t)$ has not yet passed below $M + h$, i.e.,

$$\text{(1.13)} \qquad N(T) > M + h.$$

The function $R(N)$ is negative in the closed interval $[M + h, N_0]$, so it is less than some negative number $-V$, see Figure 9.1b. It follows then from the differential equation (1.1) that, for $0 < t < T$, $N(t)$ decreases at least at the rate V, so that

$$\text{(1.14)} \qquad N(T) \leq N_0 - VT.$$

445

Comparing (1.13) and (1.14), we conclude that $M + h \leq N_0 - VT$, and therefore

$$T \leq \frac{N_0 - M - h}{V}.$$

This shows that, for T greater than the right side, $N(T)$ must violate (1.13) and be less than $M + h$. Since h is an arbitrary positive number, and since we have already shown in (1.12) that $N(t)$ does not sink below M, this proves that $N(t)$ tends to M as $t \to \infty$.

It is extremely important, in many applications, to know not only that under our assumptions $N(t)$ tends to the zero M of R, but also at what rate $N(t) \to M$ as $t \to \infty$. We have just seen that eventually, $N(t)$ gets arbitrarily close to M, so we might as well begin clocking the behavior of N at a point $M + F$ so close to M that dR/dN, which by assumption is negative at M, is negative for all N between M and $M + F = N(0) = N_0$. Then so is the maximum of dR/dN in $[M, N_0]$, which we denote by $-H_1$, with $H_1 > 0$. The number $R'(M_1)$ appearing in (1.8) is the value of dR/dN at M_1 in $[M, N_0]$, and so $R'(M_1) \leq -H_1$. Dividing (1.9) by F and using this upper bound for $R'(M_1)$, we obtain

$$\frac{1}{F} \frac{dF}{dt} \leq -H_1.$$

We analyze the consequences of this inequality as we analyzed (1.10). Thus, we write it as

$$\frac{d}{dt} [\log F + H_1 t] \leq 0,$$

and deduce that $\log F + H_1 t$ is a decreasing function of t. Hence the bracketed expression is less at $t + T$ than at t. In symbols, for $T > 0$,

$$\log F(t + T) + H_1 t + H_1 T \leq \log F(t) + H_1 t.$$

Exponentiation of this inequality gives

(1.15) $$F(t + T) \leq F(t)e^{-H_1 T}, \qquad T > 0,$$

which shows that $F(t)$ tends to zero exponentially, while inequality (1.12) shows that $F(t)$ tends to zero no faster than exponentially. The two constants $-H_0$, $-H_1$ appearing in the exponents of the lower and upper bounds of $F(t)$ are the minimum and maximum of dR/dN in $[M, N_0]$, respectively. If N_0 is close to M, the minimum and maximum of the continuous function dR/dN differ little from $-H$, its value at M:

(1.16) $$\frac{dR}{dN}(M) = -H.$$

We can combine inequalities (1.12) and (1.15) in the somewhat loose statement that as $t \to \infty$, $N(t)$ tends to M approximately at the exponential rate $\exp\{-Ht\}$.

In our analysis we have assumed, for sake of definiteness, that $R(N_0) < 0$. In the opposite case, when $R(N_0) > 0$, it follows from an entirely analogous analysis that the function $N(t)$ increases as t increases and that, as $t \to \infty$, it tends to the smallest zero of $R(N)$ which is greater than N_0.

The remaining case, when N_0 is a zero of $R(N)$, is the simplest of all. Then $N(t) \equiv N_0$ for all t. Clearly this satisfies the differential equation (1.1), for, both sides are zero in this case. We claim that the constant solution is the only one that (1.1) has which starts at a zero M of R. For if there were another one, this alleged second solution would be $\neq M$ at some later time t_1 and would run into M as t decreases from t_1 to 0. This is impossible for the same reason which shows that $N(t)$ cannot run into a zero of R as t goes from 0 to t_1. A constant solution is called a *steady state*. Our analysis shows that *the steady states of the differential equation* (1.1) *are the zeros of the function R.*

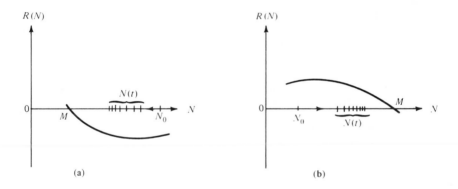

(a)

(b)

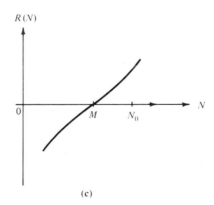

(c)

Figure 9.2

We summarize our results in

Theorem 1.1. *Suppose the function $R(N)$ is differentiable and has simple zeros. Then the differential equation (1.1) has exactly one solution with prescribed initial value $N(0) = N_0$. If N_0 is not a zero of $R(N)$, this solution is given by Formulas (1.5), (1.5)$_0$; when $R(N_0) < 0$, the solution tends to the largest zero of $R(N)$ which is less than N_0, as $t \to \infty$, and when $R(N_0) > 0$, $N(t)$ tends to the smallest zero of $R(N)$ greater than N_0 as $t \to \infty$; when $R(N_0) = 0$, the solution $N(t) = N_0$ for all t.*

Observe that a zero M of R, where dR/dN is negative, is the limit as $t \to \infty$ of solutions $N(t)$ starting at either side of M; whereas if dR/dN is positive at M, then M is *not* the limit as $t \to \infty$ of any solution, except of course the steady state $N(t) \equiv M$. We express this state of affairs by saying that *a steady state M is stable if dR/dN is negative at M, and unstable if dR/dN is positive at M;* see Figure 9.2.

EXERCISES

1.1 Find all solutions of the differential equations
  (a) $dN/dt = N^2$
  (b) $dN/dt = -N^2$.
  (c) Show that in case (a) solutions whose initial values are negative tend to zero, but at a rate much slower than exponential.

1.2 Show that the differential equation

$$\frac{dN}{dt} = \sqrt{N}$$

is satisfied by both functions $N(t) = 0$ and $N(t) = t^2/4$. Doesn't this contradict Theorem 1.1, according to which two solutions with the same initial value agree for all t?

1.3 Deduce from Formula (1.3) that if R is differentiable and $R(M) = 0$, then $Q(N)$ tends to ∞ as $N \to M$. Deduce from this that if $R(N_0) \neq 0$, then $R(N(t)) \neq 0$ for all t.

9.2 Growth and fluctuation of population

The Arithmetic of Population and Development

The rate of population growth is itself impeding efforts at social and economic development. Take the case, for example, of a developing country which has achieved an annual increase in its gross national product of five percent—a very respectable effort indeed, and one which few countries have been able to maintain on a continuing basis. Its population is increasing at 3 percent annually. Thus, its per capita income is increasing by 2 percent each year, and will take 35 years to double, say from $100 per year to $200. In the meantime, its population will have almost tripled, so that greatly increased numbers of people will be living at what is still only a subsistence level.

448

Reduction of the rate of population growth is not a *sufficient* condition for social and economic development—other means, such as industrialization, must proceed along with such reduction—but it is clear that it is a *necessary* condition without which the development process is seriously handicapped.

Dr. John Maier
Director for Health Sciences
The Rockefeller Foundation

In this section we shall study the growth of population of a single and of several species living in the same environment. The growth rate of a population is related to the birth and death rates. The basic equation governing the growth in time t of a single population of size $N(t)$, is

$$(2.1) \qquad \frac{dN}{dt} = B - D,$$

where B is the *birth rate*, D is the *death rate* for the total population. What do B and D depend on? They certainly depend on the age distribution within the population; a population with a high percentage of old members will have a higher death rate and lower birth rate than a population of the same size, but with a low percentage of old members. Yet in this section we shall disregard this dependence of birth and death rates on age distribution. The results we shall derive are quantitatively relevant in situations where the age distribution turns out to change fairly little in time.

If we assume, in addition, that the basic biological functions of the individual in the population are unaffected by the population size, then it follows that *both birth rate and death rate are proportional to the population size*. The mathematical expression of this idea is

$$(2.2) \qquad B = cN, \qquad D = dN,$$

where c, d are constants. Substituting this into (2.1) leads to

$$(2.1)' \qquad N' = aN,$$

where $a = c - d$, and prime is the usual notation for derivative. The solution of this equation is

$$(2.3) \qquad N(t) = N_0 e^{at},$$

where $N_0 = N(0)$ is the initial population size. For positive a this is the celebrated—and lamented—Malthusian law of population explosion.

If the population grows beyond a certain size, the sheer size of the population will *depress* the birth rate and *increase* the death rate. We summarize this as

$$\frac{dN}{dt} = aN - \text{effect of overpopulation},$$

where a denotes the constant entering Equation (2.1)'.

449

How can we quantify the effect of overpopulation? Let us assume that the *effect of overpopulation is proportional to the number of encounters between members of the population*, and that these encounters are *by chance*, i.e., are due to individuals bumping into each other without premeditation. For each individual, the number of encounters is proportional to the population size; therefore, the total number of such encounters is proportional to the *square* of the population. So the effect of overpopulation is to depress the rate of population growth by bN^2, b some positive number. The resulting growth equation is

$$(2.4) \qquad \frac{dN}{dt} = aN - bN^2, \qquad a, b > 0.$$

This equation was introduced into the theory of population growth by Verhulst.

Equation (2.4) is a special instance of Equation (1.1) discussed in Section 9.1,

$$(2.5) \qquad \frac{dN}{dt} = R(N),$$

with R of the form

$$(2.6) \qquad R(N) = aN - bN^2, \qquad a, b > 0.$$

The solution of the general equation is given by Formulas (1.5), $(1.5)_0$, in terms of Q defined in Formula (1.3). We now determine Q in case (2.6), and perform the inversion of Q indicated in Formula (1.5).

According to (1.3),

$$(2.7) \qquad \frac{dQ}{dN} = \frac{1}{R(N)}.$$

To find Q we have to carry out an integration. For this purpose it is convenient to expand $1/R$ in partial fractions, i.e., write it in the form

$$(2.8) \qquad \frac{1}{R(N)} = \frac{1}{aN - bN^2} = \frac{1}{a}\left[\frac{1}{N} - \frac{1}{N - (a/b)}\right].$$

It is easy to verify this identity directly; the general theory of partial fraction expansions is described in the appendix to Chapter 2. Substituting (2.8) into (2.7) leads to

$$\frac{dQ}{dN} = \frac{1}{a}\left[\frac{1}{N} - \frac{1}{N - (a/b)}\right];$$

this relation is satisfied by

$$Q(N) = \frac{1}{a}\log\frac{N}{N - (a/b)}.$$

450

Substituting this value of Q into (1.4), we get

$$\frac{1}{a} \log \frac{N}{N - (a/b)} = t + c.$$

Multiply by a and exponentiate both sides; the result is

(2.9) $$\frac{N}{N - (a/b)} = e^{at + ac} = e^{at}k,$$

where $k = \exp\{ac\}$. The value of k is easily determined by setting $t = 0$ in (2.9). If N_0 denotes the initial value of N, we get

(2.10) $$\frac{N_0}{N_0 - (a/b)} = k.$$

We can determine N as function of t from (2.9): multiplying by $N - (a/b)$ we obtain

$$N = k\left(N - \frac{a}{b}\right)e^{at}.$$

This is easily solved for N:

$$N(t) = \frac{a/b}{1 - e^{-at}/k}.$$

Substituting $1/k = 1 - a/(bN_0)$ from (2.10), we find, after multiplying top and bottom by b, that

(2.11) $$N(t) = \frac{a}{b - [b - (a/N_0)]e^{-at}}.$$

Many interesting properties of $N(t)$ can be deduced from this formula by mere inspection.

Theorem 2.1. *Assume that the initial value $N(0) = N_0$ of N is positive and $\neq a/b$. If $N_0 > a/b$, then $N(t) > a/b$ for all t and decreases as time increases. If $N_0 < a/b$, then $N(t) < a/b$ for all t and increases as time increases. In both cases, $N(t) \to a/b$ as $t \to \infty$.*

These findings are in complete agreement with Theorem 1.1 of Section 9.1. For, according to that theorem, every solution of Equation (2.5) tends to the nearest steady state, provided that steady state is stable. For the equation at hand, R is given by (2.6):

$$R(N) = aN - bN^2.$$

The steady states are the zeros of R, in this case

$$N = 0 \quad \text{and} \quad N = a/b.$$

The derivative of R is

$$\frac{d}{dN} R(N) = a - 2bN;$$

so its values at the zeros of R are

$$\frac{dR}{dN}(0) = a \quad \text{and} \quad \frac{dR}{dN}(a/b) = -a.$$

Since a is positive, we conclude that both zeros are simple, and that dR/dN is positive at $N = 0$, negative at $N = a/b$. Therefore, according to Section 9.1, the zero 0 is unstable and the zero a/b is stable, and all solutions with initial value $N_0 > 0$ tend to the stable steady state a/b as t tends to ∞. This is exactly what we found in Theorem 2.1 by studying an explicit formula for all solutions. It is gratifying that properties of solutions can be deduced directly from the differential equation which they satisfy without help from an explicit formula for solutions. Indeed, there are very few differential equations whose solutions can be described by an explicit formula.

The result we have just obtained, that all solutions of Equation (2.4) tend to a/b as t tends to ∞, has great demographic significance, for it predicts the eventual steady state of any population which can be reasonably said to be governed by an equation of form (2.4). In order to make practical use of this theoretical prediction, we have to estimate the parameters a and b. This can be done either by relating a and b to measurable quantities, or by fitting the population history observed over the past to a function of form (2.11), and estimating the parameters a and b as the ones that give the best fit. In order to do that, we have to know something about the shape of solutions $N(t)$ of the differential equation (2.4). The next result gives valuable information about the shape of graphs of solutions $N(t)$, and is obtained not by looking at Formula (2.11) for solutions, but by studying the differential equation directly.

Differentiating (2.4) with respect to t, we get

$$N'' = aN' - 2bNN'.$$

Substituting into the right side the expression for N' given in (2.4) we obtain the relation

(2.12) $$N'' = (a - 2bN)(a - bN)N.$$

The right side of (2.12) is a function of N which is

$$\text{positive for} \quad N < \frac{a}{2b},$$

(2.13) $$\text{negative for} \quad \frac{a}{2b} < N < \frac{a}{b},$$

$$\text{positive for} \quad \frac{a}{b} < N.$$

A function whose second derivative is positive was called *convex* in Section 3.8, and one whose second derivative is negative was called *concave*. We saw there that the graph of a convex function curves upward, and the graph of a concave function curves downward. So we deduce from (2.12) and (2.13) that *the graph of $N(t)$ is convex whenever $N(t)$ is less than $a/2b$ or greater than a/b, and concave when $N(t)$ lies between $a/2b$ and a/b;* see Figure 9.3.

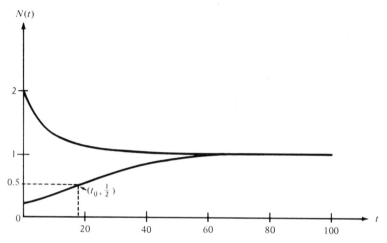

Figure 9.3   Solutions of $N' = aN - bN^2$ for initial values $N_0 = 0.2$ and 2 and parameters $a = b = 0.75$.

We can verify these facts by plotting the graphs of some typical functions $N(t)$. In Figure 9.3 we have drawn such graphs, choosing $a = b = 0.75$ and taking two different initial values, $N_0 = 2$ and $N_0 = 0.2$. Observe that both functions $N(t)$ tend to $a/b = 1$, as asserted in Theorem 2.1. The lower graph, starting at $N_0 = 0.2$, is convex for $t \leq t_0$ and concave for $t > t_0$. $N(t_0) = a/2b = 0.5$, in accordance with our observations above. The point $(t_0, 0.5)$, separating the convex and concave portions of the graph of $N(t)$, is called a *point of inflection*.

The lower curve in Figure 9.3 is called an S-curve because its shape resembles the letter S. Such S-shaped population curves have been observed in diverse situations, such as the growth of population in various countries (see Figure 9.7), as well as the proliferation of fruit flies in a fixed environment.

Our analysis shows that *at the point of inflection the population size is exactly half of the steady state population.* If a population under observation has already passed the point of inflection, one can predict the eventual size of the steady state population as double the population at inflection time.

We now return to the basic equation (2.1) of population growth and again we assume that the *death rate is proportional to the population size.* This amounts to assuming that death is due to "natural" causes, and not due to one member of the population eating the food needed by another member, or due to one member eating another. On the other hand, we *challenge the assumption that birth rate is proportional to population size.* This assumption holds true for extremely primitive organisms, such as amoeba, which reproduce by dividing. It is also true of well organized species, such as humans, who seek out a partner and proceed to produce a biologically or socially determined number of offspring. But there are important classes of organisms whose reproductive sophistication falls between those of the amoeba and humans, who need a partner for reproduction but must rely on chance encounters for meeting a mate. The expected number of encounters is proportional to the *product* of the number of males and females. If these are equally distributed in the population, the number of encounters—and so the birthrate—is proportional to N^2. The death rate on the other hand is proportional to the population size N. Since the rate of population growth is the difference between birth rate and death rate, the equation governing the growth of such population is

$$(2.14) \qquad \frac{dN}{dt} = bN^2 - aN, \quad a, b > 0.$$

Equation (2.14) is of the form (2.5) with

$$R(N) = bN^2 - aN, \quad a, b > 0.$$

This function has two zeros, 0 and a/b. The derivative of R is

$$\frac{dR}{dN} = 2bN - a;$$

so its value at the zeros of R is

$$\frac{dR}{dN}(0) = -a \qquad \text{and} \qquad \frac{dR}{dN}(a/b) = a.$$

Since a is positive, it follows that both zeros are simple, and that dR/dN is negative at $N = 0$, positive at $N = a/b$. Therefore, according to Section 9.1, the zero 0 is stable and the zero a/b is unstable; *all solutions with initial value $N_0 < a/b$ tend to 0 as $t \to \infty$.*

This stability of 0 is the stability of death; what we have discovered by our simple analysis is a very interesting and highly significant threshold effect: *Once the population size N_0 drops below the critical size a/b, the population tends to extinction.* This notion of a critical size is important for the preservation of a species. A species is classified *endangered* if its current size is perilously close to its critical size.

We now turn to a situation involving *two* species, where one species feeds on nourishment whose supply is ample, and the other species feeds on the first species. We denote the population size of the two species by N and P,

N denoting the *prey*, P denoting the *predators*. Both N and P are functions of t, and their growth is governed by differential equations of the form (1.1). The initial task is to choose suitable functions B and D describing the birth rates and death rates of each species.

We assume that the two species encounter each other by chance, at a rate proportional to the product of the size of the two populations. If we assume that the principal cause of death among the first species is due to being eaten by a member of the second species, then the death rate for the first species is proportional to the product NP. We assume that the birth rate for the predator is proportional to the population size P, and that the portion of the young that survive is proportional to the available food supply N. Thus the effective birthrate is proportional to NP. Finally we assume that the birth rate of the first species and the death rate of the second species are proportional to the size of their respective populations. So the equations governing the growth of these species are

$$\frac{dN}{dt} = aN - bNP,$$

(2.15)

$$\frac{dP}{dt} = -cP + hNP.$$

Here a, b, c and h are all positive constants. These equations were first set down and analyzed, independently, by Volterra and by Lotka. Lotka's work on this and other population models is described in his book *Elements of Physical Biology*, originally published in 1925 and reprinted by Dover, New York, in 1956. The work of Volterra, inspired by the fluctuations in the size and composition of the catch of fish in the Adriatic, appeared in *Cahier Scientifique*, Vol. VII, Gauthier-Villars, Paris, 1931, under the romantic title "Leçons sur la théorie mathématique de la lutte pour la vie" (Lessons on the mathematical theory of the fight for survival). It is reprinted in his collected works published by the Accademia Lincei, Rome.

The first order of business is to show that these laws of growth, and knowledge of the initial population size, are sufficient to determine the size of both populations at all future times. We formulate this as a

Uniqueness theorem 2.2. *A solution is uniquely determined for all time by the specification of the initial values N_0, P_0 of N and P, respectively.*

A proof of this proposition is outlined in Exercise 2.4.

We now study the solutions of (2.15). First we look at the simplest kind of solutions, the *steady states*, where both P and N are constants. These are the values where the right sides of (2.15) are zero:

(2.16) $$aN - bNP = 0, \qquad -cP + hNP = 0.$$

Both equations are satisfied by $N = 0, P = 0$; this is a "trivial" solution, corresponding to no population. It is easy to verify that the only other solution of (2.16) is

(2.17)
$$P = \frac{a}{b}, \qquad N = \frac{c}{h}.$$

In order to investigate nonconstant solutions, let us take the case that both species are present, i.e., $P > 0$, $N > 0$. We divide Equations (2.15) by N and P respectively, obtaining

$$\frac{1}{N}\frac{dN}{dt} = a - bP,$$

(2.18)

$$\frac{1}{P}\frac{dP}{dt} = hN - c.$$

We multiply the first equation by $hN - c$, the second equation by $bP - a$ and add. The sum of the right sides is 0, so we get the relation

$$\left(h - \frac{c}{N}\right)\frac{dN}{dt} + \left(b - \frac{a}{P}\right)\frac{dP}{dt} = 0.$$

Using the chain rule we rewrite this relation as

(2.19)
$$\frac{d}{dt}[hN - c\log N + bP - a\log P] = 0.$$

We introduce the abbreviations H and K through

(2.20) $\qquad H(N) = hN - c\log N, \qquad K(P) = bP - a\log P.$

Note that (2.19) says the derivative of $H(N) + K(P)$ with respect to time is zero; so we conclude from the fundamental theorem of calculus

Theorem 2.3. *For any solution* $N(t)$, $P(t)$ *of the system of Equations* (2.15), *the quantity*

(2.21)
$$H(N) + K(P)$$

is independent of t. *Here* H *and* K *are defined by* (2.20).

The constancy of the sum (2.21) is strongly reminiscent of the law of conservation of energy in mechanics and can be used, like the law of conservation of energy, to gain qualitative and quantitative information about solutions. For that purpose the following useful facts about the functions H and K are contained in

Theorem 2.4

    (a) $H(N)$ *and* $K(P)$ *tend to* ∞ *as* N, P *tend to 0 or* ∞.
    (b) *Both functions* $H(N)$ *and* $K(P)$ *are convex.*
    (c) $H(N)$ *reaches its minimum at* $N = c/h$, $K(P)$ *at* $P = a/b$.

PROOF. Recall that all four constants h, c, b, and a are positive. As N, P tend to 0, $\log N$ and $\log P$ tend to $-\infty$; so it follows from the definitions (2.20) of H and K that both tend to ∞ as N, P tend to zero. Since N tends to ∞ faster than $\log N$, $H(N) = hN - c \log N$ tends to ∞ as $N \to \infty$, and similarly, $K(P)$ tends to ∞ as $P \to \infty$. This completes the proof of part (a).

To prove part (b) we differentiate twice each function H and K given by Formula (2.20). We obtain

$(2.22)_N$
$$\frac{dH}{dN} = h - \frac{c}{N}, \qquad \frac{d^2 H}{dN^2} = \frac{c}{N^2}$$

and

$(2.22)_P$
$$\frac{dK}{dP} = b - \frac{a}{P}, \qquad \frac{d^2 K}{dP^2} = \frac{a}{P^2}.$$

Clearly, both $d^2 H/dN^2$ and $d^2 K/dP^2$ are positive. This proves the convexity of H, K, and completes part (b).

A convex function which tends to ∞ as its argument tends either to 0 or to ∞ has exactly one minimum. This minimum is where the first derivative of the function vanishes. Using Formulas $(2.22)_N$ and $(2.22)_P$, we see that these occur as indicated in part (c). □

It follows from part (c) that the minimum of the conserved quantity

$$H(N) + K(P)$$

occurs at $N = c/h$, $P = a/b$. Note that this is the steady state (2.17) of the system of equations (2.15).

Theorem 2.5

(a) *Neither species can become extinct; i.e., there is a positive lower bound for each population, $N(t)$ and $P(t)$ throughout the whole time history.*

(b) *Neither species can proliferate ad infinitum; i.e., there is an upper bound for each population throughout its time history.*

(c) *The steady state is neutrally stable in the following sense: if the initial state N_0, P_0 is near the steady state $(c/h, a/b)$, then $N(t)$, $P(t)$ stays near the steady state throughout the whole time history.*

PROOF. All three results follow from the conservation law contained in Theorem 2.3 and the properties of the conserved quantity stated in Theorem 2.4. According to part (a) of Theorem 2.4, $H(N)$ and $K(P)$ tend to ∞ as N and P respectively, tend to 0 or ∞. This shows that the sum $H(N) + K(P)$ tends to ∞ if either N or P tend to 0 or ∞. Since this is incompatible with the constancy of $H + K$, we conclude that neither $N(t)$ nor $P(t)$ can approach 0 or ∞. This proves parts (a) and (b) of Theorem 2.5.

Part (c) asserts that a solution $N(t)$, $P(t)$ which is initially close to the steady state remains close to the steady state. The reason for this is the combination of the following circumstances. The sum $H(N) + K(P)$ remains constant; therefore each term can grow only to the extent that the other diminishes. If P_0 is near a/b, $K(P_0)$ is near the minimum value of the function K; therefore $K(P)$ cannot diminish much. So $H(N)$ cannot grow much. Similarly $K(P)$ cannot grow much beyond its initial value. This shows that both $H(N)$ and $K(P)$ are trapped near their minimum values. But then N and P are trapped near c/h and a/b, the unique points where the minima are reached. Now we spell out the details.

Suppose the initial values N_0, P_0 differ little from c/h, a/b, respectively. Then, by continuity, $H(N_0)$ and $K(P_0)$ differ little from $H(c/h)$, $K(a/b)$:

(2.23)
$$H(N_0) < H(c/h) + s,$$
$$K(P_0) < P(a/b) + s,$$

s a small quantity. Let N, P denote the values of the two populations at some later time. According to the conservation law,

$$H(N) + K(P) = H(N_0) + K(P_0).$$

Using inequalities (2.23) we deduce that

(2.24) $$H(N) + K(P) \leq H(c/h) + K(a/b) + 2s.$$

On the other hand, since H achieves its minimum at c/h, K at a/b, it follows from (2.24) that

$$H(N) + K(a/b) < H(c/h) + K(a/b) + 2s,$$
$$H(c/h) + K(P) < H(c/h) + K(a/b) + 2s.$$

These can be rewritten as

(2.25)
$$H(N) < H(c/h) + 2s,$$
$$K(P) < K(a/b) + 2s.$$

Since H achieves its only minimum at c/h, any point N where the value of H is near the minimum must be near c/h:

$$\left| N - \frac{c}{h} \right| \quad \text{is small for all } t > 0.$$

Similarly

$$\left| P - \frac{a}{b} \right| \quad \text{is small for all } t > 0.$$

This completes the proof of part (c) of Theorem 2.5.   □

The next result is both interesting and surprising.

Theorem 2.6. *Every time history is periodic; i.e., for every solution $N(t)$, $P(t)$ of Equations (2.15) there is a time T such that*

$$N(T) = N(0), \qquad P(T) = P(0).$$

T is called the *period* of this particular time history. Different time histories have different periods.

PROOF. We write Equations (2.15) as

$(2.26)_N$
$$\frac{dN}{dt} = N(a - bP)$$

$(2.26)_P$
$$\frac{dP}{dt} = P(hN - c)$$

and conclude, from the monotonicity criterion that N and P are increasing or decreasing functions of t, depending on whether the right sides of Equation (2.26) are positive or negative. The sign of the right side of $(2.26)_N$ is the sign of $a/b - P$, while the sign of the right side of $(2.26)_P$ is the same as the sign of $N - c/h$. Therefore

$(2.27)_N$
$$N(t) \begin{cases} \text{increases when } P < \dfrac{a}{b} \\[2mm] \text{decreases when } P > \dfrac{a}{b} \end{cases}$$

and

$(2.27)_P$
$$P(t) \begin{cases} \text{decreases when } N < \dfrac{c}{h} \\[2mm] \text{increases when } N > \dfrac{c}{h}. \end{cases}$$

We shall, using the two relations (2.27) trace qualitatively the time histories of $N(t)$ and $P(t)$, very much the way we have traced, in Section 7.2, the ups and downs of sine and cosine (refer to Figure 9.4a).

The initial values N_0, P_0 may be chosen arbitrarily. For sake of definiteness we choose $N_0 < c/h$, $P_0 < a/b$. N starts to increase, P to decrease, until time t_1, when N reaches c/h. At t_1, P starts to increase, N continues to increase until time t_2 when P reaches a/b. Then N starts to decrease, P continues to increase until time t_3 when N reaches c/h. P starts to decrease, N continues to decrease until time t_4 when P reaches a/b. N starts to increase, P continues to decrease until time t_5 when N reaches c/h.

We claim that at time t_5 the value of P is the same as at time t_1. To convince the reader of this, we appeal to the conservation law of Theorem 2.3. Denoting the values of N and P at t_1 and t_5 by subscripts 1 and 5, we conclude from the conservation law that

(2.28)
$$H(N_5) + K(P_5) = H(N_1) + K(P_1).$$

459

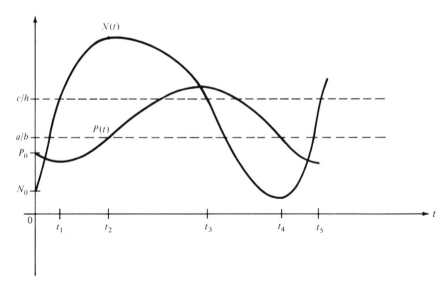

Figure 9.4a

The times t_1 and t_5 were so chosen that both N_1 and N_5 are c/h. It follows then from (2.28) that

$$K(P_5) = K(P_1).$$

By Theorem 2.4, $K(P)$ decreases monotonically in the interval $0 \leq P \leq a/b$. Since t_1 and t_5 were so chosen that both P_1 and P_5 are $< a/b$, it follows therefore from $K(P_5) = K(P_1)$ that

$$P_5 = P_1.$$

We have already seen that $N_5 = N_1$. It follows from the uniqueness theorem (2.2) that the time history of $N(t)$, $P(t)$ after t_5 is a mere repetition of its time history after t_1. Therefore the periodicity claimed in Theorem 2.6 is established, the period being $T = t_5 - t_1$. $\qquad\square$

As remarked before, each solution is periodic, but different solutions have different periods. It is quite remarkable that the following quantities are the same for all solutions.

Theorem 2.7. *The average values of P and N over a period are the same for all solutions, and equal their steady state values a/b and c/h.*

PROOF. We shall use form (2.18) of the equations:

$$\frac{1}{N}\frac{dN}{dt} = a - bP, \qquad \frac{1}{P}\frac{dP}{dt} = hN - c.$$

We integrate both equations from 0 to T, where T is the period of the solution in question. Using the chain rule we get

$$\log N(T) - \log N(0) = \int_0^T \frac{1}{N}\frac{dN}{dt}\,dt = \int_0^T (a - bP)dt,$$

$$\log P(T) - \log P(0) = \int_0^T \frac{1}{P}\frac{dP}{dt}\,dt = \int_0^T (hN - c)dt.$$

Since T is the period of N and P, the left sides are zero; so we obtain the relations

$$0 = aT - b\int_0^T P(t)dt, \qquad 0 = h\int_0^T N(t)dt - cT.$$

Dividing the first equation by bT, the second by hT, we get

$$\frac{1}{T}\int_0^T P(t)dt = \frac{a}{b}, \quad \text{and} \quad \frac{1}{T}\int_0^T N(t)dt = \frac{c}{h}.$$

The expressions on the left are the average values of P and N over a period, those on the right are their steady state values. This proves Theorem 2.7. □

This result contains several interesting features; we mention two. The constants a, b, c, h in Equations (2.15), which determine the steady state a/b, c/h, have nothing to do with the initial values P_0, N_0 of our populations. So it follows from Theorem 2.7 that the average values of P and N are independent of the initial values. Thus, if we were to increase the initial population N_0, for example by stocking a lake with fish, this would not affect the average size of $N(t)$ over a period, but would only lead to bigger oscillations in the size of $N(t)$.

For another application, suppose we introduce fishing into the model. Assuming that the catch of predator and prey is proportional to the number of each, fishing diminishes each population at a rate proportional to the size of that population. Denoting by f the constant of proportionality, we have the following modification of the basic Equations (2.15):

$$\frac{dN}{dt} = aN - bNP - fN,$$

$$\frac{dP}{dt} = -cP + hNP - fP.$$

We may write these equations in the form

(2.15)$_f$

$$\frac{dN}{dt} = (a - f)N - bPN,$$

$$\frac{dP}{dt} = -(c + f)P + hNP,$$

461

and observe that they differ from (2.15) only in that the coefficient a of N in the first has been replaced by $a - f$, and the coefficient $-c$ of P in the second has been replaced by $-(c + f)$. According to Theorem 2.7, applied to $(2.15)_f$, the average values of P and N are $(a - f)/b$ and $(c + f)/h$, respectively. In other words, increased fishing depresses the average population of predators, but increases the average population of edible fish. During the first World War, the Italian fishing industry reported a marked increase in the ratio of sharks to edible fish. Since less fishing was done during that war than before, this observation is consistent with Volterra's surprising result.

It is instructive to picture the time histories $N(t)$, $P(t)$ (see Figure 9.4a) of population change as curves in the P, N-plane. Periodicity means that these curves close; Figure 9.4b shows the graph in the P, N-plane corresponding to the qualitative histories sketched in Figure 9.4a.

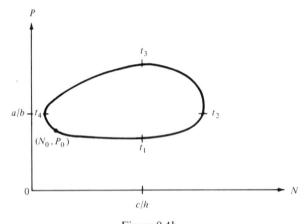

Figure 9.4b

These closed curves can be determined without solving the differential equations (2.15). According to the conservation law of Theorem 2.3, on each curve,

(2.29) $$H(N) + K(P) = \text{constant},$$

where the functions H and K are given by Formula (2.20). The value of the constant can be determined from the initial condition

$$H(N_0) + K(P_0) = \text{constant}.$$

We cannot expect to write down the history of one of the populations, say P, as a single function of the other, since in a closed curve, like that in Figure 9.4b, one value of N may correspond to two values of P. However, we shall explain in Exercise 2.5 how Equation (2.29) can be solved for P by considering separately the intervals where P is monotonic; there $K(P)$ may be inverted. The argument presented there will lead to two solutions $P_-(N)$ and $P_+(N)$ of (2.29), one convex and the other concave.

The representation of time histories of population changes as curves in the N, P-plane is analogous to representing the motions of vibrating particles as curves in the x, v-plane, as explained in Section 8.7.

Equations (2.15) can, of course, be solved approximately by numerical methods. The solutions whose graphs are given in Figures 9.5 and 9.6 were obtained by the numerical method described in Exercise 7.2, Chapter 8. These graphs and curves bear out what has been proved in the text, that solutions of (2.15) are periodic in time.

In more complicated models, numerical computations are indispensable. They not only provide numerical answers which cannot be found any other way, but often reveal patterns of behavior amenable to mathematical analysis.

To conclude we point out several simplifications that were made in the models presented in this section:

(i) We have neglected to take into account the *age distribution* of the population. Since birth rate and death rate are sensitive to this, our models are deficient and would not describe correctly population changes accompanied by shifts in the population in and out of the child bearing age. This phenomenon is particularly important in demography, the study of human populations.
(ii) We have assumed that the population is *homogeneously distributed* in its environment. In many cases this is not so; the population distribution changes from location to location.

In problems such as the geographic spread of epidemics, and the invasion of the territory belonging to one species by another, the interesting phenomenon is precisely the change in population size as function of time and location. Population sizes which depend on age and location as well as time are prototypes of functions of several variables. The calculus of functions of several variables is the natural language for the formulation of laws governing the growth of such populations. This subject will be taken up in Volume II. For the moment we offer some references for further reading:

George I. Bell, Predator–prey equations simulating an immune response, *Mathematical Biosciences*, *16*, 291–314 (1973).
Frank C. Hoppenstaedt, *Mathematical Theories of Population Demography, Genetics and Epidemics*, Vol. 20, Conference Board of the Mathematical Sciences, SIAM, Philadelphia, Pa., April 1975.

Some critical readers were probably disturbed throughout this section that we have treated population size as a differentiable function of t, whereas, in fact, population changes by whole numbers, and so is not even a continuous function of t. Our defense is that these models are just models, i.e., approximations to reality, where some less essential features are sacrificed for the

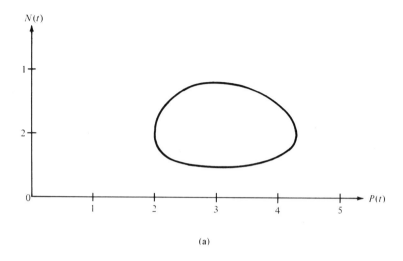

(a)

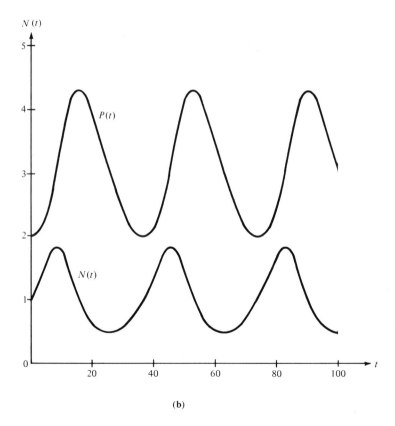

(b)

Figure 9.5   Solutions $N(t)$, $P(t)$ of Equations (2.15) with initial values
$N_0 = 1$, $P_0 = 2$, and parameters $a = 0.3$, $b = c = h = 1$.

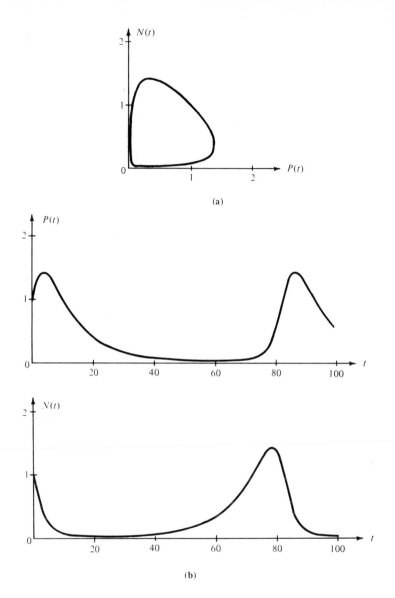

Figure 9.6   Solutions $N(t)$, $P(t)$ of Equations (2.15) with initial values $N_0 = P_0 = 1$ and parameters $a = b = 0.3, c = h = 1$.

sake of simplicity. The point we are making is that the *continuous is sometimes simpler than the discrete since it allows us to use the powerful notions and tools of calculus.* Analogous simplifications are made when dealing with matter, e.g., in applying calculus to such physical quantities as pressure or density as functions of time or space. After all, according to the atomic theory of matter these functions too change discontinuously.

EXERCISES

2.1 Derive a formula for solutions of the differential equation (2.14),

$$\frac{dN}{dt} = bN^2 - aN, \quad a, b > 0,$$

using the method described in Section 9.1 for solving Equation (1.1). Verify, using this formula for the solutions, that if the initial value N_0 is less than a/b, then $N(t)$ tends to 0 as t tends to ∞.

2.2 Verify by a direct calculation that the function $N(t)$ defined by Formula (2.11) has the property that at the point of inflection, defined as the point where $d^2N(t_0)/dt^2 = 0$, $N(t_0) = (\frac{1}{2})N(\infty)$. [*Hint*: To simplify the algebra, show first that by stretching N and t, N can be put in the form $N(t) = 1/(k + e^{-t})$.]

2.3 Get hold of population statistics (see, for example, Figure 9.7), and investigate to what extent it is true that the population size at the inflection point is half of the steady state population size.

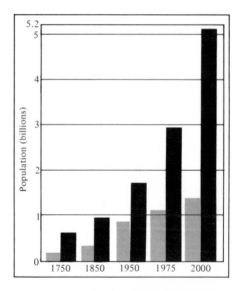

Figure 9.7   World population increase since 1750 is charted for developed countries (light) and underdeveloped countries (dark). Classification as developed or underdeveloped is according to economic and demographic differences now prevailing. Data for the year 2000 are based on a United Nations projection that assumes slowly ebbing growth rates.

2.4 Let p and n be functions of t which satisfy the following differential equations:

$$(2.30) \qquad n' = f(p), \qquad p' = g(n),$$

where $'$ denotes differentiation with respect to t, and f, g are differentiable functions.
(a) Let n_1, p_1 and n_2, p_2 be two pairs of solutions of (2.30). Show that the differences

$$n_1 - n_2 = m, \qquad p_1 - p_2 = q$$

satisfy the inequalities

$$(2.31) \qquad |m'| \le K|q|, \qquad |q'| \le K|m|,$$

where K is an upper bound for the absolute value of the derivatives of the functions f and g.
(b) Deduce from (2.31) that

$$(2.32) \qquad mm' + qq' \le 2K|m||q| \le K(m^2 + q^2).$$

(c) Define E to be

$$E = \tfrac{1}{2}m^2 + \tfrac{1}{2}q^2.$$

Deduce from (2.32) that

$$(2.33) \qquad E' \le 2KE.$$

(d) Deduce from (2.33) that $\exp\{-2Kt\}E$ is a decreasing function of t. Deduce from this that, if $E(0) = 0$, then $E(t) = 0$ for all $t > 0$. Show that this implies that two solutions n_1, p_1 and n_2, p_2 of (2.30) which are equal at $t = 0$ are equal forever after.
(e) Deduce the uniqueness theorem (2.2) for solutions of Equation (2.15) by writing them in form (2.18) and introducing $n = \log N$, $p = \log P$ as new variables. Show that in terms of these variables, (2.18) is of form (2.30).

2.5 Consider Equation (2.29):

$$(2.34) \qquad H(N) + K(P) = \text{constant}.$$

Suppose that H and K are convex functions of their arguments, and that $K(P)$ is a decreasing function of P for $P < P_0$, and an increasing function for $P > P_0$. These are precisely the properties stated in Theorem 2.4 that we have verified for the functions H and K given by (2.20) with $P_0 = a/b$.
(a) Show that solutions of (2.34) where $P > P_0$ can be described by expressing P as function of N. Show that solutions of (2.34) where $P < P_0$ can be described similarly. Denote these functions by $P_+(N)$ and $P_-(N)$.
(b) Let $P(N)$ be either of the functions $P_+(N)$ or $P_-(N)$. Show by differentiating (2.34) twice that

$$\frac{dH}{dN} + \frac{dK}{dP}\frac{dP}{dN} = 0$$

and

$$(2.35) \qquad \frac{d^2H}{dN^2} + \frac{d^2K}{dP^2}\left(\frac{dP}{dN}\right)^2 + \frac{dK}{dP}\frac{d^2P}{dN^2} = 0.$$

(c) Solve (2.35) for $\dfrac{d^2P}{dN^2}$:

$$\frac{d^2P}{dN^2} = - \frac{\dfrac{d^2H}{dN^2} + \dfrac{d^2K}{dP^2}\left(\dfrac{dP}{dN}\right)^2}{\dfrac{dK}{dP}}.$$

Deduce from this formula and the information given about H and K that $P_+(N)$ is a concave function and $P_-(N)$ is convex.

9.3. Mathematical theory of chemical reactions

In high school chemistry we studied the concept of a *chemical reaction*: it is the formation of one or several compounds called the *products* of the reaction out of one or several compounds or elements called *reactants*. Familiar examples are:

(3.1) $$2\,H_2 + O_2 \longrightarrow 2\,H_2O.$$

In words: 2 molecules of hydrogen and 1 molecule of oxygen form 2 molecules of water.

(3.2) $$H_2 + I_2 \longrightarrow 2\,HI.$$

In words: 1 molecule of hydrogen and 1 molecule of iodine form 2 molecules of hydrogen iodide.

A chemical reaction may require energy or may release energy in the form of heat; the technical terms are *endothermic* and *exothermic*. Familiar examples of reactions that release energy are the burning of coal or oil, and, more spectacularly, the burning of an explosive. In fact, the whole purpose of these chemical reactions is to garner the energy they release; the products of these reactions are uninteresting, in fact they can be a severe nuisance, namely pollution. In the chemical industry, the desired commodity is the product of the reactions or of a series of reactions.

The above description of chemical reactions deals with the phenomenon entirely in terms of its initial and final states. In this section we shall study *time histories* of chemical reactions. This branch of chemistry is called *reaction kinetics*. An understanding of kinetics is essential in the chemical industry, because many reactions necessary in certain production processes must be set up so that they occur in the right order within specified time intervals. Similarly, the kinetics of burning must be understood in order, for example, to design an efficiently functioning gasoline engine. The effect of fluorcarbons on depletion of ozone in the stratosphere must be judged by computing the rates at which various reactions involving these molecules occur. Last but not least, reaction kinetics is a valuable experimental tool for studying the structure of molecules.

In this section we shall describe the kinetics of fairly simple reactions; in particular, with those where both, reactants and products, appear as gases. Furthermore we assume that all components are homogeneously distributed in the vessel in which the reaction takes place; that is, we assume that the concentration, temperature and pressure of all components at any given time are the same at all points in the vessel. A few remarks will be made at the end of this section about other cases.

The *concentration* of a reactant measures the number of molecules of that reactant present per unit volume. Note that if two components in a vessel have the same concentration, then that vessel contains the same number of molecules of each component.

In what follows we shall denote different molecules as well as atoms, ions, and radicals, which play important roles in chemical reactions, by different capital letters such as A, B, C, etc., and we shall denote their concentrations by the corresponding lower case letter* such as a, b, c, etc. These concentrations change with time. The *rates* at which they change, i.e., the derivatives of the concentrations with respect to time, are called the *reaction rates*. A basic principle of reaction kinetics says that the reaction rates are completely determined by pressure, temperature, and the concentrations of all components present. Mathematically this can be expressed by specifying the rates as functions of pressure, temperature, and concentrations; then the laws of reaction kinetics take the form of differential equations:

$$(3.3) \qquad \frac{da}{dt} = f(a, b, \ldots ; T, p), \qquad \frac{db}{dt} = g(a, b, \ldots ; T, p), \ldots$$

where f, g, etc. are functions specific to each particular reaction. In the simple reactions considered here, we suppress the dependence of f, g, ... on the temperature T and the pressure p. The determination of these functions is the task of the theorist and experimenter. We shall start with some theoretical observations; of course, the last word belongs to the experimentalist.

The products of a chemical reaction are built out of the same basic components as the reactants, i.e., the same nuclei and the same number of electrons, but the components are just arranged differently. In other words, the chemical reaction is the process by which the rearrangement of the basic components occurs. One can think of this process of rearrangement as a continuous distortion, starting with the original component configuration and ending up with the final one. There is an *energy* associated with each transient configuration; the initial and the final states are *stable*, which means that energy is at a local minimum in those configurations. It follows that during a continuous distortion of one state into the other, energy increases until it reaches a peak and then decreases as the final configuration is reached. There are many paths along which this distortion can take place; the reaction is channelled mainly along the path where the peak value is minimum.

---

* In the chemical literature, the concentration of molecule A would be denoted by [A].

The difference between this minimum peak value of energy and the energy of the initial configuration is called the *activation energy*. It is an energy barrier that has to be surmounted for the reaction to take place.

This description of a chemical reaction as rearrangement in one step is an oversimplification; it is applicable to only a minority of cases, called *elementary reactions*. In the great majority of cases the reaction is *complex*, meaning that it takes place in a number of stages which lead to the formation of a number of intermediate states. The intermediate states—atoms, free radicals, and activated states—may disappear when the reaction is completed. The transitions from initial state to intermediate state, from one intermediate state to another, and from intermediate states to final state are all elementary reactions. So a complex reaction may be thought of as a network of elementary reactions.

We now study the rate of an elementary reaction of form

$$A_2 + B_2 \longrightarrow 2AB,$$

where one A_2 molecule consisting of two A atoms and one B_2 molecule consisting of two B atoms combine to form two molecules of the compound AB. The reaction takes place only if the two molecules collide and are energetic enough* to supply the activation energy needed for the reaction. The frequency with which this happens is proportional to the *product* of the concentrations of A_2 and B_2 molecules, i.e., is equal to

(3.4)   *kab.*

Here a and b denote these concentrations and k is the *rate constant*. Equation (3.4) is called the *law of mass action*.

Denote the concentration of the reaction product AB at time t by $x(t)$. Since the rate of formation of AB is given by (3.4), x satisfies the differential equation

$$(3.5) \qquad\qquad \frac{dx}{dt} = kab.$$

Denote by a_0 and b_0 the initial concentrations of A_2 and B_2. Since each molecule of A_2 and B_2 make 2 molecules of AB,

$$a(t) = a_0 - \frac{x(t)}{2}, \qquad b(t) = b_0 - \frac{x(t)}{2}.$$

Substituting this into (3.5) yields the differential equation

$$(3.5)' \qquad\qquad \frac{dx}{dt} = k\left(a_0 - \frac{x}{2}\right)\left(b_0 - \frac{x}{2}\right).$$

---

* The kinetic energies of the molecules in a vessel are not uniform, but are distributed according to a maxwellian probability distribution (see Section 6.4). Therefore, some molecules always have sufficient kinetic energy to react when they collide.

This equation is of the form (1.1), with x in place of N:

(3.6) $$\frac{dx}{dt} = R(x), \qquad R(x) = k\left(a_0 - \frac{x}{2}\right)\left(b_0 - \frac{x}{2}\right).$$

If the initial concentration of AB is zero,

$$x(0) = x_0 = 0.$$

Since $R(0)$ is positive, the solution $x(t)$ of (3.6) with initial value $x(0) = 0$ starts to increase. According to Theorem 1.1, this solution of (3.6) tends to that zero of $R(x)$ to the right of $x = 0$ which is nearest to $x = 0$. The zeros of $R(x)$ are $x = 2a_0$ and $x = 2b_0$; the one nearest to zero is the smaller of the two. We denote it by x_∞:

(3.7) $$x_\infty = \min[2a_0, 2b_0].$$

It follows then that as $t \to \infty$, $x(t)$ tends to x_∞.

Observe that the quantity x_∞ defined by (3.7) is the largest amount of AB that can be made out of the given amounts a_0 and b_0 of A_2 and B_2. Therefore our result shows that as $t \to \infty$, one or the other of the reactants gets completely used up.

Our second observation is that although $x(t)$ tends to x_∞ as $t \to \infty$, $x(t)$ never reaches x_∞; so, strictly speaking, the reaction goes on forever. However when the difference between $x(t)$ and x_∞ is so small that it makes no practical difference, the reaction is practically over. We show how to estimate the time required for the practical completion of the reaction.

We have shown in Section 9.1 that $N(t)$ approaches the steady state M at an exponential rate. Specifically, we have shown in (1.15) that the deviation $F(t) = M - N(t)$ satisfies the inequality

(3.8) $$F(t + T) \le F(t)e^{-HT}, \quad T > 0, \quad H > 0.$$

If $N(t)$ is near M, H can be taken to be

(3.9) $$H = -\frac{dR}{dN}(M).$$

Let us say, somewhat arbitrarily, but with some justification, that the reaction is practically over if $F(t)$ is decreased by a factor of 1000. Now $e^{-7} < 10^{-3}$, so it follows from (3.8) that

$$F(t + T) < 10^{-3}F(t)$$

if

$$HT = 7.$$

Thus the reaction is practically over after time $T = 7/H$.

To find the time T in our case, we must compute H for R given by (3.6). According to (3.9), $-H$ is the value of the derivative,

$$\frac{dR}{dx} = \frac{k}{2}(x - a_0 - b_0),$$

at the zero M of R, that is, at

$$M = x_\infty = \min[2a_0, 2b_0].$$

If $a_0 < b_0$, $x_\infty - a_0 - b_0 = a_0 - b_0 < 0$, and if $b_0 < a_0$, $x_\infty - a_0 - b_0 = b_0 - a_0 < 0$; so by (3.9) we obtain

$$H = -\frac{dR}{dx}(x_\infty) = \frac{k}{2}|a_0 - b_0|.$$

So a reasonable practical estimate for the reaction time is

$$T = \frac{14}{k|a_0 - b_0|}.$$

Notice that the reaction time becomes very large if a_0 and b_0 are nearly equal. If $a_0 = b_0$ our formula gives infinite reaction time, which is nonsense as it stands, but very correctly alerts us that something goes wrong with our setup. What goes wrong is that when $a_0 = b_0$, then

$$R(x) = k\left(a_0 - \frac{x}{2}\right)^2$$

has a *double zero* at $x = 2a_0$. This violates the assumption for the theory developed in Section 9.1 that all zeros of R are simple. In fact we saw in part (c) of Exercise 1.1 that in such cases $N(t)$ tends to M not exponentially but at a much slower rate.

There is an equally good chemical reason why the reaction proceeds to completion much more slowly when the ingredients a_0 and b_0 are so perfectly balanced that they get used up simultaneously. If there is a shortage of both kinds of molecules, a collision leading to a reaction is much less likely than when there is a shortage of only one kind, but an ample supply left of the other kind of molecules.

Equation (3.5) for the reaction rate is not quite correct, for it only takes into account the forward reaction of A_2 and B_2 combining to form AB. There is also a reverse reaction of AB dissociating into A_2 and B_2. We assume that the mechanism which brings this dissociation about is the collision of sufficiently energetic AB molecules, therefore the rate of dissociation is proportional to x^2 and so equal to

(3.10) $$rx^2,$$

where r is the rate constant for the reverse reaction. The rate at which $x(t)$ changes is the difference of the birth rate or rate of formation (3.4) and the death rate or rate of destruction (3.10); in symbols:

(3.11)′
$$\frac{dx}{dt} = k\left(a_0 - \frac{x}{2}\right)\left(b_0 - \frac{x}{2}\right) - rx^2.$$

This equation is analogous to the equations in Section 9.2 expressing the population growth rate as the difference between the birth and death rates.

Equation (3.11)′ is still of general form (1.1), with

(3.11)
$$R(x) = \left(\frac{k}{4} - r\right)x^2 - k\frac{a_0 + b_0}{2}x + ka_0 b_0.$$

$R(0) = ka_0 b_0$ is positive, so a solution of (3.11)′ starting at $x(0) = 0$ increases and tends, according to Theorem 1.1 of Section 9.1, to the smallest positive zero of $R(x)$, which we denote again by x_∞:

$$R(x_\infty) = 0, \quad x_\infty > 0, \quad R(x) > 0 \quad \text{for } x \text{ in } [0, x_\infty).$$

The expression (3.11) for $R(x)$ is a quadratic in x which, for $k > 4r$, has two positive roots. They can be found by the quadratic formula; the smaller of the two is

(3.12)
$$x_\infty = \frac{a_0 + b_0 - \sqrt{(a_0 - b_0)^2 + (16a_0 b_0 r/k)}}{1 - (4r/k)}.$$

In laboratory work, a_0 and b_0 are at the experimenter's disposal; x_∞ is measured at the conclusion of the experiment. Thus Formula (3.12) can be used to measure the ratio of the reverse rate to the forward rate constant, r/k.

Formula (3.12) shows that the final concentration, x_∞, of the product depends only on the ratio, not on the absolute magnitudes of the rate constants r and k. This makes perfectly good sense from the point of view of chemical kinetics, since x_∞ is the equilibrium concentration for which the rates of forward and reverse reactions just balance, and this clearly depends only on the ratio of these rates.

Theorem 3.1. x_∞ *is a decreasing function of* r/k.

PROOF. Let us abbreviate the ratio r/k by s. By definition, x_∞ is the smaller positive zero of Equation (3.11). After dividing by k we can write (3.11) in the form

(3.13)
$$\frac{1}{k}R(x) = R_s(x) = \left(a_0 - \frac{x}{2}\right)\left(b_0 - \frac{x}{2}\right) - sx^2.$$

For every fixed x except $x = 0$, the derivative dR_s/ds is $-x^2 < 0$, so $R_s(x)$ is a *decreasing* function of s for every x except $x = 0$; $R_s(0)$ is independent of s. Thus if $s_1 < s_2$, the graph of $R_{s_1}(x)$ lies above the graph of $R_{s_2}(x)$ as shown on Figure 9.8. It is immediately evident from this figure that the smaller zero of $R_{s_1}(x)$ is greater than the smaller zero of $R_{s_2}(x)$. This completes the proof of Theorem 3.1. □

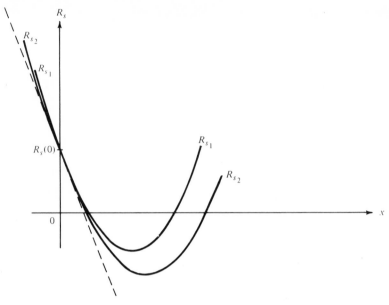

Figure 9.8

Another way of showing the decreasing dependence of x_∞ on $s = r/k$ is to differentiate x_∞, as given by Formula (3.12), with respect to s, and to verify that dx_∞/ds is negative. This is suggested in Exercise 3.3. However the geometric argument presented above is easier, as well as more general, since it does not rely on an explicit formula.

We have seen that, for $s = 0$, $x_\infty = \min(2a_0, 2b_0)$; this means that at the conclusion of the reaction, one or the other of the reactants is totally used up. It follows from Theorem 3.1 that when r is positive, $x_\infty(s) < \min(2a_0, 2b_0)$, which means that when $r > 0$ a certain amount of both reactants is left; the larger the value of r, the larger the reactant population at completion of the reaction. Since larger r means faster reverse reaction, this mathematical conclusion is plausible from the point of view of reaction kinetics.

We consider a typical complex reaction such as the spontaneous decomposition of some molecule A, for example N_2H_4. The decomposition occurs in two stages; the first stage is the formation of a population of *activated molecules* B, followed by the spontaneous split-up of the activated molecules. The mechanism for the formation of activated molecules B is through collision of two sufficiently energetic A molecules. The number of these per unit time in a unit volume is proportional to a^2, the square of the concentration of A. There is also a reverse process of *deactivation*, due to collisions of activated and nonactivated molecules; the number of these per unit time in a unit volume is proportional to the product ab of the concentrations of A and B. There is, finally, a spontaneous decomposition of B molecules into the end products which we do not specifically compute in the text (however, see Exercise 3.1). The number of these per unit time in a unit volume is proportional to the concentration of B. If we denote the rate constant of the formation of activated molecules by k, that of the reverse process by r, and that of the spontaneous dissociation by d, we get the following rate equations:

$(3.14)_a$
$$\frac{da}{dt} = -ka^2 + rab$$

$(3.14)_b$
$$\frac{db}{dt} = ka^2 - rab - db.$$

What can we say about the solutions of this system of differential equations? We claim the following:

Theorem 3.2

   (i) *The concentrations $a(t)$, $b(t)$ remain positive for all $t > 0$.*
   (ii) *$a(t)$ and $b(t)$ tend to 0 as t tends to ∞.*

PROOF. These facts will be deduced directly from Equations (3.14). As a first step we add the equations, obtaining

(3.15)
$$\frac{d}{dt}(a + b) = -db.$$

It follows from (3.15) by the monotonicity criterion that as long as b is positive, the sum $a + b$ decreases.

Let us take the case of practical interest where initially both A and B are present:

$$a(0) = a_0 > 0, \qquad b(0) = b_0 > 0.$$

Then, by continuity, both will be positive at least for a time interval $0 \le t \le T$; their sum, by (3.15), is decreasing as long as $b > 0$, so

$$0 < a + b < a_0 + b_0, \quad \text{for } 0 \le t \le T.$$

Since $b > 0$,

$$a < a_0 + b_0.$$

Dividing $(3.14)_a$ by a, we get

(3.16)
$$\frac{1}{a}\frac{da}{dt} = -ka + rb;$$

the right side is greater than $-ka$ which, in turn, is greater than $-k(a_0 + b_0)$ since $a < a_0 + b_0$ and $k > 0$. Thus

(3.17)
$$\frac{1}{a}\frac{da}{dt} \ge -k(a_0 + b_0).$$

The left side is the derivative of log a, so (3.17) is equivalent to

$$\frac{d}{dt}[\log a + k(a_0 + b_0)t] \geq 0,$$

and this implies, by the monotonicity criterion, that

$$\log a(t) + k(a_0 + b_0)t \geq \log a_0.$$

Since the exponential function is an increasing function, we deduce

$(3.18)_a$
$$a(t) \geq a_0 e^{-k(a_0 + b_0)t}.$$

We obtain a lower bound for $b(t)$ in a similar fashion. In Equation $(3.14)_b$, the right side is diminished if the positive term ka^2 is omitted and if, instead of rab, the even larger term $r(a_0 + b_0)b$ is subtracted. This yields the inequality

$$\frac{db}{dt} > -[r(a_0 + b_0) + d]b.$$

We divide by b and employ the same reasoning that led from (3.17) to $(3.18)_a$ to deduce

$(3.18)_b$
$$b(t) > b_0 e^{-[r(a_0 + b_0) + d]t}.$$

Inequalities $(3.18)_a$ and $(3.18)_b$ were derived under the assumption that the initial concentrations were positive; they show that the concentrations remain positive for all $t > 0$. This proves part (i) of Theorem 3.1.

To prove part (ii) we integrate Equation (3.15) from 0 to T and use the fundamental theorem of calculus:

$$\int_0^T -db(t)dt = \int_0^T \frac{d}{dt}(a + b)dt = a(T) + b(T) - (a_0 + b_0).$$

This may be rewritten as

(3.19)
$$d\int_0^T b(t)dt = a_0 + b_0 - [a(T) + b(T)].$$

We integrate also Equation (3.16) obtaining

(3.20)
$$\int_0^T \frac{1}{a}\frac{da}{dt}dt = \log\frac{a(T)}{a_0} = -k\int_0^T a(t)dt + r\int_0^T b(t)dt.$$

The abbreviations

(3.21)
$$\int_0^T a(t)dt = A(t), \qquad \int_0^T b(t)dt = B(T)$$

enable us to write (3.19) and (3.20) in the simpler forms

(3.19)′
$$dB(T) = a_0 + b_0 - [a(T) + b(T)],$$

(3.20)′
$$\log\frac{a(T)}{a_0} = -kA(T) + rB(T).$$

We saw in part (i) that $a(T) + b(T)$ is positive and decreasing; from this and (3.19)', it follows that B(T) is increasing but bounded from above by $a_0 + b_0$. Since a bounded increasing function tends to a limit (a consequence of the monotone convergence theorem in Section 1.6) we conclude that B(T) has a limit B_∞ which does not exceed the bound $a_0 + b_0$:

$$(3.22) \qquad \lim_{T \to \infty} B(T) = B_\infty, \quad 0 < B_\infty \leq a_0 + b_0.$$

Since $a(t) > 0$ for all t, A(T) is an increasing function of T: We claim that A(T) tends to ∞ as T tends to ∞. We shall prove this by showing that an assumption to the contrary leads us to conclude that $a(t)$, the positive integrand in $A(T) = \int_0^T a(t)dt$, tends to a nonzero constant, contradicting our assumption that A(T) has a finite limit as $T \to \infty$. For suppose A(T) does not tend to ∞ as $T \to \infty$; then A(T), being increasing and bounded, would tend to some finite limit A_∞ as T tends to ∞. Then the right side of (3.20)' would tend to a finite limit, namely $-kA_\infty + rB_\infty$; this is also the limit of $\log[a(T)/a_0]$, the left side of (3.20)'. This shows that $a(T)$ itself tends to the finite, nonzero limit $a_0 \exp\{-kA_\infty + rB_\infty\}$ as T tends to ∞. But then A(T), the integral of $a(t)$ from 0 to T, surely tends to ∞ as T tends to ∞. So we have shown that $A(T) \to \infty$ as $T \to \infty$. But then it follows that the right side of (3.20)' tends to $-\infty$. Then so does the left side, $\log[a(T)/a_0]$; this proves that $a(T)$ tends to 0 as T tends to ∞.

The following argument shows that $b(T)$ tends to zero: It is a consequence of (3.15) that $a(T) + b(T)$ is a decreasing function. Since it is positive, it tends to some limit. We have just shown that $a(T)$ tends to zero, therefore $b(T)$ tends to some limit. We claim that this limit is zero. For, if it were a positive number, then B(T), the integral of b from 0 to T, would surely tend to ∞, contrary to what we have already shown. This completes the proof of part (ii) of Theorem 3.2. $\qquad \square$

We have shown that A(T), the integral of $a(t)$ from 0 to T tends to ∞ as T tends to ∞, whereas B(T), the integral of $b(t)$ from 0 to T, tends to some finite value as $T \to \infty$. It follows that, on the average, $b(t)$ is much smaller than $a(t)$. Let us venture to guess that their ratio $b(t)/a(t)$ tends to zero on the average:

$$(3.23) \qquad \frac{1}{T} \int_0^T \frac{b(t)}{a(t)} dt \to 0$$

as $T \to \infty$. Let us divide Equation (3.16) by a:

$$(3.16)' \qquad \frac{1}{a^2} \frac{da}{dt} = -k + r\frac{b}{a}.$$

The left side is, according to the chain rule, the derivative of $-1/a$. Integrate both sides of (3.16)' with respect to t from 0 to T and apply the fundamental theorem to the left side:

$$-\frac{1}{a(T)} + \frac{1}{a(0)} = -kT + r \int_0^T \frac{b(t)}{a(t)}\,dt.$$

Divide both sides by $-T$; we get

$$\frac{1}{Ta(T)} = \frac{1}{Ta_0} + k - \frac{r}{T}\int_0^T \frac{b(t)}{a(t)}\,dt.$$

It follows from (3.23) that the last term on the right tends to 0 as $T \to \infty$; so we conclude that

(3.24)
$$\frac{1}{Ta(T)} \to k \quad \text{as } T \to \infty.$$

Thus, as $T \to \infty$, $a(T)$ tends to 0 at the rather slow rate

$$\frac{1}{kT}.$$

The above result hinged on the hypothesis (3.23), which is strongly suggested by our previous analysis, but is not a strict logical consequence of it. However (3.23) can be rigorously deduced by a more elaborate argument outlined in Exercise 3.2.

The behavior indicated by (3.24) applies only for large values of T. For other times the behavior of $a(t)$ is quite different. We now present a somewhat intuitive discussion of the behavior of $a(t)$ and $b(t)$ at early time. We shall study the case, of considerable practical interest, when the rate constant d of dissociation is considerably smaller than the reverse rate constant r. More precisely, *we assume that d is very much smaller than ra_0*. Let us take the case that the initial concentration b is zero. The rate at which b develops is given by the right side of Equation (3.14)$_b$:

(3.25)
$$ka^2 - rab - db.$$

From our assumption that d is small compared to ra_0, it follows that db is small compared to rab, as long as a is not much smaller than a_0. This is certainly true for small values of t. Also, for small values of t, the second term rab in (3.25) is smaller than the first term, ka^2. Thus the third term db in (3.25) is unimportant until a has decreased and b has increased so far that $ka^2 - rab$ is of the same size as db. We now calculate the solution during this initial phase. Dropping the term $-db$ in (3.14)$_b$ leads to the following simplified form of Equations (3.14):

$$\frac{da}{dt} = -ka^2 + rab$$

$$\frac{db}{dt} = ka^2 - rab.$$

Adding these equations we obtain

$$\frac{da}{dt} + \frac{db}{dt} = 0,$$

which implies that $a + b = $ constant. The value of that constant is a_0, so

$$b = a_0 - a.$$

Setting this into the first equation, we get

$$\frac{da}{dt} = -ka^2 + ra(a_0 - a)$$

$$= -(k + r)a^2 + ra_0 a.$$

This equation is again of form (1.1) with

(3.26) $$R(a) = -(k + r)a^2 + ra_0 a.$$

$R(a_0)$ is negative, so $a(t)$ starts to decrease. According to Theorem 1.1, $a(t)$ tends to the largest zero of $R(a)$ to the left of a_0. The zeros of $R(a)$ are 0 and $ra_0/(k + r)$; so during this first phase, $a(t)$ approaches $ra_0/(k + r)$. Since during this phase $a + b = a_0$, it follows that $b(t)$ approaches $a_0 - ra_0/(k + r)$ $= ka_0/(k + r)$.

The first phase comes to an end and the second phase begins when $a(t)$ is so near $ra_0/(k + r)$ and $b(t)$ to $ka_0/(k + r)$ that $ka^2 - rab$ becomes comparable to db. Since db is very much smaller than rab, it follows that, in the second phase, ka^2 and rab are very much bigger than their difference. Introduce the quantity

(3.27) $$p = ka - rb.$$

According to the above remark, p is very much smaller than ka.

Let us denote the sum of a and b by c:

(3.28) $$c = a + b.$$

We can express b in terms of p and c by multiplying (3.28) by k and subtracting from it (3.27). After dividing by $k + r$ we get

(3.29) $$b = \frac{kc - p}{k + r}.$$

Since, in phase 2, p is very much smaller than ka, and since $c > a$, p is even smaller compared to kc, and so it follows from (3.29) that $kc/(k + r)$ is a very good approximation to b. We can use this approximation on the right in Equation (3.15) to get

$$\frac{dc}{dt} = -\frac{kd}{k + r} c.$$

The solution of this equation is

$$c(t) = c_0 \exp\left\{-\frac{kd}{k+r}t\right\}.$$

Since b is well approximated by $kc/(k+r)$ and since $c_0 = a_0$,

$$(3.30)_b \qquad b(t) = \frac{ka_0}{k+r}\exp\left\{-\frac{kd}{k+r}t\right\};$$

since $a + b = c$, $a = c - b$; so

$$(3.30)_a \qquad a(t) = \frac{ra_0}{k+r}\exp\left\{-\frac{kd}{k+r}t\right\}.$$

This is a good approximation to the solution as long as the assumptions of phase 2 hold. As soon as $a(t)$ becomes appreciably smaller than a_0, phase 2 is over.

By now we have strayed pretty far from the straight and narrow path of strict mathematical deductions and have embarked on approximations. It is time to check our speculations against some solid numerical calculations.

With the rate constants

$$k = 0.2, \qquad r = 0.1, \qquad d = 0.001$$

Table 3.30

t	a	b	b/a	$(ka^2 - rab)/db$
0	1.00000	0.00000	0.00000	—
5	0.55946	0.43914	0.78494	86.60270
10	0.44115	0.55492	1.25790	26.02573
15	0.39054	0.60263	1.54307	11.56430
20	0.36464	0.62544	1.71522	6.05421
25	0.35006	0.63686	1.81927	3.47760
30	0.34131	0.64242	1.88225	2.13520
35	0.33574	0.64476	1.92041	1.39142
40	0.33201	0.64528	1.94357	0.96398
45	0.32934	0.64472	1.95763	0.71280
50	0.32730	0.64353	1.96617	0.56311
75	0.32061	0.63424	1.97822	0.35302
100	0.31525	0.62388	1.97896	0.33512
200	0.29512	0.58365	1.97766	0.33336
300	0.27630	0.54601	1.97616	0.33336
400	0.25869	0.51080	1.97455	0.33337
500	0.24222	0.47786	1.97285	0.33337
750	0.20553	0.40450	1.96808	0.33339
1000	0.17448	0.34241	1.96250	0.33341

and initial concentrations

$$a_0 = 1, \qquad b_0 = 0,$$

the system (3.14) of differential equations was solved by computer using the method outlined in Exercise 7.2 of Chapter 8. The results are shown in Table 3.30 and Figures 9.9a and b. Table 3.30 shows that $ka^2 - rab$ and db become comparable at about $t = 40$; this marks the end of phase 1. The computed values of a and b are

$$a(40) = 0.332, \qquad b(40) = 0.645$$

in reasonably good agreement with the theoretical values

$$a = \frac{ra_0}{k + r} = \frac{1}{3}, \qquad b = \frac{ka_0}{k + r} = \frac{2}{3}.$$

According to Formulas $(3.30)_a$ and $(3.30)_b$, the ratio $b(t)/a(t)$ is constant in phase 2, its value being equal to r/k. Table 3.30 indicates that for $40 < t < 1000$ the value $b(t)/a(t)$ is nearly constant, the constant being 1.97. This is in excellent agreement with $r/k = 0.2/0.1 = 2$, and indicates that the time interval 40 to 1000 belongs to phase 2. To further verify this we set in Formulas $(3.30)_a$, $(3.30)_b$ the values given above and $t = 10^3$. We get

$$a(1000) = \frac{0.1}{0.3} \exp\left\{ - \frac{10^{-3}0.2}{0.3} 10^3 \right\}$$

$$= \tfrac{1}{3} e^{-2/3} \cong \tfrac{1}{3}(0.51)$$

$$\cong 0.1715,$$

$$b(1000) = \frac{0.2}{0.3} e^{-2/3} \cong 0.3423,$$

gratifyingly close to the exact values determined by the numerical solution.

We have graphed the concentrations $a(t)$, $b(t)$ in the b, a-plane in Figure 9.9a, which dramatically shows the two phases. Figure 9.9b shows the time histories of $a(t)$ and $b(t)$.

We conclude with some general observations. The rates at which chemical reactions proceed depend very much on the temperature at which they take place. The higher the temperature, the faster the reaction; that is why we apply heat to promote a reaction. It follows that rate constants are temperature dependent. Statistical physics suggest that this temperature dependence takes the following form:

(3.31) $$k = k_0 \exp\left\{ - \frac{E}{RT} \right\}.$$

Here T is absolute temperature, E the activation energy, R a universal constant, and k_0 a factor depending on the components of the reaction

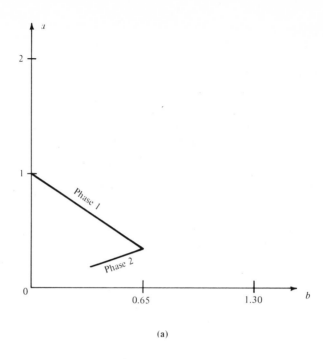

(a)

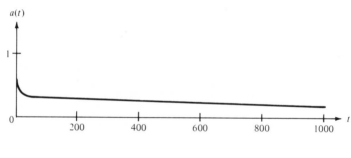

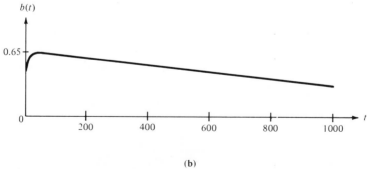

(b)

Figure 9.9   Solutions $a(t)$, $b(t)$ of Equations (3.14) with initial values $a_0 = 1$, $b_0 = 0$ and rate constants $k = 0.2$, $r = 0.1$ and $d = 0.001$.

and varying very slowly with T. Notice that according to Formula (3.31), k is an increasing function of T and a decreasing function of E.

The appearance of the exponential function in Formula (3.31) is related to the property of the exponential function discussed in Exercise 5.7 of Chapter 6.

The reactions we have studied in this section were assumed to be homogeneous, i.e., they take place simultaneously at all points of the vessel containing them. There are many reactions of great practical interest which stray far from homogeneity; the explosion of fuel in the cylinder of an internal combustion engine is such an example. In such reactions, concentrations must be described as functions of time and space. We shall study the laws of propagation of such reactions in Volume II.

Finally, we again call the reader's attention to the striking similarity between the differential equations governing the evolution of concentrations of chemical compounds during reaction and the laws governing the evolution of animal species interacting with each other.

EXERCISES

3.1 Using the numerical method described in Exercise 7.2 of Chapter 8, compute the concentration of the end product formed by spontaneous dissociation of activated molecules given the rate constants $f = r = 0.1, d = 0.001$, and the initial concentrations $a_0 = 1, b_0 = 0$ for the rate equations (3.14).

3.2 In this exercise we lead the reader to investigate in further detail the behavior of solutions of Equation (3.14) as t tends to ∞. We recall the equations:

$$\frac{da}{dt} = -ka^2 + rab$$

$$\frac{db}{dt} = ka^2 - rab - bd.$$

(a) Show that the quantity

(3.32) $$p = b/a^2$$

satisfies the differential equation

(3.33) $$\frac{dp}{dt} = k - dp + (2k - r)ap - 2r(ap)^2.$$

(b) Show that the inequality

(3.33)' $$\frac{dp}{dt} < k - [d - (2k - r)a]p$$

is satisfied.

(c) Show that if $2k < r$, the right side of (3.33)' is negative when $p > k/d$.
(d) Show that if $2k > r$, the right side of (3.33)' is negative when $a < d/2(2k - r)$ and $p > 2k/d$.

(e) We have seen in part (ii) of Theorem 3.2 that $a(t)$ tends to 0 as $t \to \infty$. Deduce from this and inequality (3.33)' that for all t large enough

(3.34) $$p(t) < 2k/d.$$

(f) Deduce (3.23) from (3.34)

(g) Deduce from (3.33), (3.34), and the fact that $a(t)$ tends to 0 as $t \to \infty$ that

$$\lim_{t \to \infty} p(t) = k/d.$$

3.3 In this exercise the reader is led to prove by calculus that x_∞, defined by Formula (3.12), is an increasing function of r.

(a) Show that, by introducing the abbreviations

$$(a_0 - b_0)^2 = c, \qquad 4a_0 b_0 = d, \qquad 4r/k = s,$$

Formula (3.12) can be rewritten as follows:

(3.35) $$x_\infty(s) = \frac{\sqrt{c + d} - \sqrt{c + ds}}{1 - s}.$$

(b) Show that (3.35) can be rewritten as

(3.35)' $$x_\infty(s) = \frac{f(1) - f(s)}{1 - s},$$

where

(3.36) $$f(s) = \sqrt{c + ds}.$$

Show that

(3.37) $$\frac{dx_\infty}{ds} = \frac{f(1) - f(s) - (1 - s)\dfrac{df(s)}{ds}}{(1 - s)^2}.$$

(c) Show, using the mean value theorem, that the right side of (3.37) is negative if $df(s)/ds$ is a decreasing function of s.

(d) Show that, for f defined by (3.36), $df(s)/ds$ is a decreasing function of s.

(e) Conclude that $x_\infty(s)$ is a decreasing function of s.

P.1 The bisection method for finding a zero
of a function

The program below is designed to locate a zero of a continuous function F defined on an interval $[x_1, x_2]$ at whose endpoints F has opposite signs. We describe the basic step of the program, then add some comments.

The interval $[x_1, x_2]$ is divided into two equal parts by its midpoint

$$x_{\text{mean}} = \frac{x_1 + x_2}{2},$$

and $F(x_{\text{mean}})$ is evaluated. If $|F(x_{\text{mean}})| \leq$ epsilon, the program ends; if $|F(x_{\text{mean}})| >$ epsilon, then F has opposite signs at the endpoints of either $[x_1, x_{\text{mean}}]$ or of $[x_{\text{mean}}, x_2]$. The program selects the subinterval where F changes sign by testing if the product

$$F(x_1) \times F(x_{\text{mean}})$$

is negative or positive. It picks the next interval according to the rule.

$$[x_1, x_2]_{\text{new}} = \begin{cases} [x_1, x_{\text{mean}}] & \text{if } F(x_1) \cdot F(x_{\text{mean}}) < 0 \\ [x_{\text{mean}}, x_2] & \text{if } F(x_1) \cdot F(x_{\text{mean}}) > 0, \end{cases}$$

and then repeats the basic step of evaluating F at the midpoint, etc. At each step, the length of the interval is halved. The program ends when the length of the interval is less than some prescribed value epsilon. Thus, unless $|F(x_{\text{mean}})| \leq$ epsilon at some earlier stage, the program ends after n steps where n is sufficiently large to satisfy

(P.1) $$(x_2 - x_1)2^{-n} < \text{epsilon}$$

485

Flow Chart for BISECT

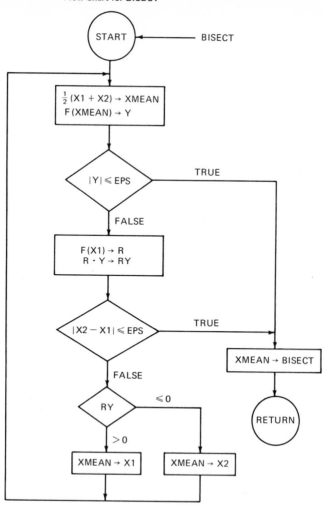

```
      SUBROUTINE BISECT(F,X1,X2,EPS,ROOT)
   2  XMEAN=0.5*(X1+X2)
      Y=F(XMEAN)
      IF(ABS(Y).LE.EPS)GO TO 1
      R=F(X1)
      RY=R*Y
      IF(ABS(X2-X1).LE.EPS)GO TO 1
      IF(RY)3,3,4
   3  X2=XMEAN
      GO TO 2
   4  X1=XMEAN
      GO TO 2
   1  ROOT=XMEAN
      RETURN
      END
```

486

that is,

$$n \cong \log_2 \frac{x_2 - x_1}{\text{epsilon}}.$$

How accurately has our program performed its task? To answer this question, we distinguish two cases:

(i) the program ends when $|F(x_{\text{mean}})| \leq \text{epsilon}$,
(ii) the program ends after n steps, where n satisfies (P.1).

Suppose the subroutine used to evaluate F is such that it gives $F(x)$ within epsilon of its precise value. Then, in case (i), the program has located

(i) a point x where $|F(x)| < 2 \times \text{epsilon}$,
   and in case (ii), the program has located
(ii) a point x within epsilon of a true zero of F. A flow chart for this algorithm follows.

The input is

the function F,
the interval $[x_1, x_2]$.

The output of the program is the zero of F in $[x_1, x_2]$.
Note. The notation

$$a \rightarrow b$$

means the value of b is replaced by the current value of a.

EXERCISE

P1.1 Try to write a program which finds all zeros of F, in some approximate sense.

P.2 A program for locating the maximum of a unimodal function

This program is designed to find the interior point of an interval S where a function F, belonging to a certain restricted class of functions defined on S, reaches its maximum. *The class of functions* defined on S for which this program is designed *consists of continuous functions that increase on S until they reach their maximum and decrease thereafter*, see Figure P.1. We shall call such functions *unimodal on S*.

Given an interval S and a unimodal function F, the basic step performed by the program is a selection of a subinterval S_{new} of S, half as long as S, with respect to which F is again unimodal; that is, F increases on S_{new} until it reaches its maximum, then decreases.

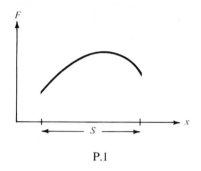

P.1

Denote the length of the given interval by $2h$, and its midpoint by x_{mean},

$$S = [x_{mean} - h, x_{mean} + h].$$

Now mark also the midpoints $x_{mean} - h/2$ and $x_{mean} + h/2$ of the left and right half of S, see Figure P.2. The maximum of F is in one of the three overlapping intervals of length h

(i) $[x_{mean} - h, x_{mean}]$,   (ii) $[x_{mean} - \frac{1}{2}h, x_{mean} + \frac{1}{2}h]$

(iii) $[x_{mean}, x_{mean} + h]$.

The following tests determine which interval contains the maximum of the unimodal function F on S:

If $F(x_{mean} - h/2) > F(x_{mean})$,   then max F is in (i),

if $F(x_{mean} + h/2) > F(x_{mean})$,   then max F is in (iii),

if neither inequality holds,   then max F is in (ii).

The next interval, selected by these tests to be either (i) or (ii) or (iii), contains the maximum of F and is half as long as S. Its length is redesignated as $2h_{new}$, its midpoint as $(x_{mean})_{new}$, and the basic step is repeated. The program ends when the interval containing the maximum of F is narrowed down to a length less than a prescribed quantity $2 \times$ epsilon, and the midpoint of that interval then lies within epsilon of the point where F reaches its maximum.

Note that the interval selected by our program is always such that the function F has its maximum at an interior point, not at an endpoint, and that F's property of being unimodal is therefore preserved at each stage.

The input to the program is the function F over $[x_1, x_2]$; the output is the approximate value x near which F reaches its maximum.

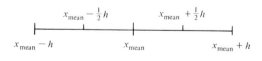

P.2

Flow Chart for MAX

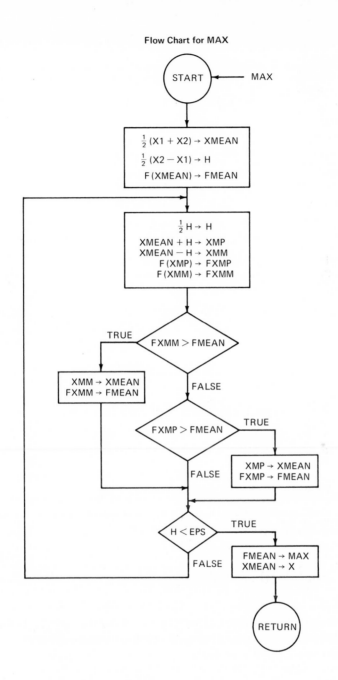

489

```
   REAL FUNCTION MAX(F,X1,X2,EPS,X)
   XMEAN=0.5*(X1+X2)
   H=0.5*(X2-X1)
   FMEAN=F(XMEAN)
6  H=H*0.5
   XMP=XMEAN+H
   XMM=XMEAN-H
   FXMP=F(XMP)
   FXMM=F(XMM)
   IF(FXMM.GT.FMEAN)1,4
1  XMEAN=XMM
   FMEAN=FXMM
   GO TO 2
4  IF(FXMP.GT.FMEAN)3,2
3  XMEAN=XMP
   FMEAN=FXMP
2  IF(H.LT.EPS)GO TO 5
   GO TO 6
5  MAX=FMEAN
   X=XMEAN
   RETURN
   END
```

EXERCISES

P2.1 Write a computer program, called MIN, based on MAX, for finding, approximately, the point where a unimodal* function reaches its minimum.

P2.2 Write a computer program to determine approximately the point where a unimodal function reaches its extremum. The output should indicate the *type* of extremum.

P.3 Newton's method for finding a zero of a function

Newton's method is described in Section 3.10. The main step starts with any given approximation x_{old} to a zero of F and constructs an improved new approximation x_{new} by the formula

(P.2) $$x_{new} = x_{old} - \frac{F(x_{old})}{F'(x_{old})}.$$

* Here we modify our previous definition to read: A function is unimodal on S if it is continuous and decreases on S until it reaches its minimum and increases thereafter.

490

Flow Chart for NEWTØN

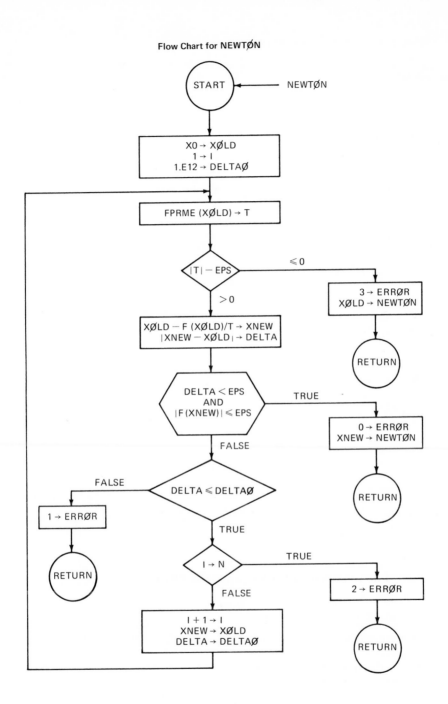

```
  REAL FUNCTION NEWTON(F,FPRME,X0,EPS,N,ERROR)
  INTEGER ERROR
  XOLD=X0
  I=1
  DELTAO=1.E12
3 T=FPRME(XOLD)
  IF(ABS(T)-EPS)1,1,2
2 XNEW=XOLD-F(XOLD)/T
  DELTA=ABS(XNEW-XOLD)
  IF((DELTA.LT.EPS).AND.ABS(F(XNEW)).LE.EPS)GO TO 5
  IF(DELTA.LE.DELTAO)GO TO 6
  ERROR=1
  RETURN
6 IF(I.GT.N)GO TO 7
  I=I+1
  XOLD=XNEW
  DELTAO=DELTA
  GO TO 3
7 ERROR=2
  RETURN
5 NEWTON=XNEW
  ERROR=0
  RETURN
1 ERROR=3
  NEWTON=XOLD
  RETURN
  END
```

The program then generates, using (P.2), a sequence x_1, x_2, etc. of (hopefully) better and better approximations to the zero. The difference of two successive ones is denoted by delta:

$$\text{delta}_i = |x_i - x_{i-1}|, \quad i = 1, 2, \ldots, N.$$

The program can fail to find a zero in one of 3 ways:

1. The difference between successive approximations fails to decrease; i.e., for some i, $1 \le i \le N$,

$$\text{delta}_i > \text{delta}_{i-1}$$

   where delta_0 is taken to be very large, equal to 10^{12}.
2. The sequence fails the convergence test; i.e., for *all* i, $1 \le i \le N$,

$$\text{delta}_i > \text{epsilon}.$$

3. Division by $F'(x_i)$ cannot be performed because $F'(x_i)$ is too small for some i, $1 \le i \le N$.

492

Whenever any of these failures occurs, the program sets the output called ERROR equal to 1, 2, or 3. On the other hand if the program is successfully completed, ERROR is set equal to 0 and the last value of x_i is taken as an approximate zero, under the name NEWTON.

The *inputs* into the program are:

The functions F and F'.

An initial guess x_0 for the root.

A small number epsilon used to test the convergence of the sequence of
   approximations $x_1, x_2, \ldots, x_N$.

An integer N which limits the total number of iterations.

Caution. The result of this program should not be used without checking ERROR to see whether the program was successful or not.

EXERCISE

P3.1 Write a program for finding the zero of a function by combining the Bisection and
   Newton programs. Start with Bisection, continue with Newton, with a provision
   to return to Bisection in case of failure.

P.4 Simpson's rule

This program, written as a subroutine, implements Simpson's rule for approximating integrals. The method is explained in Section 4.6. Over the interval $[a, b]$ Simpson's rule yields

$$\text{(P.3)} \qquad \int_a^b f(x)dx \cong \frac{b-a}{6}\{f(a) + 4f(m) + f(b)\},$$

where $m = (a + b)/2$. In this program the total interval of integration $[x_1, x_2]$ is broken up into N equal parts; the function is evaluated at the mid- and end points of all the subintervals so that the expressions in braces of Formula (P.3) can be formed. Multiplication by $(b - a)/6 = (x_2 - x_1)/6N$ is carried out out after all the parts

$$f(x_j) + 4f\left(\frac{x_j + x_{j+1}}{2}\right) + f(x_{j+1})$$

are added.

The *inputs* to the program, specified by the user, are:

the function F to be integrated;

the limits x_1 and x_2 of integration; and

N, the number of subdivisions.

The output of the program, called ANS, is an approximation to the integral.

Flow Chart for SIMPSON

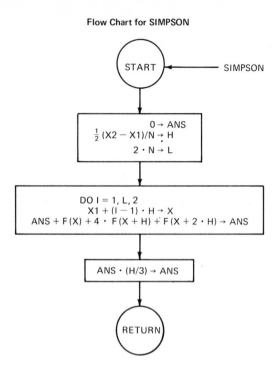

```
SUBROUTINE SIMPSON(F,X1,X2,N,ANS)
ANS=0.
H=0.5*(X2−X1)/FLOAT(N)
L=2*N
DO 1 I=1,L,2
X=X1+FLOAT(I−1)*H
1 ANS=ANS+F(X)+4.*F(X+H)+F(X+2.*H)
ANS=(H/3.)*ANS
RETURN
END
```

EXERCISES

P4.1 Write a program for Simpson's rule which incorporates an option to first divide the interval of integration into two parts. Such a program would be useful for selecting disjoint intervals, $[x_1, x_2] = [x_1, a] + [a, x_2]$, in which a more efficient integration can be achieved by using two stepsizes $(a - x_1)/N_1$ and $(x_2 - a)/N_2$; for example, if the OSC of the function is very small over one disjoint interval but large over the other.

P4.2 Write a program which doubles the value of N until the approximations I_N satisfy the convergence criterion

$$|I_{2N} - I_N| < \text{epsilon.}$$

494

P.5 Evaluation of log x by integration

This program evaluates an approximation to the log function, using the definition of log x as the integral

$$(P.4) \qquad \log x = \int_1^x \frac{dt}{t} \, ;$$

see Section 5.2. When $x < e$, the program evaluates the integral by Simpson's rule applied to a subdivision of $[1, x]$ into 10^3 equal parts.

To find log z for $z > e$, the program calculates $z/e, z/e^2, \ldots$ up to z/e^n where n is the first integer such that $z/e^n < 1$. Thus the program finds an n

Flow Chart for Natural Logarithm (LN)

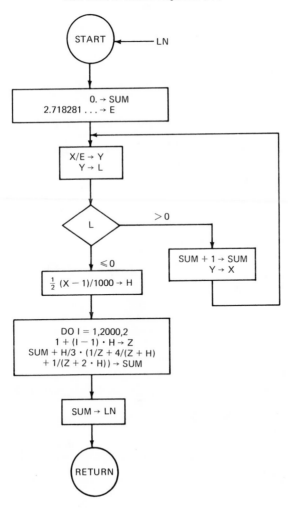

```
  REAL FUNCTION LN(X)
  SUM=0.
  E=2.718281828459045
4 Y=X/E
  L=Y
  IF(L)2,2,3
3 SUM=SUM + 1.
  X=Y
  GO TO 4
2 H=0.5*(X-1.)/1000.
  DO 5 I=1,2000,2
  Z=1.+(I-1)*H
5 SUM=SUM+(H/3.)*(1./Z+4./(Z+H)+1./(Z+2.*H))
  LN=SUM
  RETURN
  END
```

such that $z \geq e^{n-1}$, and $z < e^n$. Then the program sets $x = z/e^{n-1} < e$ and evaluates log x as described. Set $m = n - 1$; if $x = z/e^m$, then

$$\log x = \log z - m, \qquad \text{so} \qquad \log z = \log x + m.$$

Thus the program reduces the job of finding the log of a number $z > e$ to that of finding the log of a number $x < e$.

The input is x; the output is log x.

EXERCISES

P5.1 Write a program for evaluating log x which employs the Simpson subroutine. Choose the number of subdivisions N sufficiently small so that the step size h does not fall below $1/1000$.

P5.2 Compare the efficiency and accuracy of this log program with the one proposed in Exercise 3.2 of Section 5.3.

P.6 Evaluation of e^x using the Taylor series

At the end of Section 5.3, we derived the Taylor approximation of order n, Equation (3.24), to the exponential function

(P.5) $$E_n(x) = 1 + x + \frac{x^2}{2!} + \frac{x^3}{3!} + \cdots + \frac{x^n}{n!}.$$

This program uses (P.5) to approximate e^x, with a maximum permissible $n = 100$.

496

Flow Chart for e^x (E)

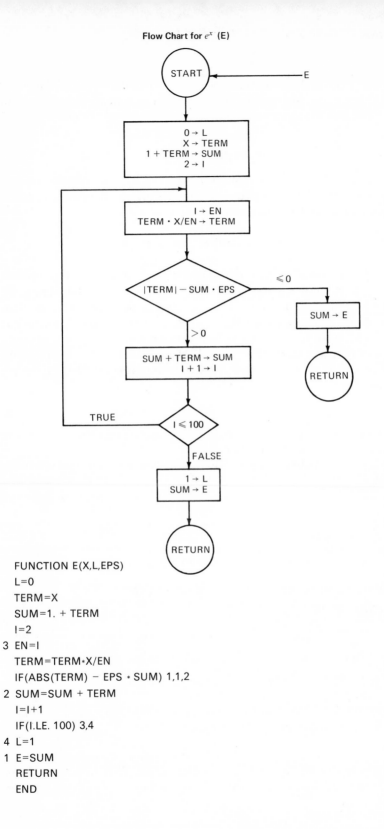

```
    FUNCTION E(X,L,EPS)
    L=0
    TERM=X
    SUM=1. + TERM
    I=2
3   EN=I
    TERM=TERM*X/EN
    IF(ABS(TERM) − EPS * SUM) 1,1,2
2   SUM=SUM + TERM
    I=I+1
    IF(I.LE. 100) 3,4
4   L=1
1   E=SUM
    RETURN
    END
```

497

The basic step in the program is the evaluation of the i-th term in the sum (P.5). The terms $x^i/i!$ are computed successively; each term is obtained from the previous one according to the formula

$$\frac{x^{i+1}}{(i+1)!} = \frac{x^i}{i!} \frac{x}{i+1}.$$

Before being added onto the sum on the right in (P.5), each term is tested for size. If a term $x^n/n!$ is less than epsilon times the accumulated sum E_{n-1}, it is discarded as relatively too small, for then

$$\frac{1}{E_{j-1}} \frac{x^j}{j!} < \text{epsilon} \quad \text{for all } j > n,$$

and the program declares the accumulated sum, called E, to be the approximation for e^x. If a term is greater than epsilon times the accumulated sum, the term is added on, and the basic step is repeated. If there is no term less than epsilon times the accumulated sum before $i = 100$ is reached, an output called L is set equal to 1; otherwise L is set equal to zero.

The user of this program should check what the value of L is before going on. The inputs are x, and the convergence tolerance epsilon. The outputs are e^x and the flag L.

EXERCISE

P6.1 Test the rate of convergence of (P.5) by finding out how I, the number of terms in the sum, depends upon x. Fix the value of epsilon and choose several values of x in the intervals
(a) $0 < x < 1$;
(b) $1 \le x < 2$;
(c) $0 > x > -1$;
(d) $-2 > x \ge -1$.

P.7 Evaluation of sin x and cos x using the Taylor series

The Taylor series for the sine function was derived in Section 7.3, Equation (3.27):

(P.6) $$\sin x = x - \frac{x^3}{3!} + \frac{x^5}{5!} - \cdots.$$

This program uses (P.6) to calculate an approximation to sin x.
The first step makes use of the periodicity of sine:

$$\sin x = \sin y$$

whenever x and y differ by an integer multiple of 2π. To reduce the problem of finding the sine of any angle to that of finding the sine of an angle between

Flow Chart for SIN(X), COS(X)

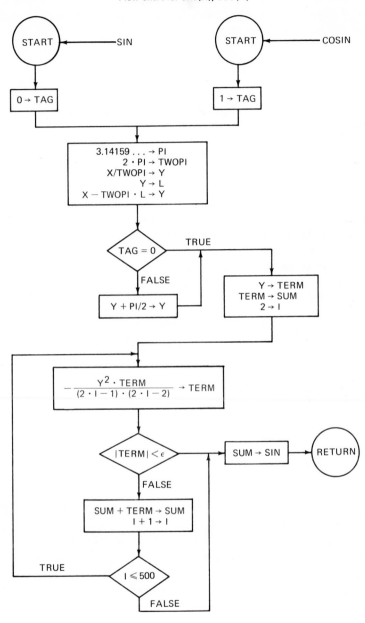

499

```
      FUNCTION SIN(X)
      INTEGER TAG
      TAG=0
3     PI=3.141592653589793
      TWOPI=2.*PI
      Y=X/TWOPI
      L=Y
      Y= X- TWOPI * FLOAT(L)
      IF(TAG.EQ.0)1,2
2     Y=Y+PI/2.
1     TERM=Y
      SUM=TERM
      I=2
7     RI=I
      TWCRI=2.*RI
      TERM=-TERM*Y*Y/((TWCRI-1.)*(TWCRI-2.))
      IF(ABS(TERM)-1.E-12)15,15,6
6     SUM=SUM+TERM
      I=I+1
      IF(I.LE.500)GO TO 7
15    SIN=SUM
      RETURN
      ENTRY COSIN
      TAG=1
      GO TO 3
      END
```

zero and 2π, we set

$$y = x - 2\pi \left[\frac{x}{2\pi}\right], \qquad \left[\frac{x}{2\pi}\right] = \text{greatest integer} \leq \frac{x}{2\pi}.$$

In Fortran, the greatest integer less than a real number z is obtained by storing a type real variable, say Z into a type integer variable such as L. The number y obtained this way lies between 0 and 2π.

The basic step is the computation of the i-th term in the series (P.6). As with the exponential function, the terms are computed successively, each term obtained from the previous one by the formula

$$\frac{x^{2i+1}}{(2i + 1)!} = - \frac{x^{2i-1}}{(2i - 1)!} \frac{x^2}{2i(2i + 1)}.$$

Before being added onto the accumulated sum, each term is tested for size; if less than 10^{-12} in absolute value, it is discarded as insignificantly small, the accumulated sum is declared to be sin, and the program is terminated. If the term is not less in absolute value than 10^{-12}, it is added onto the accumulated sum, the index i is increased by 1 and the basic step is repeated.

500

The program has a provision to compute cos x; this is accomplished by the use of a second entry point into the function called ENTRY COSIN. When the program is entered here, a tag is set equal to 1, whereas after the main entry, the tag is set equal 0. After the tag is set, the two routes merge; the tag is tested and if found 1, the argument y is replaced by $y + \pi/2$, so that the program computes

$$\sin\left(y + \frac{\pi}{2}\right) = \sin\left(x + \frac{\pi}{2}\right) = \cos x.$$

The input is x, the output is sin x.

EXERCISES

P7.1 Given any real number x, there is a unique y which differs from x by an integer multiple of 2π, and lies in $-\pi \le y < \pi$. Write a short program which computes this value of y.

P7.2 The number of terms used by the program to obtain an approximation to sin x is the largest value of i for which $x^{2i+1}/(2i + 1)!$ is greater than 10^{-12}. The larger x is, the larger the number of terms. Find out how large i is when

(a) $x = 2\pi - 10^{-12}$;   (b) $x = \pi$.

Index

Index